D0773366

HOW THINGS WORK

THEODORE GRAY

Photographs by Nick Mann

HOW THINGS WORK

THE INNER LIFE OF EVERYDAY MACHINES

THEODORE GRAY

bestselling author of *The Elements*

Photographs by Nick Mann

Copyright © 2019 Theodore Gray

Jacket design by Katie Benezra
Jacket photographs by Nick Mann
Jacket copyright © 2019 by Hachette Book Group, Inc.

Hachette Book Group supports the right to free expression and the value of copyright.

The scanning, uploading, and distribution of this book without permission is a theft of the author's intellectual property. If you would like permission to use material from the book (other than for review purposes), please contact permissions@hbgusa.com. Thank you for your support of the author's rights.

Black Dog & Leventhal Publishers
Hachette Book Group
1290 Avenue of the Americas
New York, NY 10104

www.hachettebookgroup.com
www.blackdogandleventhal.com

First Edition: October 2019

Black Dog & Leventhal Publishers is an imprint of Perseus Books, LLC, a subsidiary of Hachette Book Group, Inc. The Black Dog & Leventhal Publishers name and logo are trademarks of Hachette Book Group, Inc.

The publisher is not responsible for websites (or their content) that are not owned by the publisher.

The Hachette Speakers Bureau provides a wide range of authors for speaking events.
To find out more, go to www.HachetteSpeakersBureau.com or call (866) 376-6591.

Additional copyright/credits information is on page 254.

Print book interior design by Matthew Cokeley

Library of Congress Cataloging-in-Publication Data
Names: Gray, Theodore W., author. | Mann, Nick, photographer.
Title: How things work: the inner life of everyday machines / Theodore Gray; photographs by Nick Mann.
Description: New York, NY: Black Dog & Leventhal Publishers,
Hachette Book Group, 2019. | Includes index.
Identifiers: LCCN 2019009572| ISBN 9780316445436 (hardcover) | ISBN 9780316445450 (ebook)
Subjects: LCSH: Technology—Popular works. | Technology—Pictorial works.
Classification: LCC T47 .G738 2019 | DDC 621.8—dc23
LC record available at https://lccn.loc.gov/2019009572

Printed in China

Contents

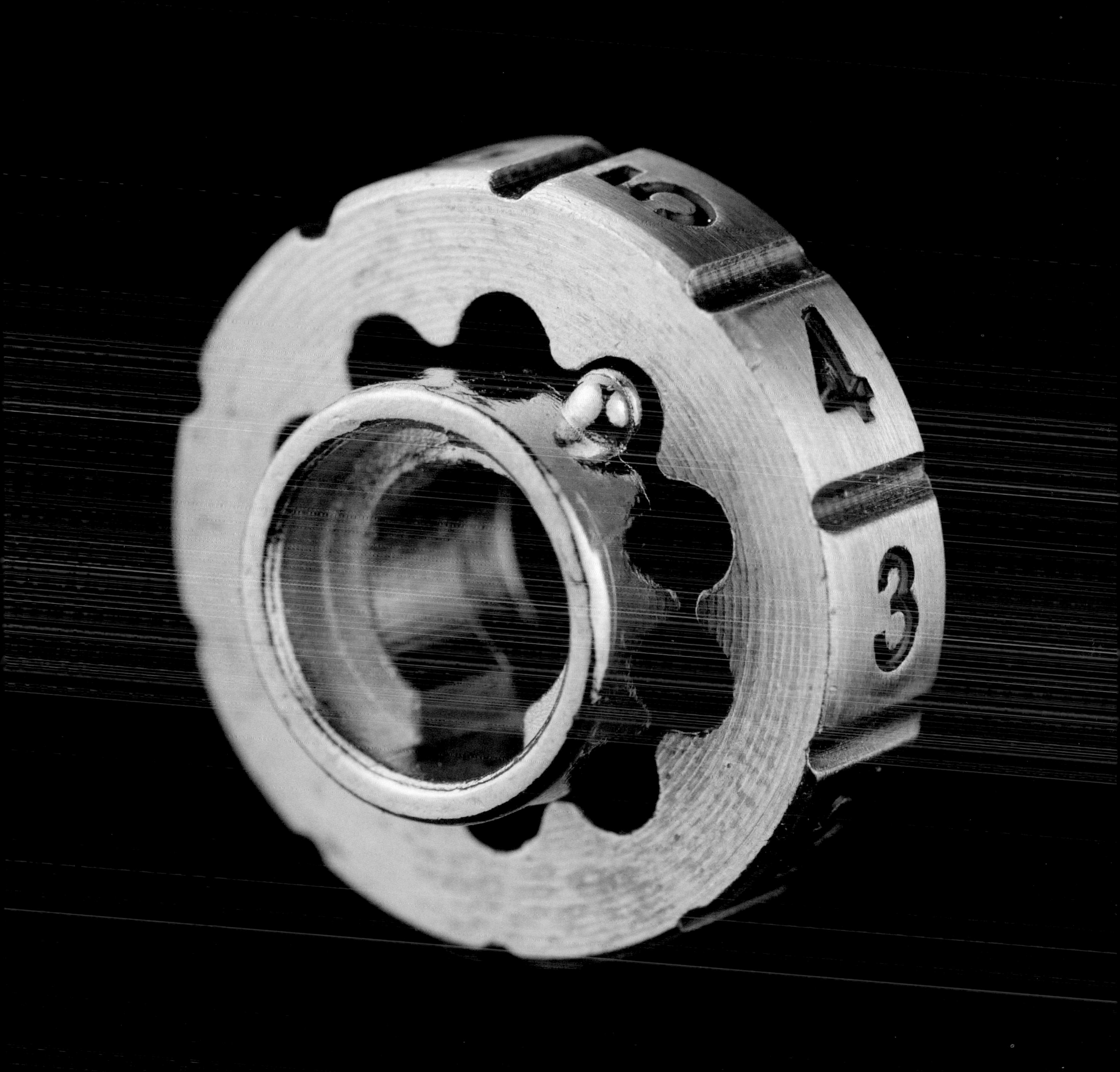
5
4
3

COMFORT IN THINGS

PEOPLE ARE MESSY. They're complicated and unpredictable. They can hurt you, and sometimes they hurt inside. Machines are not like that. They are what they are. They don't lie or cheat or turn the screw just when they know it will hurt the most. (Except printers. Printers are the psychopaths of machines.) Machines follow the rules. Even if you don't understand those rules at first, once you learn them, they will never change. They will remain true and the same, forever and always. This is especially true of machines you make yourself.

Other people's machines, complex manufactured things, can be frustrating and difficult to master. (Hence the existence of shooting ranges in Las Vegas where you can bring your printer and shoot at it with pistols, assault rifles, sawed-off shotguns . . . you know, whatever it takes.) But a thing you've made yourself is an open book. As you design it and bring it into existence, the machine reveals its nature to you one step at a time. In the end, it exists because of you, and you will always know it better than anyone else. You know not only its final form, but all the roads not taken, the other shapes and ways of being it could have embraced. When a manufactured thing breaks, it can be nearly impossible to fix, but when a machine you made breaks, well, you made it once, you can make it again.

A life built around making things is a good life. When you make things, they become a big part of who you are. And, as you make them, you sometimes find that they in turn are making you.

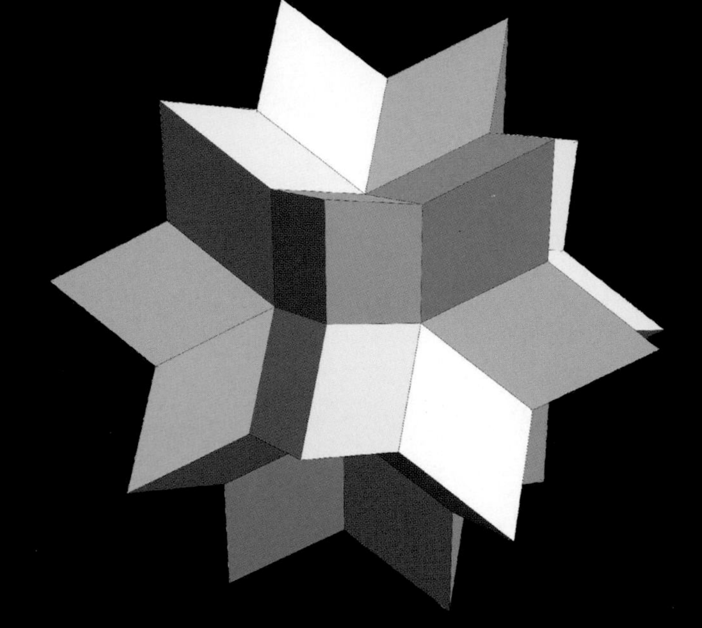

FOR MANY YEARS I worked on creating a computer program called *Mathematica*. Many other people worked on it too, but for several years, at the beginning, the Notebook user interface portion of it was mine and mine alone. I found great satisfaction in watching this creation of years come together. It was my life. (Literally: I had no other life, not even a date, for a decade.)

Even today I can see in my mind the structures inside this program, the internal logic, the chaos in a few places. I haven't looked at the actual code in many years, but if I did, I'm sure I would find many old friends, along with new ones added by the programmers who have come after me.

I still use *Mathematica* nearly every day to do my work. Yes, sometimes I curse my former self for bugs I could have fixed or features I will never see because no one else will ever care enough to add them. It was my baby, but not anymore. And, like my kids who were once small but then grew up, I have to accept my baby for what it is, not resent it for what it might have been.

Making *Mathematica* shaped me for twenty-three years. It's an important thing in the world, and I am proud of my work on it. But I'm also glad that when the chance presented itself to make something else, I took that road.

I got started collecting elements by accident, but have since spent a lot of time with them—I wrote a few books about them. Elements are good. Like kids, they are primal and raw, unique and universal. Everything is made of elements—everything we know and everything we are. And the kids, too, they are made of elements. The machines we call molecules are some of the most intricate and remarkable machines in existence. The machines called DNA and the machines called proteins are life itself to us. The time I spent with elements changed my life, and made me better. More interesting, I think.

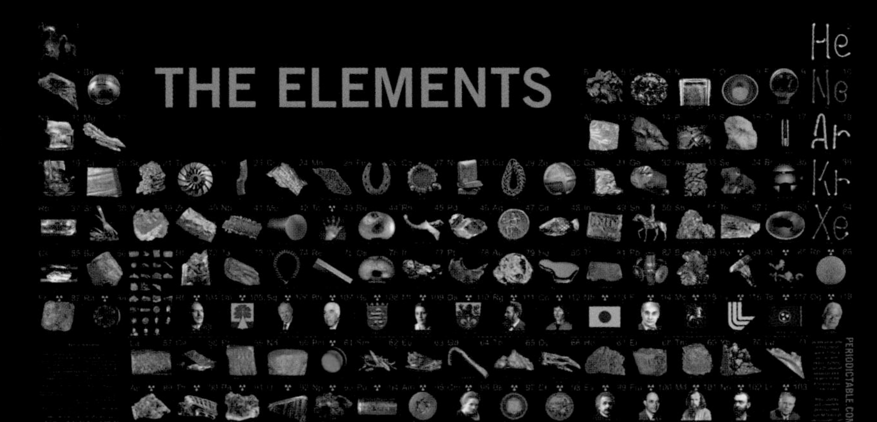

I Grew Up with Things

SOFTWARE AND ELEMENTS are great, but this book is not about software or elements. It is about the certain and comfortable world of mechanical things. I made, played with, took apart, broke, and fixed a lot of things when I was young, and in the years since. I learned a lot from these things. Most of all I learned to love them. They changed me. They were there for me in good times and bad. They kept me going, and opened worlds for me. These things are now a part of my past, but other things are very much part of my present, and yours. I have gathered many interesting things for you to see and learn about in this book. I hope that in the pages that follow, you will see the beauty I do in the inanimate, yet very alive world of things.

▼ This is what a calculator used to look like. I loved these things, and am just barely old enough to remember when a few of them were still actually being used. I've kept half a dozen of them. None of these work anymore, but I keep them anyway, and maybe someday I'll try to bring them back to life.

▶ Read about how I made this thing from scrap roof flashing on page 174.

▲ Remember LP records? I hear they are making a comeback now. This is a briefcase I made specifically for carrying records to parties in case I was ever invited to one, which of course I wasn't.

Read about this touch-keyboard on pages 250–251.

This was one of the most secret military devices of its day: the Norton bombsight. My dad got one as surplus and I spent hours figuring out how to make its gyroscopes spin. It is a sophisticated mechanical computer designed to calculate the precise moment a bomb should be dropped given the airplane's altitude and speed, and the wind speed and direction.

I made this thing. Then I dropped it.

I made several animals that my parents kept, for which I am eternally grateful.

When I was seventeen I knitted a regulation Dr. Who scarf. Yes, they published the exact row count for each color of yarn. I was working on this scarf when my dad came into my room to tell me that my sister had just died. I thought I'd always remember the exact color band I was on, but I don't. Just the sun in the window, the smell in the air, the sound of the door, and the feeling that has no name.

On Christmas break from college in Germany, I showed my grandfather Armin (see page 163) how to make detached rings on a small improvised wood lathe in his cabin high in the Swiss Alps. Don't get the wrong idea: he was a professional inventor and his small cabin in the mountains had the most finely appointed personal machine shop I have ever seen, before or since. I got a real thrill when he praised my humble baby rattle–making skills. I still remember his exact words: "Ein rechtes Lehrlingsstück," spoken in the Swiss dialect and meaning "a fine apprentice's work."

I invented this thing for listening to the vibrations of metal objects. It has a magnet mounted behind a fine wire coil, so nearby movement of anything metal will create a current in the coil, which can be picked up and amplified. It's the same idea as electric guitar pickups, but generalized into a "mechanophone" that you can point at any vibrating mechanical thing. I never actually found a good use for it back then, but during the writing of this book it turned out to be the perfect thing for the sonic scale shown on page 162! Never throw away any machine or device you've made yourself.

The Mechanisms in This Book

IN THIS BOOK you will find chapters on things with clear cases, locks, clocks, scales, and the machines needed to make a potholder from scratch. These topics may seem random—and, OK, the chapter on clear things is indeed a bit random (but totally fun, I promise).

The other chapters, however, are anything but random. They represent ancient and foundational machines that helped us measure our world, secure our place in it, and guide us into the fearfully and wonderfully made world of today. No finite list of inventions can ever be complete, and there are many more I would have liked to write about, but these are a good start. By understanding these mechanisms, you can understand a big part of how the mechanical world works.

Writing this book was an absolute joy, and I hope you get a sense of the pleasure it was to encounter and learn about each and every one of the fine things on this page and the many pages that follow.

CLEAR THINGS: The chapter on clear things may have a lot of modern inventions, but it speaks to the ancient human desire to see inside, to understand, to reveal the inner workings of a thing in your hand. The first machines were simple things with all of the moving parts immediately visible. Clear cases help restore a bit of that old clarity to modern devices.

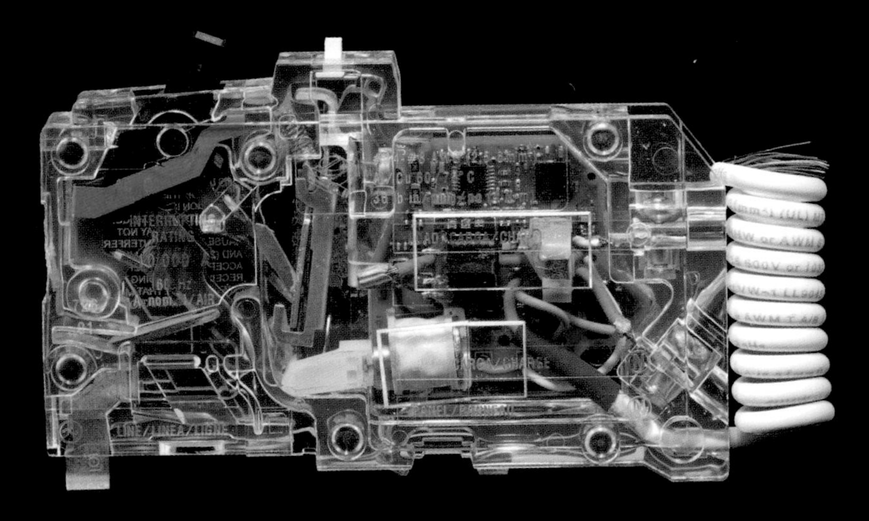

CLOCKS, the youngest category of devices in this book, are at least 3,500 years old. Of course, back then a clock was not much more than a stick in the ground: a sundial. In the centuries—stretching into millennia—since, clocks have evolved into a fantastic array of clever mechanical and electronic devices. And at a certain

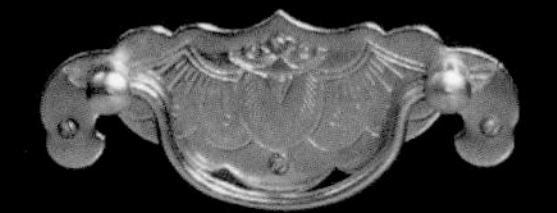

LOCKS as much as 4,000 years old have been found, making them even older than clocks. Apparently people have been trying to steal things since before there was a way to record what time the crime occurred. Secrecy is fundamental to locks. With enough information—the shape of the key, or the numbers of a combination—you can open any lock. This fact makes locks one of the first examples of information-processing equipment. They are tiny computers designed to test whether you are in possession of the secret needed to open them. In the modern age, locks have evaporated into pure information—PIN codes, computer passwords, and elaborate public-key cryptography that enables secure transactions on the internet.

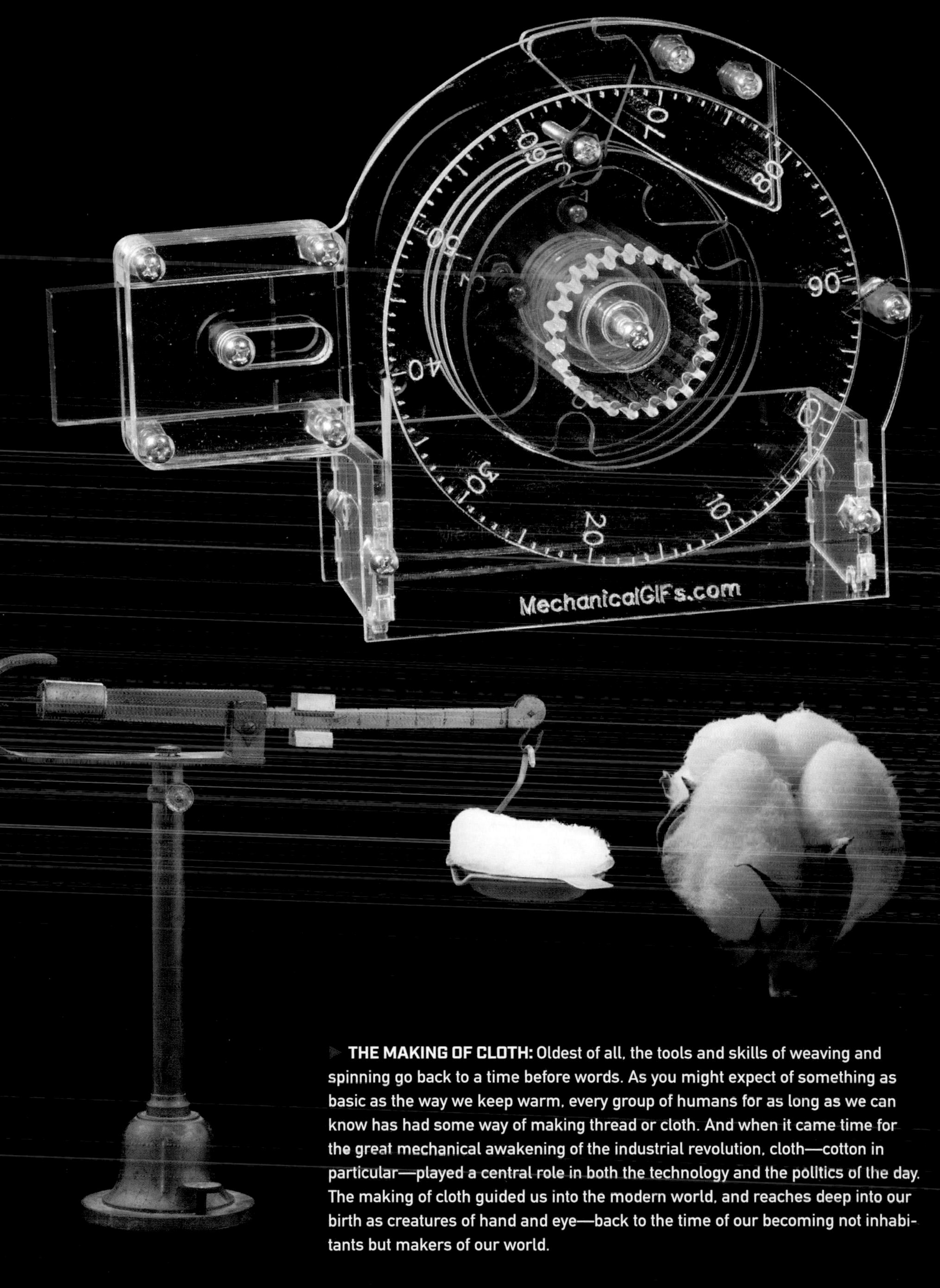

SCALES: Older even than locks are weighing scales, which date to the time of the great pyramids of Egypt, 4,500 years ago. The first and still the most important time things are weighed is when they are sold. Weight is the primary way we define how much of something we have, and therefore how much it should cost. For commerce, scales need to be honest (hard to cheat) but not especially accurate: beyond a fraction of a penny on the dollar, who really cares? But for scientific uses, scales accurate to parts per billion are needed, leading to a series of ever more clever designs. In 2018 a milestone was reached: a scale was built that is able to weigh more accurately than we are able to preserve the weight of the most carefully preserved object in the world, the Standard Kilogram in Paris—which is now no longer the standard of weight.

THE MAKING OF CLOTH: Oldest of all, the tools and skills of weaving and spinning go back to a time before words. As you might expect of something as basic as the way we keep warm, every group of humans for as long as we can know has had some way of making thread or cloth. And when it came time for the great mechanical awakening of the industrial revolution, cloth—cotton in particular—played a central role in both the technology and the politics of the day. The making of cloth guided us into the modern world, and reaches deep into our birth as creatures of hand and eye—back to the time of our becoming not inhabitants but makers of our world.

CLEAR THINGS

WHEN I WAS young I saw for the first time a picture of a telephone with a transparent case. You could look inside and see all the electronic components that made it work! My first thought was "wow, that is *so cool*," followed by "I *want* one." But then I got worried.

A clear telephone seemed obviously superior to every other possible telephone, so why weren't all telephones made with clear cases? Who could possibly want a phone that hid all the good stuff inside a pointlessly opaque case? I knew it couldn't cost more to use clear plastic rather than colored plastic. Did the people who made telephones just not realize that they could use clear plastic? Could people smart enough to make a whole telephone really be that dumb? It was at this moment that I realized, with a sense of loss, that maybe not everyone *wanted* to see inside their telephones. Maybe a lot of people cared more about the color of the case than about how the parts fit together inside it. I'm still sad about this.

I never did get a clear telephone as a kid, and now that I have one, it's too late to use it: we haven't had a landline phone connection at my house for years.

Writing a book about how things work is a great excuse to be able to collect a tremendous number of things with clear cases, which are useful for understanding how products are put together. They are also every bit as fun and cool as I imagined they would be. What I did not imagine is that many of them exist only because of prisons.

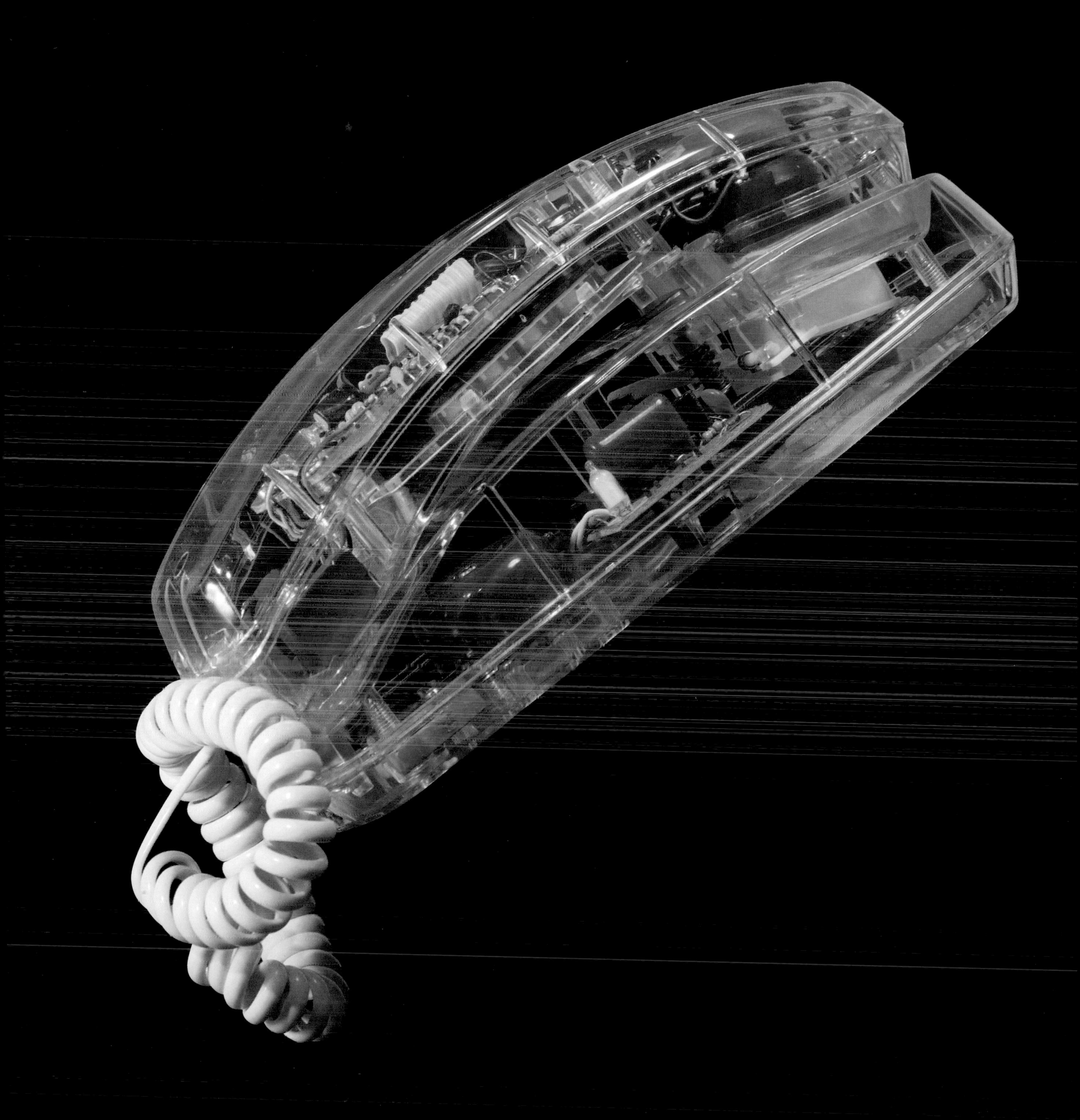

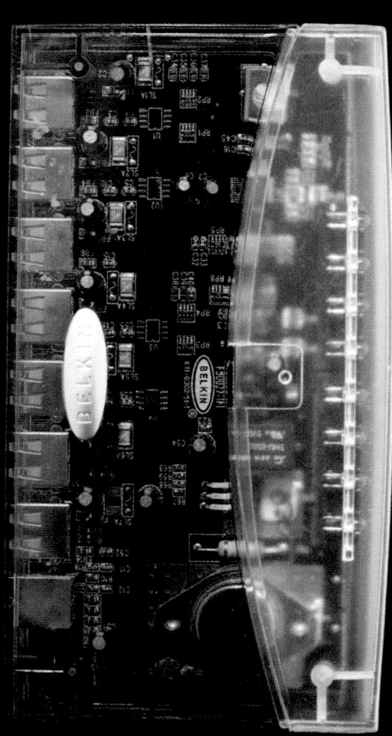

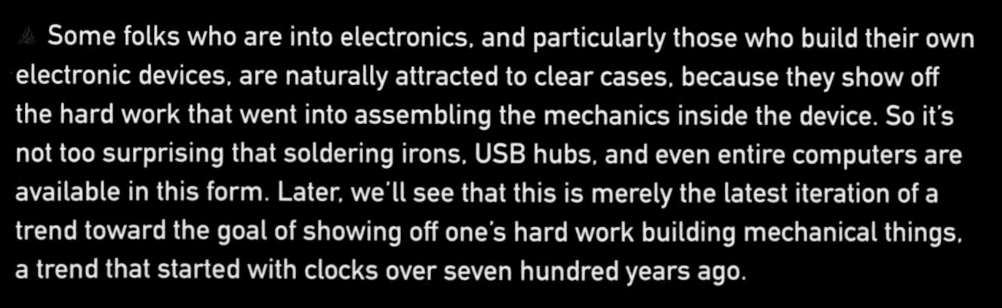

▲ Some folks who are into electronics, and particularly those who build their own electronic devices, are naturally attracted to clear cases, because they show off the hard work that went into assembling the mechanics inside the device. So it's not too surprising that soldering irons, USB hubs, and even entire computers are available in this form. Later, we'll see that this is merely the latest iteration of a trend toward the goal of showing off one's hard work building mechanical things, a trend that started with clocks over seven hundred years ago.

▼ There may be a lot of people who, for reasons beyond me, prefer opaque plastic cases, but these beauties are proof that I'm not the only one who appreciates how superior a clear case is. These are aftermarket transparent replacement cases for game machines and controllers that allow you to take off and throw away the silly colored shells and replace them with gloriously transparent ones.

▲ You can even get a transparent replacement case for your iPhone, so you can see all the circuitry inside! Well, OK, sorry, this is just a protective case with a photograph of the insides of an iPhone printed on it. But I have gotten double takes from a few people when I use it.

The Transparent Redemption

MY FIRST HINT of the existence of a vast prison-industrial complex of clear objects came when I saw a television much like these two at my friend Koatie's house some years ago. I thought she was being trendy, but she explained that it was her old prison TV (she used to be trouble). They make them this way so that it's impossible for convicts to hide drugs, knives, or other forbidden contraband inside the case. Little did I know that this TV was just the tip of the iceberg.

In fact, several brands and styles of old-school CRT (cathode-ray tube) televisions with clear cases were in circulation in prisons for many years. That big gray mass of glass that you see behind the screen is the "tube" part. It's hollow and opaque, so presumably the perfect place to hide contraband. But actually the tube is filled with a vacuum, so the minute you try to cut a hole in it to hide something, the whole television will violently implode.

A CRT television is full of dangerously high voltages, even days after it's been unplugged (because the capacitors in its power supply can hold dangerous amounts of charge for quite a while). Turning on a television with its case removed is not a good idea, so it's really neat to be able to safely see inside one while it's in operation.

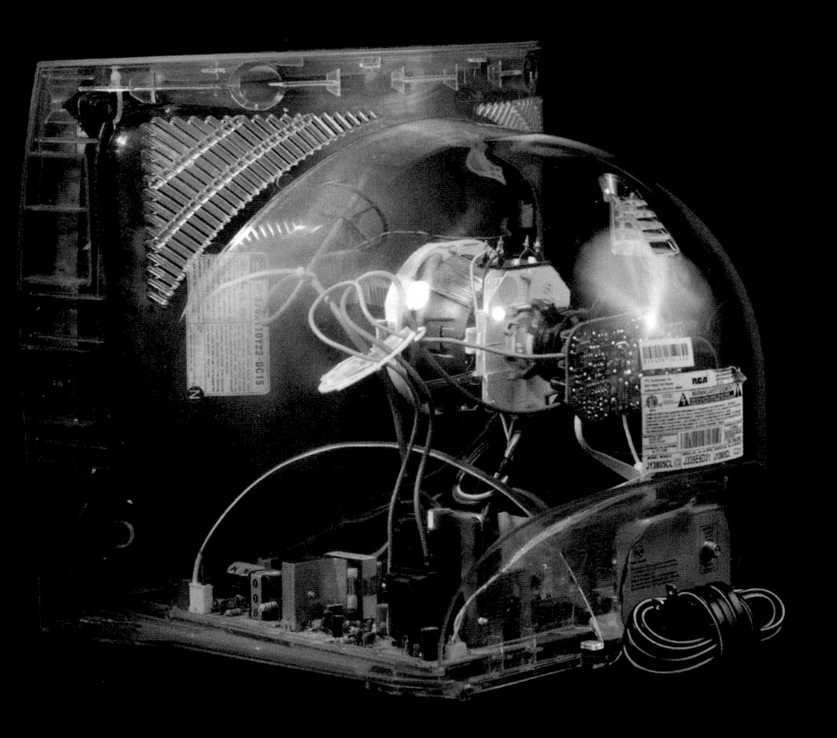

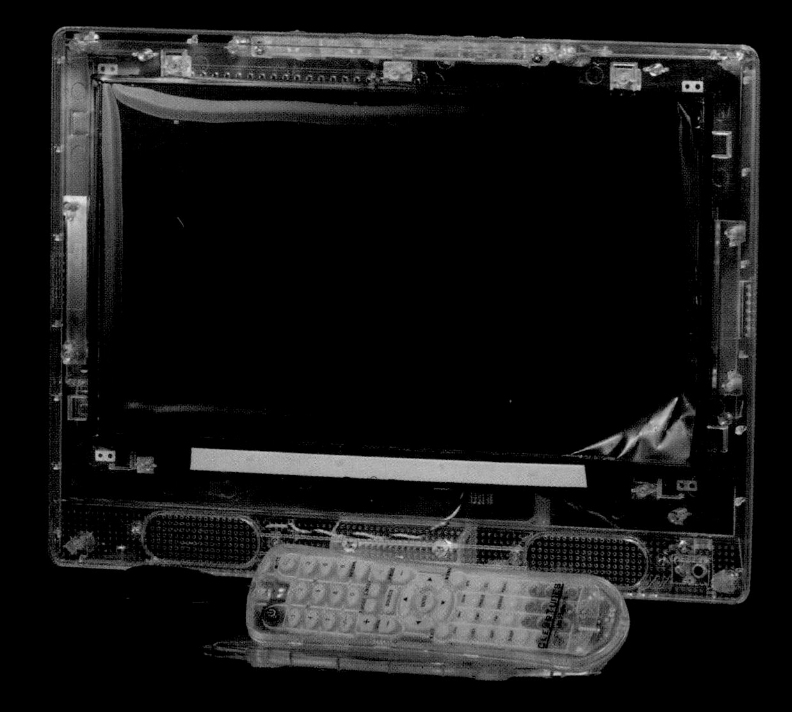

These days, of course, prison TVs are flat-screen, so there's even less room inside for hiding stuff. And where there had been magnets and high-voltage transformers, now there are only microchips and LCD (liquid crystal display) screens (see page 127 for an explanation of how these screens work).

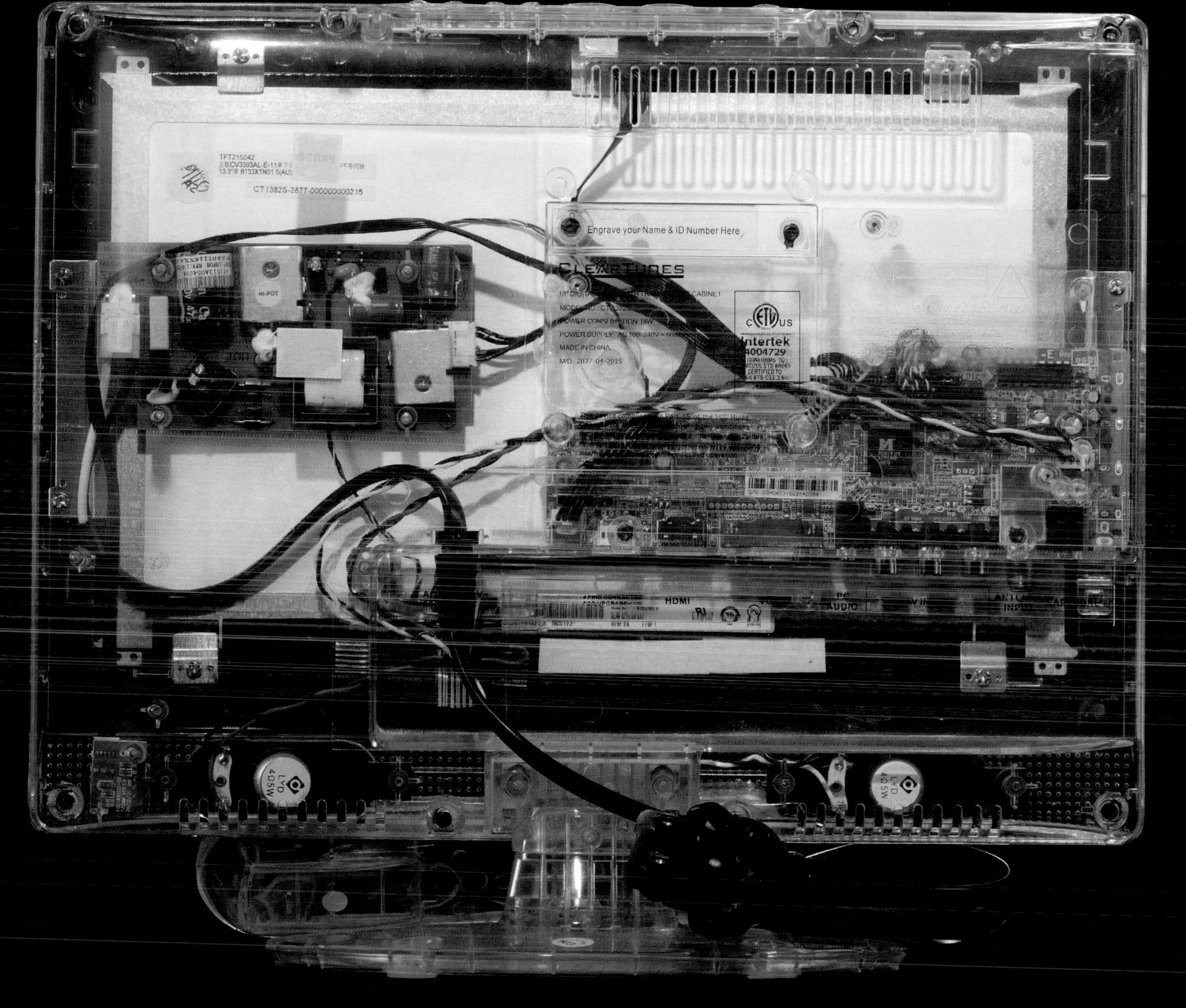

The back of the flat-screen TV on the previous page.

You can see a similar progression of technology in the wide array of prison radios over the years. The large, discrete components (capacitors, inductors, resistors, transformers, and so on) in the older models have been replaced by microchips in the newer ones. A complete radio these days can exist on a single microchip no bigger than a few breadcrumbs.

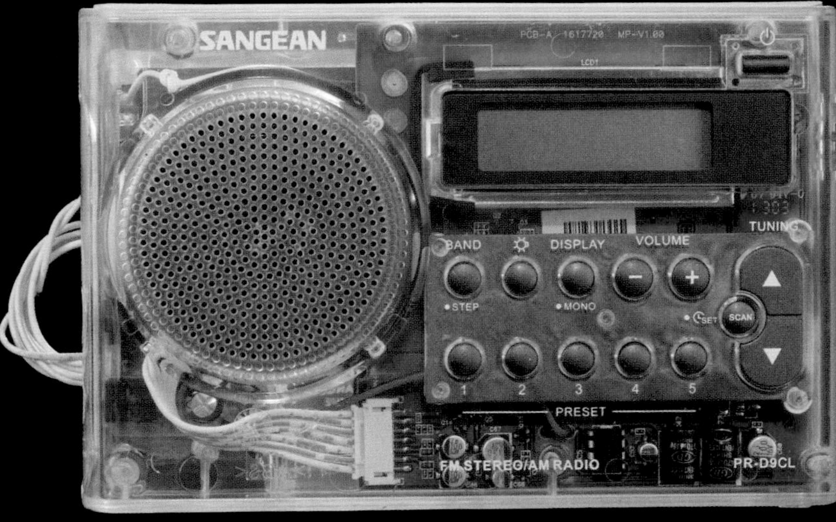

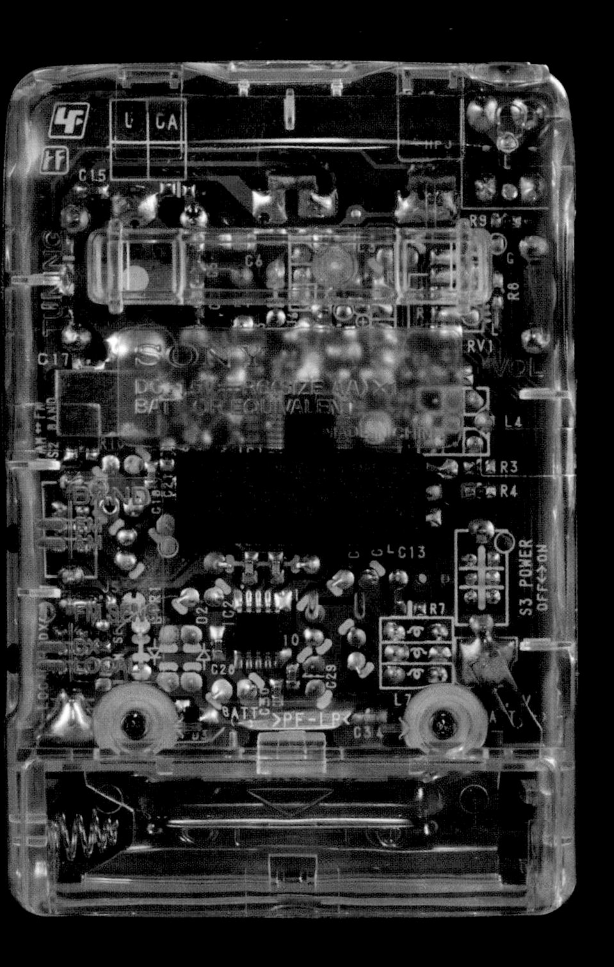

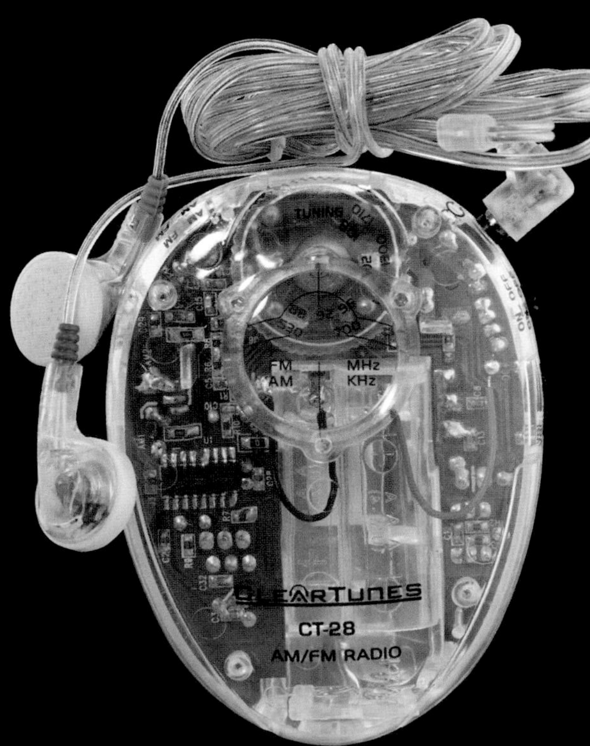

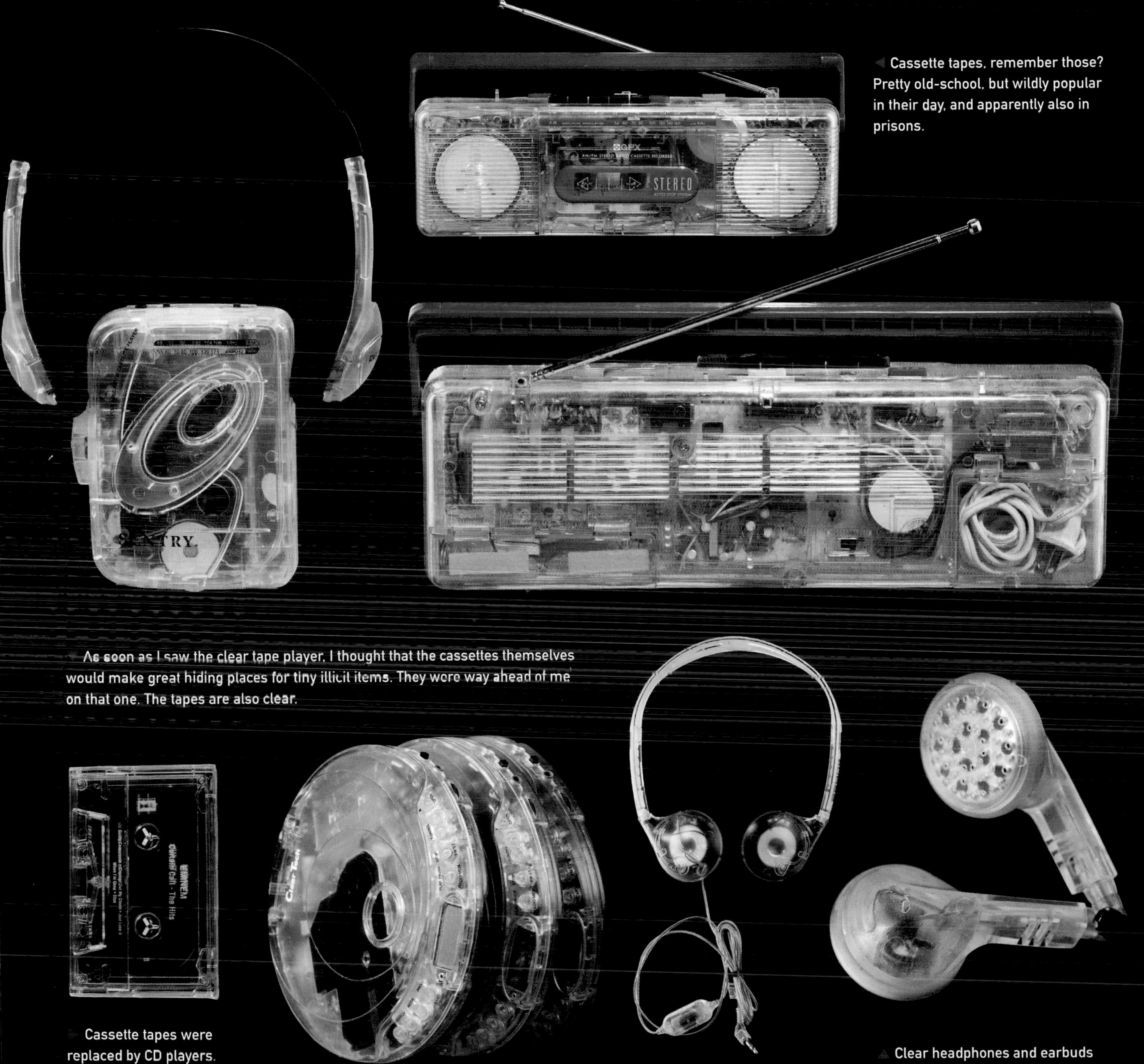

Cassette tapes, remember those? Pretty old-school, but wildly popular in their day, and apparently also in prisons.

As soon as I saw the clear tape player, I thought that the cassettes themselves would make great hiding places for tiny illicit items. They were way ahead of me on that one. The tapes are also clear.

Cassette tapes were replaced by CD players.

Clear headphones and earbuds

A CLOCKWORK RADIO

THIS IS MY all-time favorite clear object, because it combines clockwork and electronic technologies. No batteries necessary here because it employs a hand-crank, wind-up generator.

The generator creates electricity in pulses that come and go many times per second. On average there is enough energy to run the radio, but the circuitry needs steady power, not pulses. A capacitor stores up energy at the peaks of the pulses and releases it during the low points. In this way it smooths out the voltage going to the radio.

In a clever twist, the capacitor also feeds any voltage it accumulates back into the generator, causing it to push back against the force of the spring, slowing down the rate of unwinding. When the sound volume is low and the radio is using only a little power, the capacitor stays charged for quite a while and the generator spins fast only once every few seconds. But when the sound is turned up and the radio needs more power, the generator spins fast all the time.

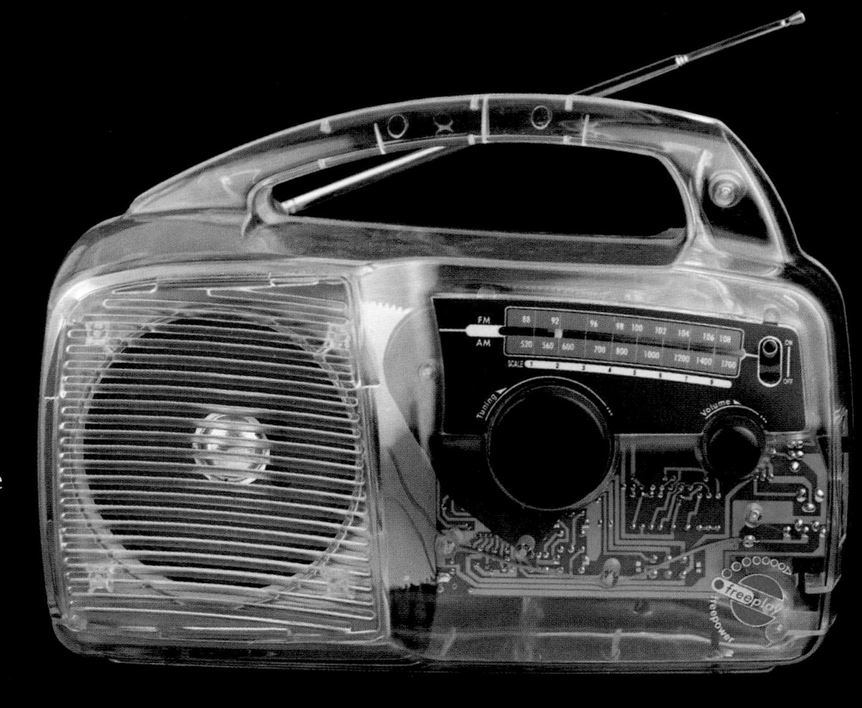

Capacitors smooth out the supply of electricity from the generator.

A handle, located behind this antenna, is turned to wind up the main spring.

An electric generator is a lot like an electric motor working in reverse. Instead of turning electrical energy into rotary motion, it turns rotary motion into electrical energy. Wires take the electricity from the generator to the circuit board, where it powers the radio.

Gears speed up the rotation of the slowly unwinding spring.

A large pulley connected by a rubber belt to a small pulley on the generator provides a final increase in rotational speed.

This powerful spiral torsion spring stores the energy you put into it by cranking. Don't ever take one of these things apart if it's wound up! It will vigorously uncoil, and by vigorously I mean slice your fingers off. The "anniversary" clock we will see on page 112 has a spring exactly like this.

THE TYPEWRITER

THIS EXAMPLE OF a clear typewriter is a fascinating hybrid of technologies from different eras: it stands halfway between computer and mechanical device. When you strike the keys, the letters go into a tiny bit of memory on a microchip inside the machine, just enough to store one line of text at a time. You can make corrections to that one line before hitting the Return key, which will trigger the line to print. That's the computer part. But then the mechanical part takes over: A wheel with lots of spokes, one for each letter and symbol, spins rapidly to position the correct spoke at the top. Then a strong and very fast electronic hammer called a solenoid whacks the spoke, which presses its raised letter onto an ink ribbon layered between the spoke and the paper. An impression of the letter is made on the page.

Although the outside world has moved on to computers, in prisons time has stood still, and clear typewriters, sometimes brand new, still circulate in systems around the country.

THE HAIR DRYER

A CLEAR HAIR dryer! Not just one but several different models! Finding these particular objects (on eBay and in prison supply catalogs) was the moment I realized that this thing with clear objects went deeper than I'd ever thought possible. If they make clear hair dryers, anything is possible.

Hair dryers are unusual in that they are the only common handheld electrical devices that consume a *lot* of power. In fact, they are typically set up so that at their highest setting, they draw the maximum amount of power that the electrical code legally allows you to pull from a standard wall socket. If manufacturers could make them to use any more power, they would.

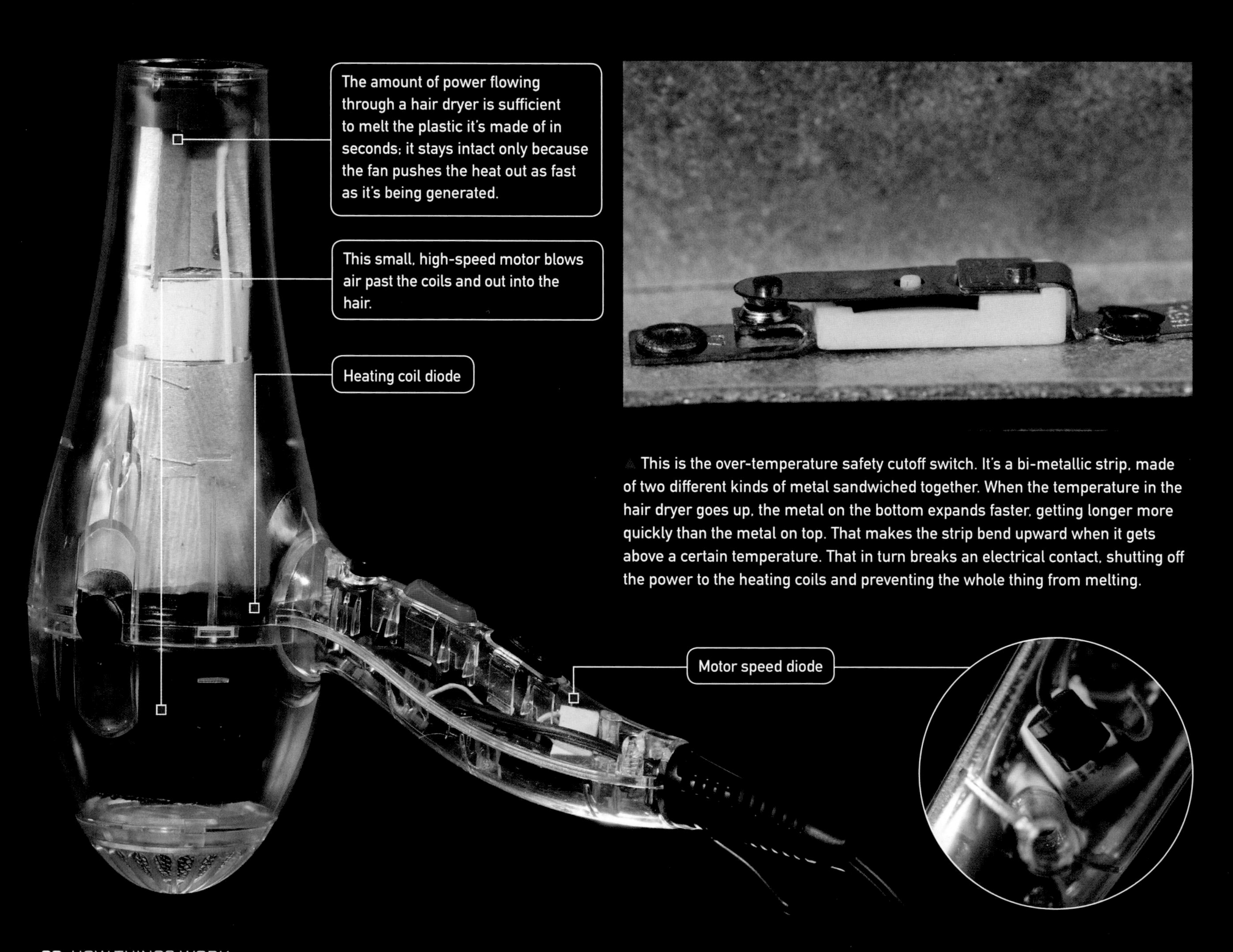

▲ This is the over-temperature safety cutoff switch. It's a bi-metallic strip, made of two different kinds of metal sandwiched together. When the temperature in the hair dryer goes up, the metal on the bottom expands faster, getting longer more quickly than the metal on top. That makes the strip bend upward when it gets above a certain temperature. That in turn breaks an electrical contact, shutting off the power to the heating coils and preventing the whole thing from melting.

This is what happens if you disable all the safety features in a cheap plastic hair dryer, including the thermal overload switch, the mica heat shield, and the fan motor. It smokes for a few seconds, then melts down around the heating coils. The thing it didn't do is catch fire, which I'm frankly a little surprised about. I've always thought it was inherently unsafe to have a hair dryer made of plastic, but now that I've tried really hard to get one to catch fire, I feel a bit better about it.

Inside this sadly opaque area are the heating coils, which turn electricity into heat. There is no clear plastic that could withstand the heat of the coils. The enclosure could in theory have been made of quartz glass, but that would have been very expensive.

There's a reason why all normal hair dryers have exactly three heat settings (cool, low, and high), and three fan speed settings (off, low, and high). Power from the wall is AC, which stands for alternating current: it switches back and forth between positive and negative voltage 60 times per second (positive for 1/120 of a second, then negative for the next 1/120 of a second, and so on). It just so happens that there is an extremely cheap, small, and efficient device, called a diode, which allows the flow of electricity in one direction and blocks it in the other.

If you make AC power flow through a diode before it gets to its destination, only half the power (represented by the shaded red area below) will make it through. The low/high power switch in the hair dryer simply patches a diode in or out of the circuit. When the diode is in, only half the power gets through—there's your low-power mode.

Anything more than two power levels is much more complicated and expensive to achieve. In fact, if you want more than two levels, you might as well go all the way to infinitely variable power, which is done with a device called a triac, found in some expensive hair dryers and in all wall switch light dimmers.

TRUE STORY. IN 1984 I spent my sophomore year of college at the University of Gottingen, Germany, living in a student dorm. Down the hall from me were a number of students from the Middle East whose hair dryer had stopped working. I "fixed" it for them in the sense of bypassing the broken thermal overload switch in hopes that the switch being blown didn't indicate a bigger problem. It did, and we all lost power about two seconds after they switched it on. Then they made a joke about taking me hostage, which I didn't understand because I didn't know the German word for "hostage" until I looked it up later that night.

Later, in an unrelated incident that I was totally not responsible for, the stereo system in the common area caught fire. I was the only one who thought it was imperative to use the available fire extinguisher immediately (which I did). I think the other students were annoyed that I got fire extinguisher powder all over the room, but I'm pretty sure I saved everyone a lot of trouble and death. I don't remember a single thing about the names or faces of the students I lived with for nearly a year, but I do remember the look on the face of the stereo's large power-filtering capacitor on fire with its guts blown out the top.

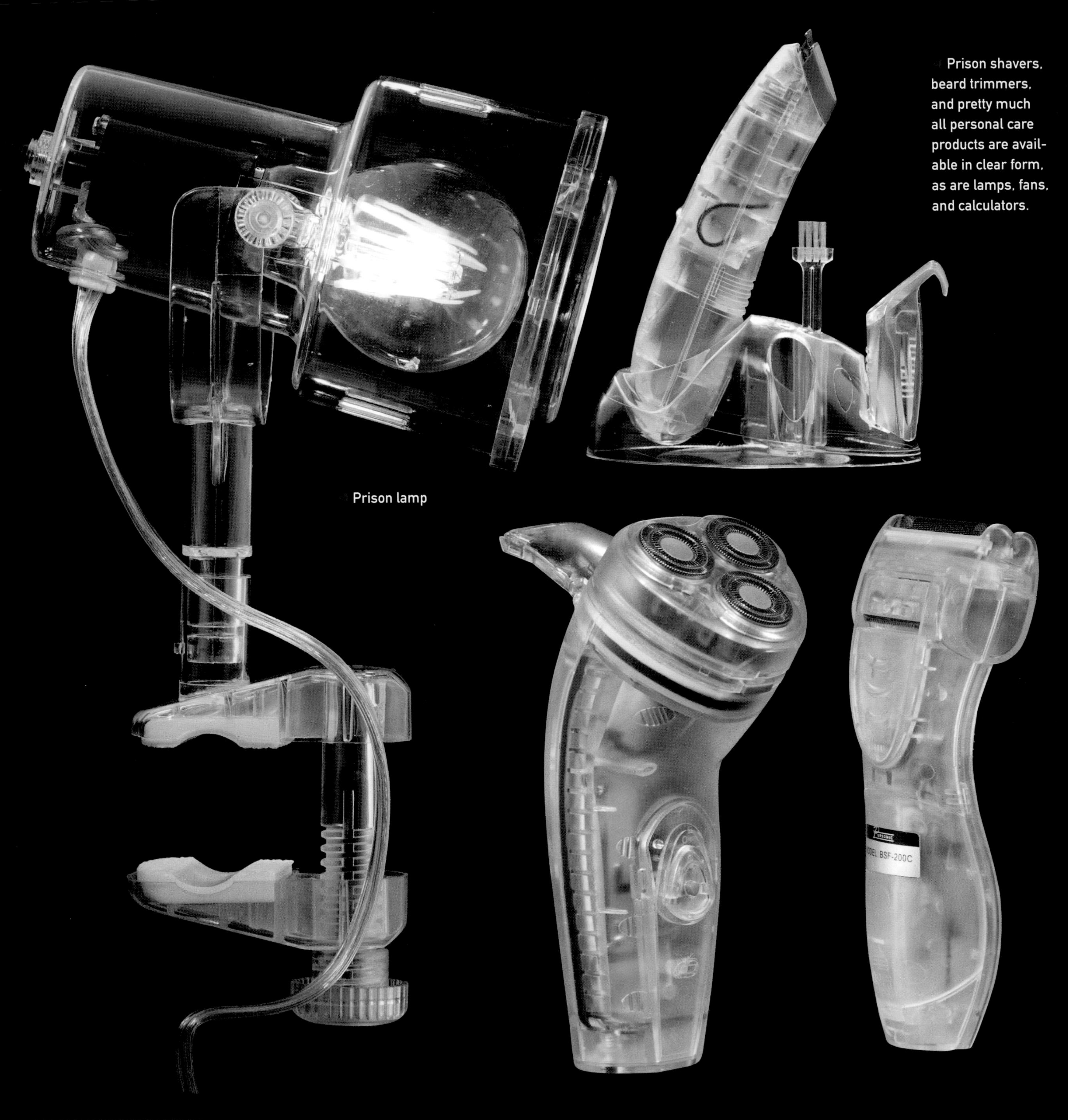

Prison shavers, beard trimmers, and pretty much all personal care products are available in clear form, as are lamps, fans, and calculators.

Prison lamp

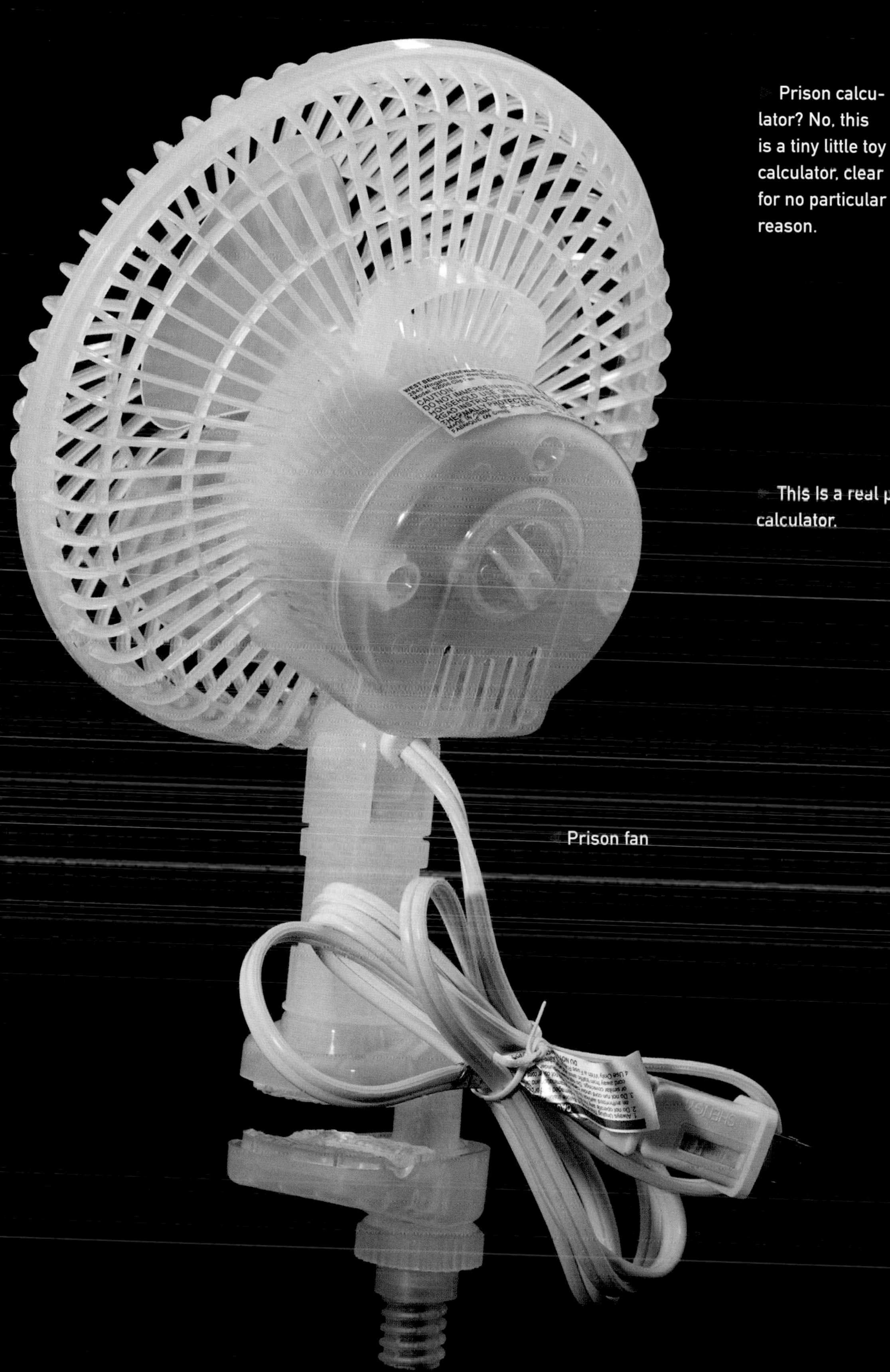

Prison fan

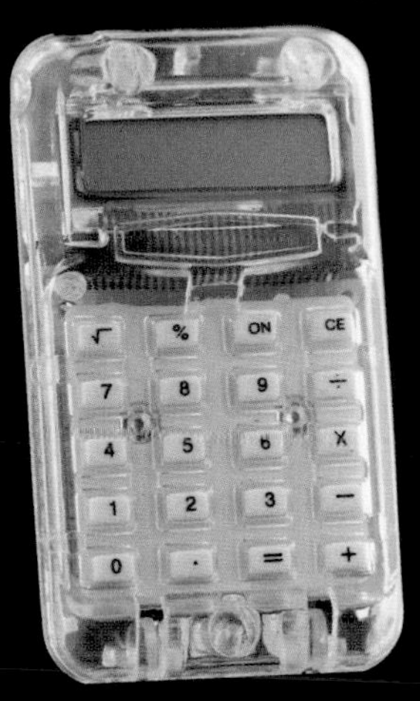

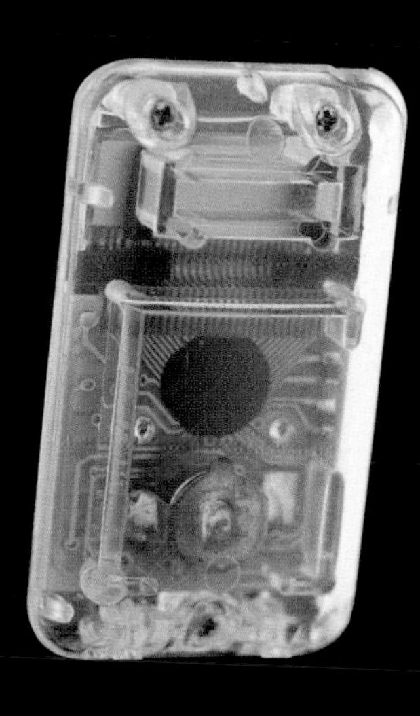

Prison calculator? No, this is a tiny little toy calculator, clear for no particular reason.

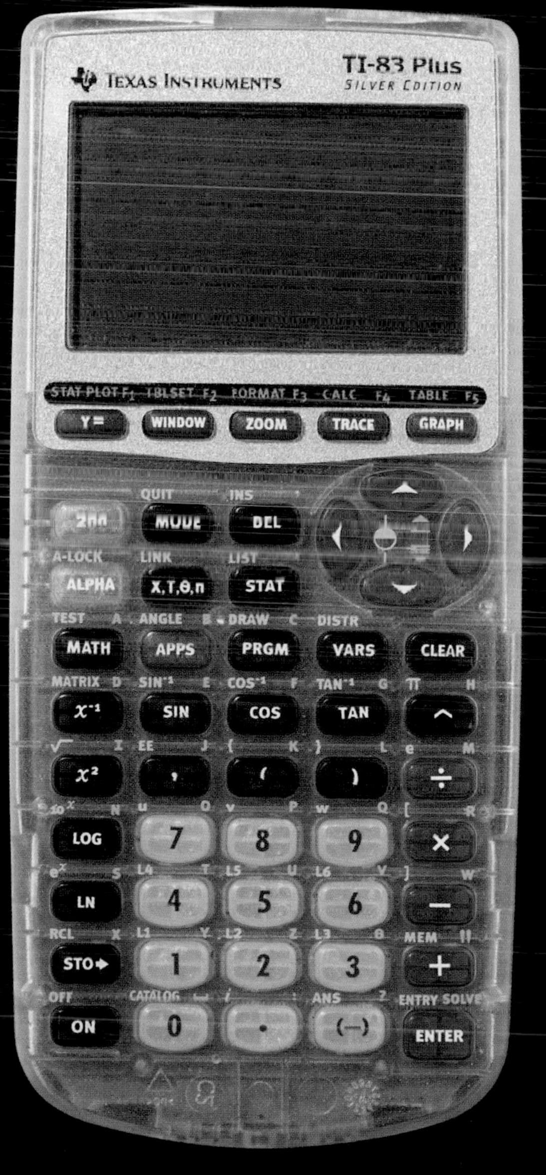

This is a real prison calculator.

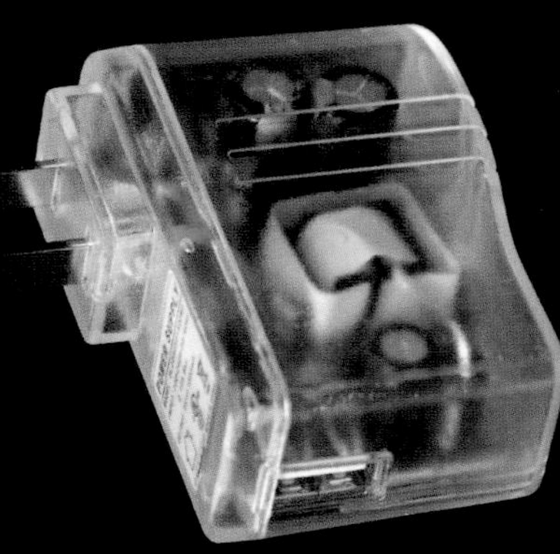

A low-security prison cell with a bunch of electronic devices requires a standard tangle of outlet strips and power adapters. The industry has responded by making these available in clear form.

Not all transparent power plugs are made for prisons. This one is a hospital-grade plug. They are made transparent so that they can be inspected to verify that the wires are solidly connected, and that there is no evidence of heat (which indicates corrosion) or insulation failure.

Circuit breakers are devices that cut off the power to an electric circuit when too much power is being drawn for the wiring to handle. These clear ones let you see the electromagnet inside that throws the switch when an overload condition occurs. I believe these are also made clear to be inspectable when used in critical life-safety situations.

▲ These clear backpacks are advertised for use in schools, which require them for pretty much the same reason clear things are required in prisons. It's a sad day when we treat our students with the same suspicion as people actually convicted of crimes.

▼ The other place innocent people are treated like criminals is, of course, in airports. Clear pouches like these are made for carrying your two-part liquid explosives through airport security. (Side note: the American system of clear pouches is a completely ineffective way of detecting potential liquid explosives, which look and x-ray just like water or shampoo. The Chinese system is superior: they have a person who opens and smells each container. People can smell nearly as well as dogs, if their nose is close enough, and we can learn to recognize—and name—hundreds of different substances.)

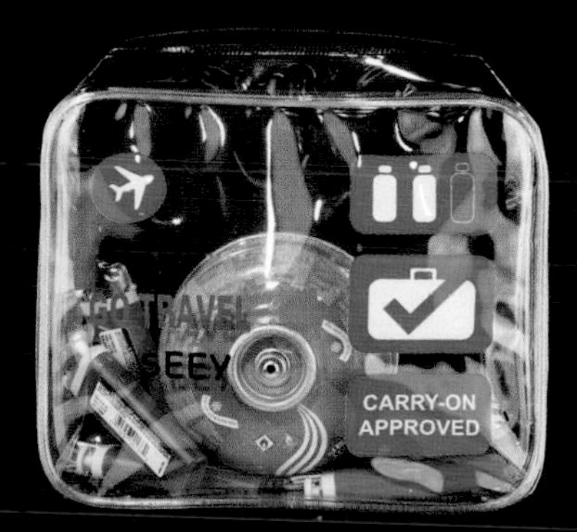

◄ Given how much the people who run prisons seem to love transparent objects, I'm sure they would require prisoners themselves to be clear, if they could get away with it.

► While I have not found any instances of prison officials advocating for clear prisoners, I did find a news report of a sheriff who got in trouble for trying to make his inmates wear transparent prison uniforms. That was apparently considered crossing a line.

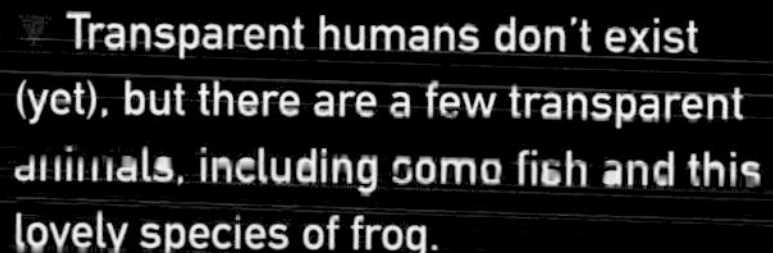

▼ Transparent humans don't exist (yet), but there are a few transparent animals, including some fish and this lovely species of frog.

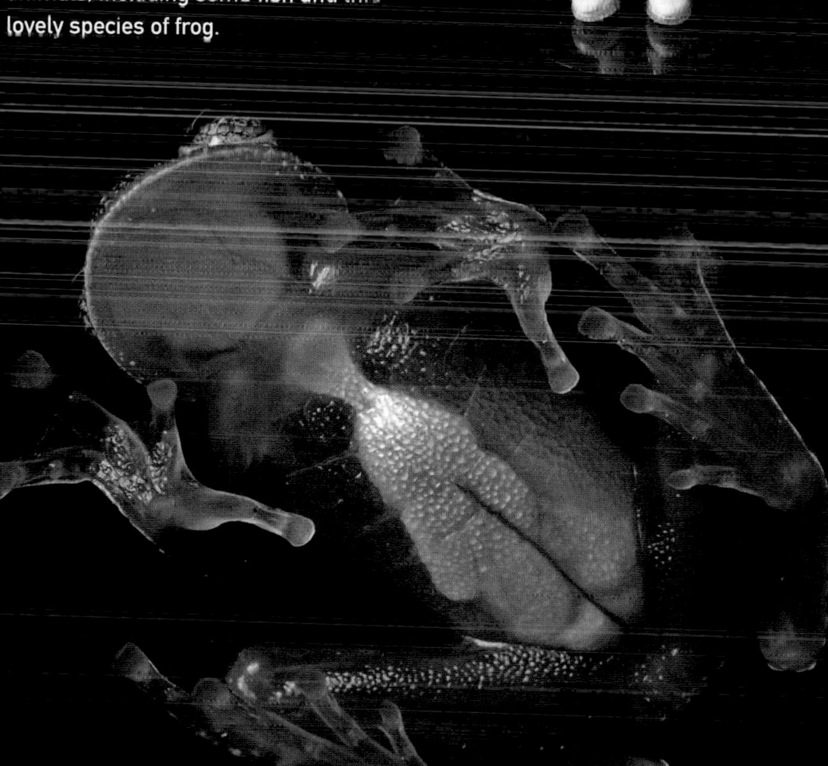

Marking Time

OK, ENOUGH ABOUT prisons: there are plenty of other clear things in the world. There's a long history of clocks that were built with their inner workings laid bare. For many centuries there were no more precise or beautiful mechanisms than the fine clocks of their day, so it's no surprise that their makers wanted to show them off. In "Clocks" you can learn all about the working parts of these beautiful old "skeleton" clocks.

I love this see-through clock because, although see-through clocks are an ancient art, this one is completely modern. First, it's made of clear acrylic instead of cut-away metal and glass. Second, it runs on electricity instead of a spring or hanging weight like clocks of yore, and third, it uses a light to project an image of the clock face on the wall.

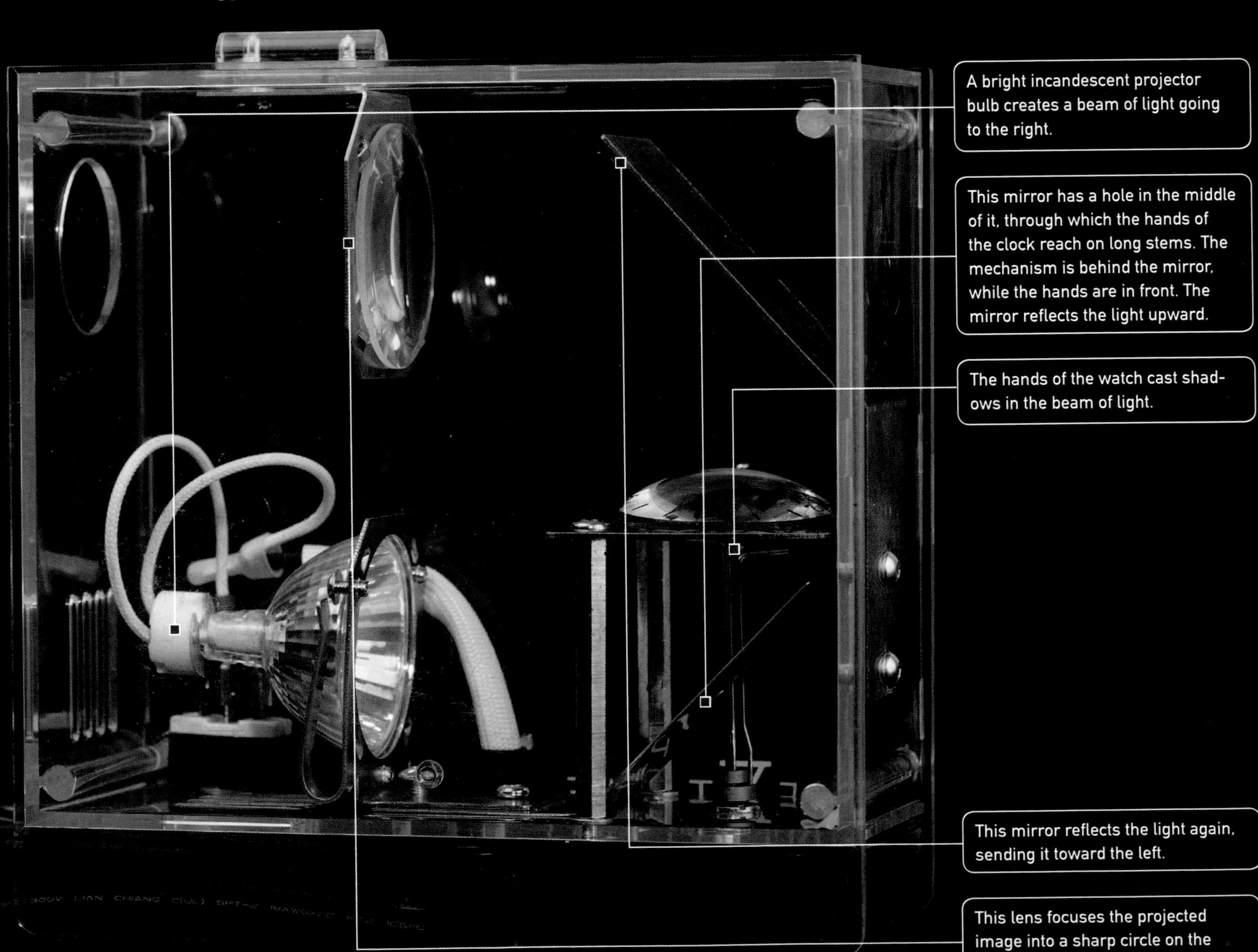

This "skeleton watch" is actually very small, maybe an inch (2.5 cm) in diameter, but it's inside a 4-inch (10 cm) solid glass sphere that acts as a magnifying glass to make it look like it fills the whole ball. It's a common type of clock sold in flea markets in China, which is where I bought this one. When it came time to fly home with it, the very nice Chinese security agent pointed to it questioningly in the x-ray of my suitcase. *It looked exactly like a cartoon bomb complete with fuse sticking out the top.* She must have seen this dozens of times before, because she didn't even bother opening the suitcase after I told her what it was.

It turns out Chinese airport security is actually pretty chill. I didn't think to ask for a picture of the bomb-clock x-ray, but when they pulled me aside again on a subsequent trip, they happily gave me permission to take a snapshot of this truly beautiful x-ray image of one of my five suitcases full of clocks. Consider it a preview of what you'll see in "Clocks." (Even more evidence of their chill: in five maxed-out suitcases of clocks, the *only* thing they wanted to see was the dark cylinder in the lower left corner of this picture. It's the heavy marble base to the self-pendulum clock on page 111.)

These two clear plastic wristwatches are quite similar but arrive at their design from opposite directions. The one on the top is clear for style reasons: it's cool to wear. The one on the bottom is a prison watch made for wearing when you're cooling your heels in the slammer.

called a U-trap in their drain, typically located directly underneath the sink basin. The drain pipe goes down from the sink, then turns around in the U-trap and heads back up for a few inches, then turns around again and heads down and off into the great beyond of the sewer. The little dip keeps a small pool of water trapped in the bottom at all times. Why? Because the great beyond of the sewer smells *really bad*. Without the trap, there would be an open path for sewer gases to come up through the drain. (This is one reason houses left alone for a long time can smell bad. After a few months the water in the drain traps will evaporate, allowing sewer smells in. The solution is to run every tap for a minute and flush every toilet at least once every few months.)

Sink traps are where most clogs occur, and with this one you can not only *see* the clog, you can turn the handle to sweep it out with a paddle. Or, if you see your wedding ring in there, turn it the other way to push the ring into position to be fished out from above.

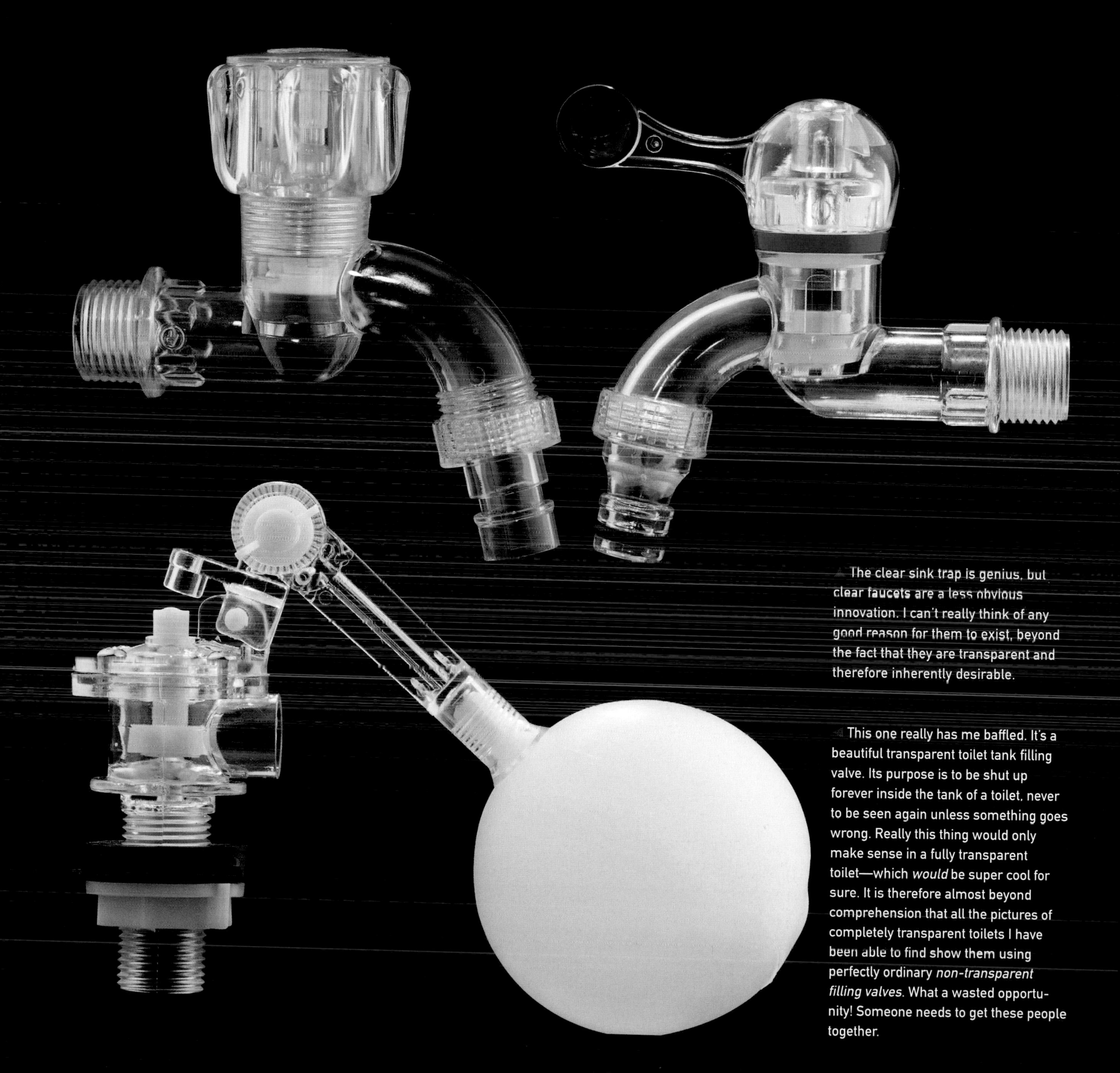

▲ The clear sink trap is genius, but clear faucets are a less obvious innovation. I can't really think of any good reason for them to exist, beyond the fact that they are transparent and therefore inherently desirable.

◄ This one really has me baffled. It's a beautiful transparent toilet tank filling valve. Its purpose is to be shut up forever inside the tank of a toilet, never to be seen again unless something goes wrong. Really this thing would only make sense in a fully transparent toilet—which *would* be super cool for sure. It is therefore almost beyond comprehension that all the pictures of completely transparent toilets I have been able to find show them using perfectly ordinary *non-transparent filling valves*. What a wasted opportunity! Someone needs to get these people together.

Transparent Everything Else

THERE ARE COUNTLESS one-off clear objects made as art pieces, or as trendy custom-made fixtures for overpriced apartments in New York City. Some of them are pretty nifty, but I'm much more interested in mass-produced clear objects, because generally a lot more thought and design work go into them. If you're planning to make millions of something and sell them for a few dollars each, you have to put serious work into optimizing the design. It's in these common objects that true genius is more often visible.

In many cases this hard and expensive design work may have been done mainly with the intention of producing ordinary non-transparent versions of the products. But a mold is a mold, and at some point the designers decided to do a production run with clear plastic instead.

Outside of the prison and plumbing worlds, the office seems to be another hotbed of clear things. Maybe people who have no window in their cubicle find comfort in at least having a window into their stapler. I'm sure the stapler, pictured here, is made from the same molds as normal black or red ones, but there's no more cost in making it clear. This one came from a dollar store where everything costs literally one dollar.

Clear hole punch

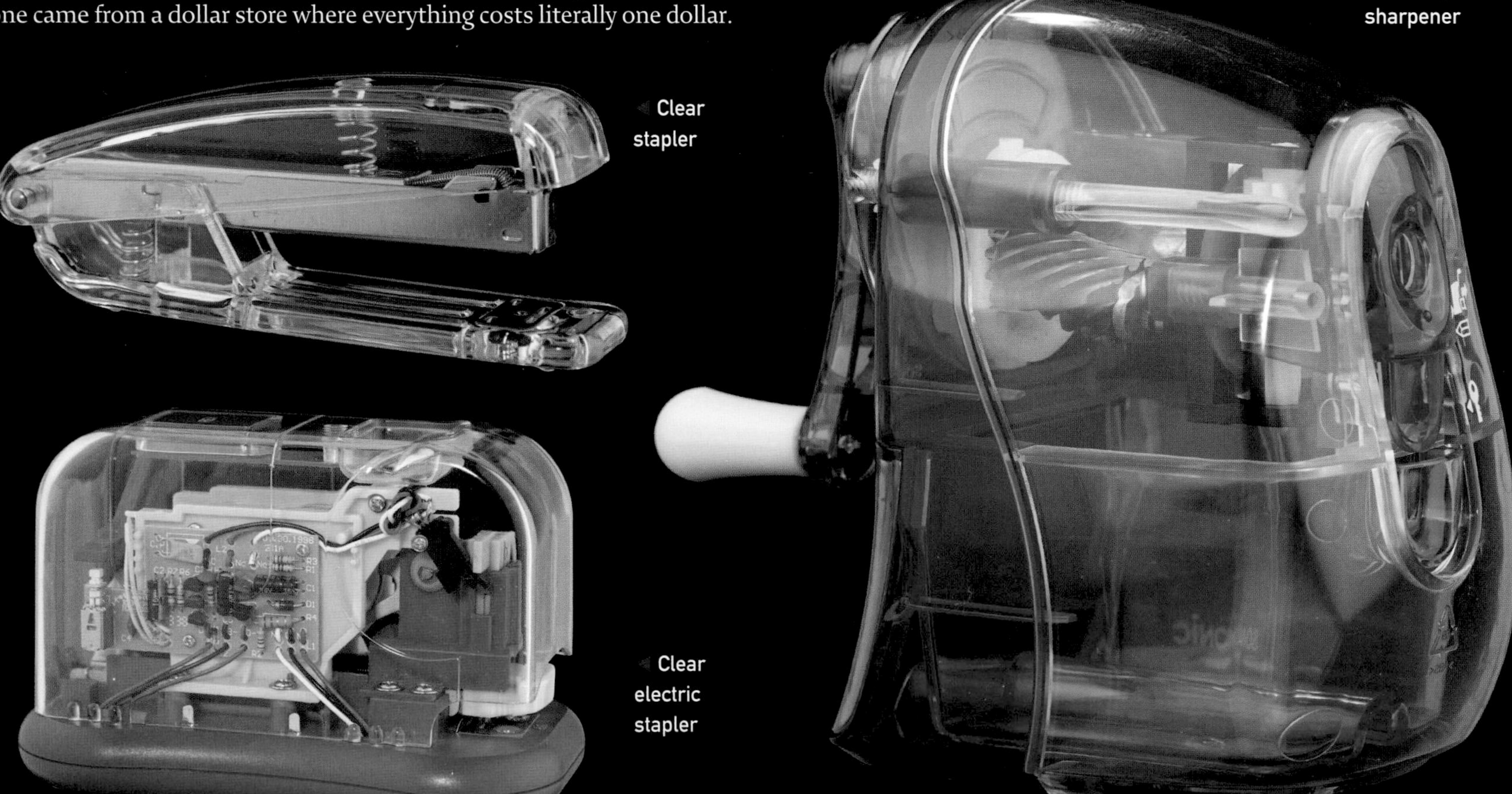

Clear stapler

Clear electric stapler

Clear pencil sharpener

▼ This clear plastic chair is surprisingly sturdy. I was afraid to sit on it at first, but now I'm just afraid to scratch it. (This is the one argument against certain clear objects. While they look great at first, once they are scratched they can look pretty ugly.)

▶ Scratching is not a problem with this *slightly* more expensive table and chair set, made of ornate cut glass of the kind you might normally expect to see in the shape of a fancy flower vase. For a time in the mid to late 1800s this stuff was popular with the wealthy families of India. An entire glass furniture industry in England was supported by the maharaja's taste for fancy glass furniture. (These priceless examples are from the Corning Museum of Glass.)

This thing bothers me because I've lived long enough and built enough "cool things" to know that everything gets dirty and then it's not cool anymore, it's just dirty. A lot of modern architecture suffers from this problem: looks great at first, then pretty soon it's just shabby. This object is seriously asking for trouble: clear acrylic looks good only when it's dust-free, and this is an [expletive deleted] *air purifier* (technically an ozone generator). Its very function in life is to suck dusty air through itself and make itself look shabby.

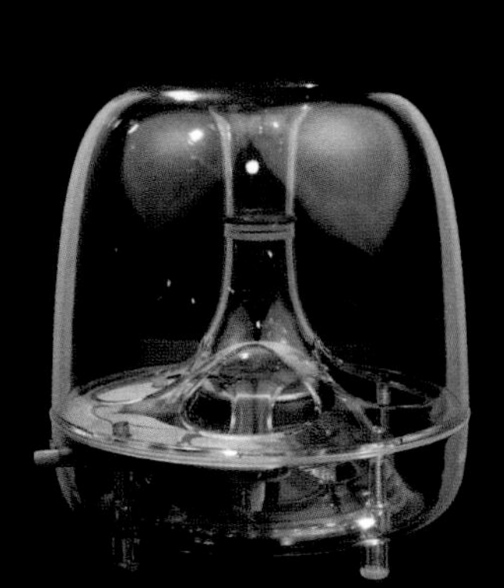

This is a transparent apple peeler. I'm including this item only because it's got such a lovely collection of visible gears. Clearly these people understood the potential of a clear case for their product.

These computer speakers from the mid-1990s have always been a favorite of mine. They sound pretty good, and you can see exactly how the woofer (low-frequency) base unit channels sound through a series of air chambers.

There seem to be two kinds of clear violins: very expensive professional models for serious, hyper-trendy musicians, and this kind, which you get on eBay for cheap. As is so often the case, the cheap version has more features, in this case a cool blue LED that lights up when you play it.

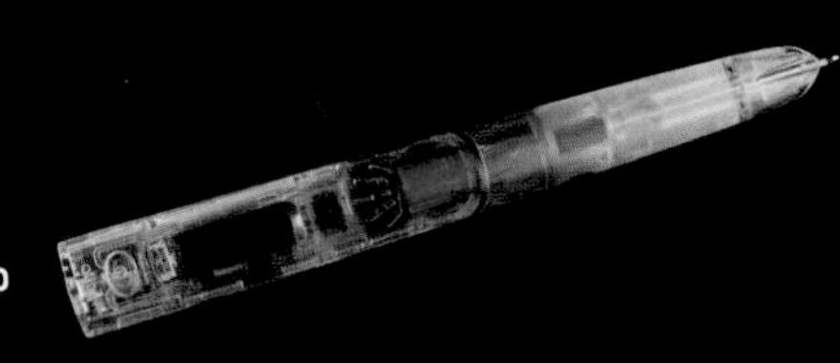

This is called, by its inventor, a "pen-top computer." That's like a laptop, except instead of being a computer on top of a lap, it's a computer on top of a pen. You write on special paper and it tracks everything you write, storing it for later transcription. You can even do things like draw a calculator and then use the pen to tap its buttons, with the answers delivered by voice. Interesting idea. Clever technology. But included here only because it's got a clear case.

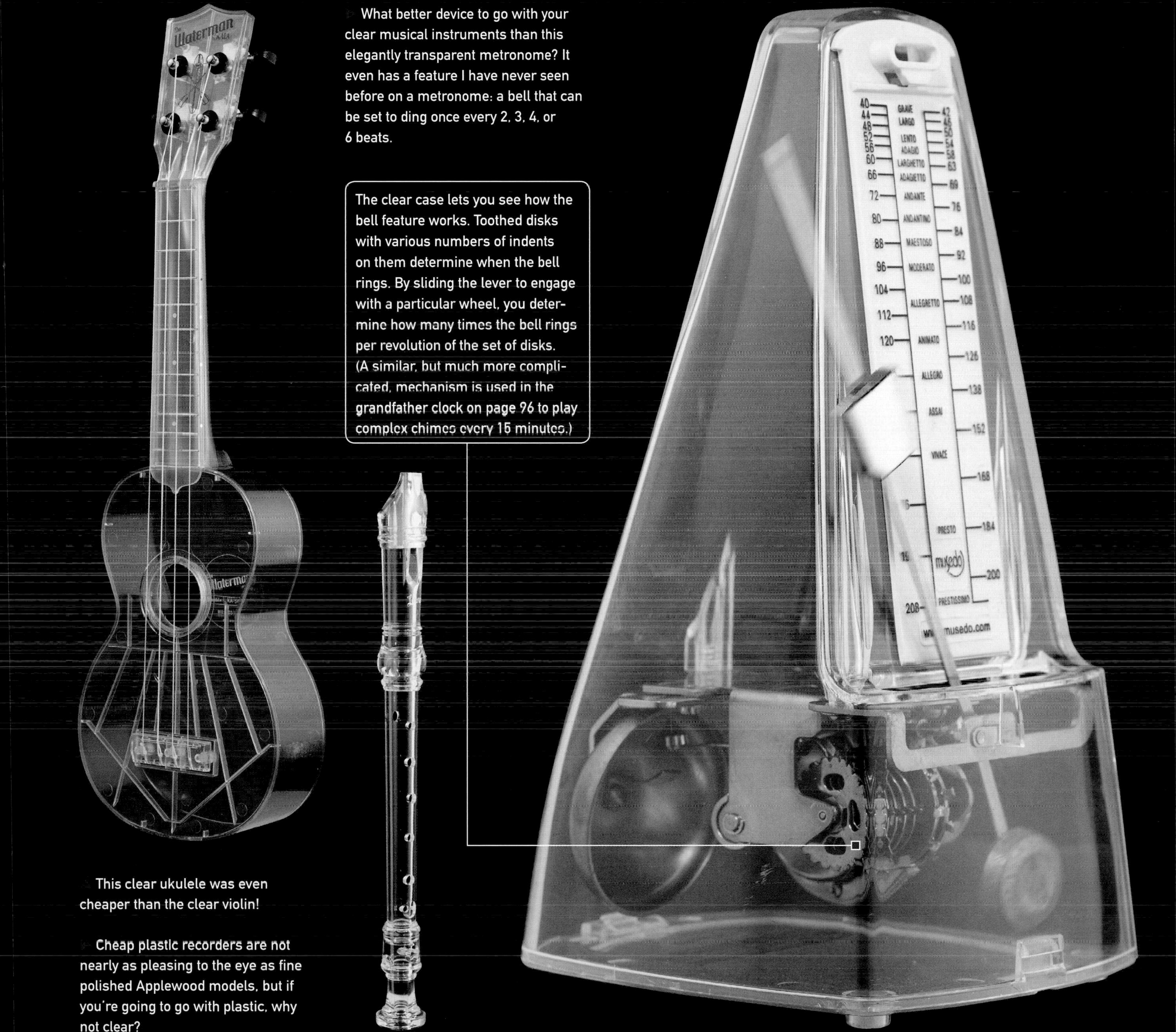

What better device to go with your clear musical instruments than this elegantly transparent metronome? It even has a feature I have never seen before on a metronome: a bell that can be set to ding once every 2, 3, 4, or 6 beats.

The clear case lets you see how the bell feature works. Toothed disks with various numbers of indents on them determine when the bell rings. By sliding the lever to engage with a particular wheel, you determine how many times the bell rings per revolution of the set of disks. (A similar, but much more complicated, mechanism is used in the grandfather clock on page 96 to play complex chimes every 15 minutes.)

This clear ukulele was even cheaper than the clear violin!

Cheap plastic recorders are not nearly as pleasing to the eye as fine polished Applewood models, but if you're going to go with plastic, why not clear?

SOME CLEAR THINGS are made as sales tools. This clear sewing machine was meant to be displayed in a sewing machine store so people could see the clever gears inside. This style of salesman's sample has become less popular now that very few machines have clever gears inside. Clever computer chips don't really have the same attraction when made clear, since the working parts are in many cases smaller than the wavelength of visible light, rendering them inherently impossible to see even with the most powerful light microscope.

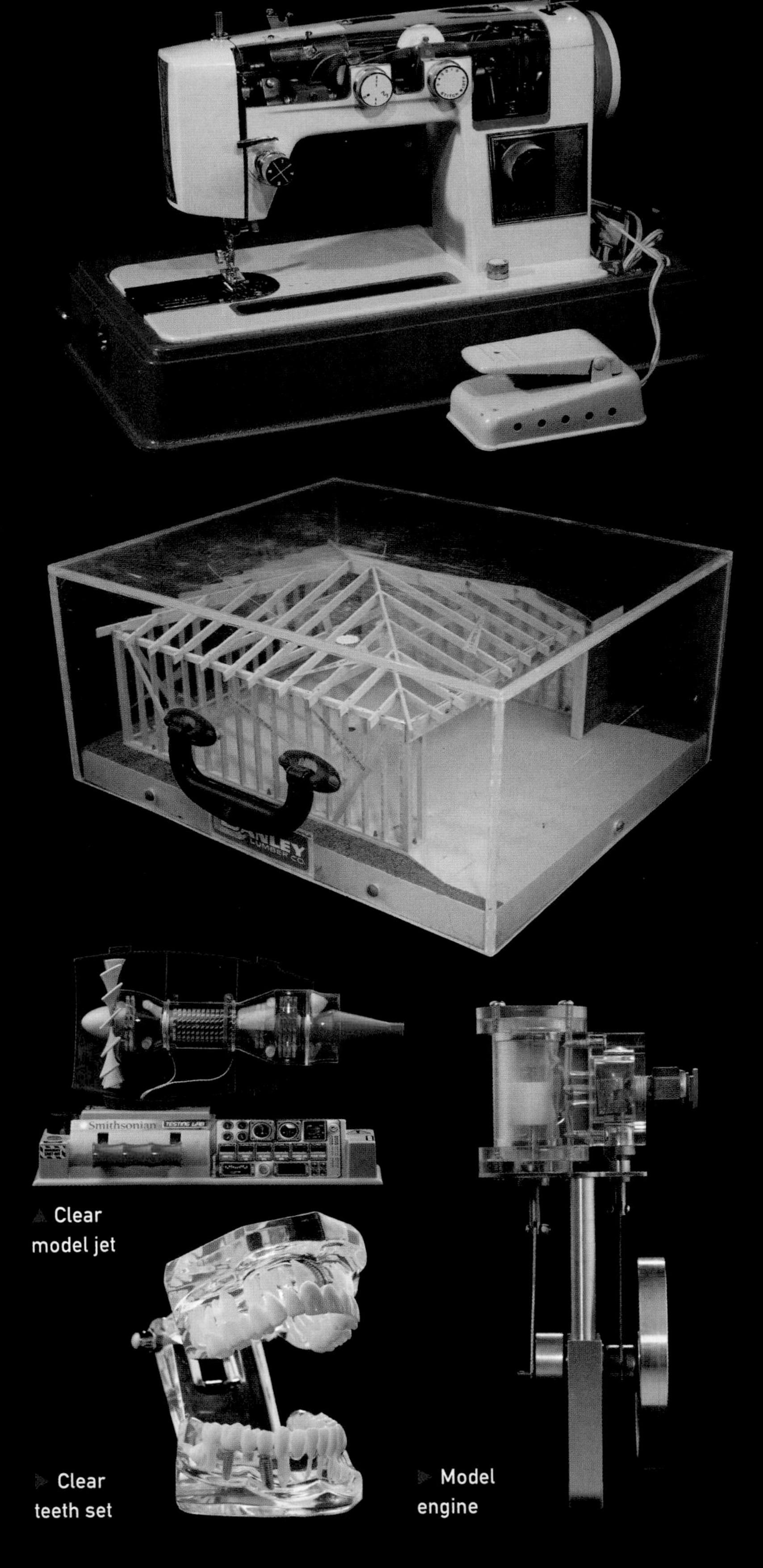

Clear sewing machine

Why would you want to wear a transparent hardhat at your construction job? Why, to show off your *fabulous* hair, of course! This one is made of polycarbonate plastic, the proper impact-resistant material for this kind of protective helmet. It is a real working hardhat.

This salesman's sample is a model of an entire garage in convenient miniature, skeletal form.

BEYOND PRACTICAL CLEAR things exists the world of *models*. These things are not practical, but are not just toys either. The lock is a practice lock designed for people who want to learn lock picking skills (for example, people who want to become professional locksmiths or bicycle thieves). The engines are educational models meant for learning about these devices. The teeth are for dentists to show patients what's in store for them when the drill comes.

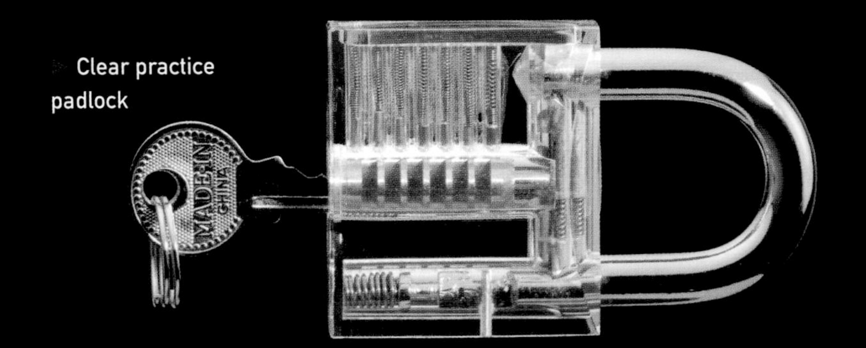

Clear practice padlock

Clear model jet

Clear teeth set

Model engine

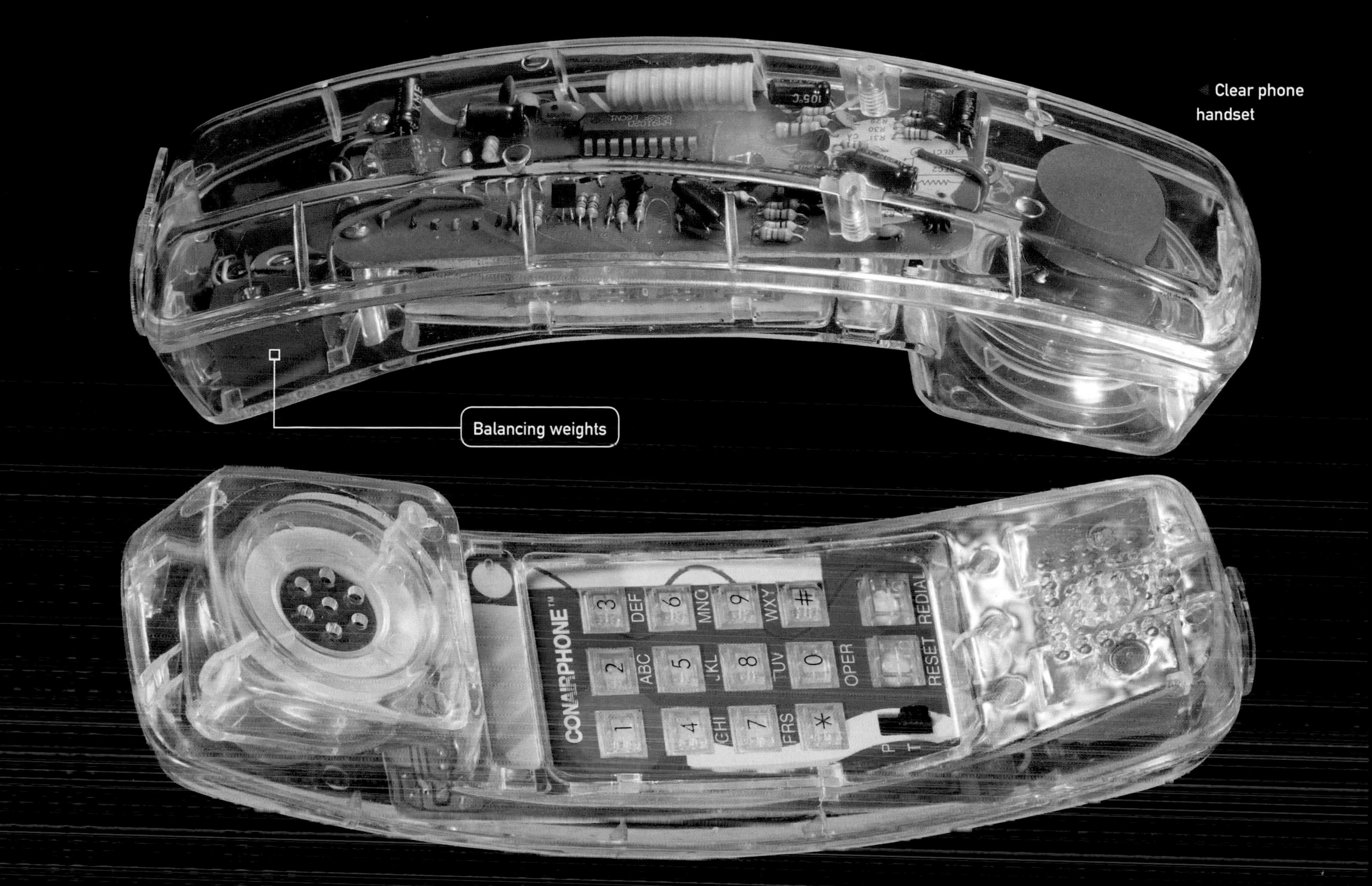

WE STARTED THIS chapter with a clear telephone, and we'll end it with the same phone, because it's a good illustration of what the rest of this book is all about: understanding the inner logic of things.

All manufactured objects, unlike living things or geological formations, are the products of human minds. Someone thought about how to make everything fit together. By studying how something is built, we can learn a lot about the person who made it. What motivated them? Were they trying to make something cheap, or good? Did they care more about how it looked, or how it felt in the hand?

See those two red angular blocks carefully packed into the bottom of the phone handset? Those are *ballast*. They are literally small lead weights, which together weigh about as much as a whole modern smartphone. Their purpose is simply to make the handpiece on this landline phone feel better balanced, more substantial, and less "plastic-y" in the hands. How priorities have changed! These days manufacturers struggle to shave off another gram or two from the weight of their ultrahigh-resolution, video-playing phones, but in a simpler time they intentionally added nearly 80 grams of deadweight to the handset because the designer wanted it to have a solid, quality feel to it.

After reading this chapter it may seem to you that you can get just about anything with a clear case, but sadly the devices made with clear cases are really only a tiny fraction of all the things in the world. Rest assured, though, that opaque things are just as interesting on the inside as transparent things. You just have to take them apart and poke around inside. And taking apart anything you can get your hands on is *highly recommended*—just make sure, before you take anything apart, that it's unplugged, doesn't have any residual charge in its capacitors, and you either know how to put it back together again, or have a younger sibling or dog you can blame the mess on.

In the following chapters I'm going to show you some of my favorite things—clear, opaque, and otherwise. I'll try to tell you how they work, why they exist, and why they matter. I hope you enjoy these things as much as I do.

LOCKS

LOCKS ARE ABOUT exclusion. They are for keeping people, or bears, out of places you don't want them. But of course there must also be a way to let in certain people, or certain bears. These two simple and contrary funtions—keeping out and letting in—are the purpose of every lock and the thing that makes them so complicated and fascinating.

In order for a lock to be effective it has to be able to differentiate who or what to let in or keep out. There are many ways to create a dividing line between the in-crowd and those left out in the cold.

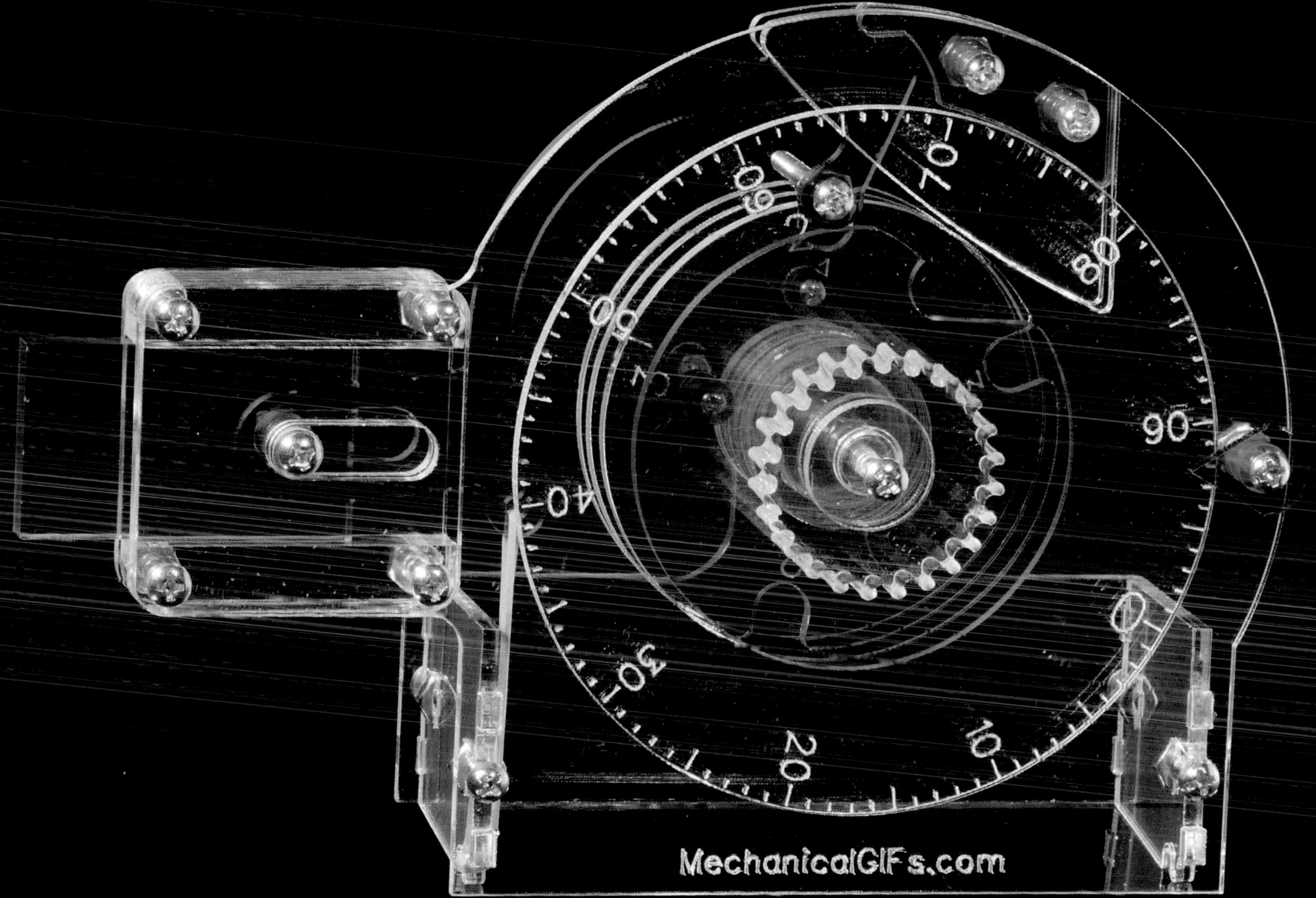
MechanicalGIFs.com

THE SIMPLEST TYPE of lock is one that functions like the cap on a childproof medicine bottle. It tries to tell the difference between adults and children by requiring a combination of knowledge ("push down while turning") and strength (you have to push and turn fairly hard). In theory, only adults have the knowledge and strength necessary to open the lock.

But strength-based locks are really not very good at their function. Everyone knows that if you, as an adult, have a childproof bottle you can't get open, it's best to find a child to open it for you. They're clever about this sort of thing, and they usually don't have arthritis.

In this same category is this childproof cabinet door lock, which has a decoy button to trick babies and toddlers who have figured out how to infiltrate regular babyproof locks. The button on the front of the case—the one that looks like it should open the lock—actually does nothing. The real buttons that open the lock are barely visible on the sides of the case. Both children and adults have trouble opening this lock if they don't know the trick.

Childproof medicine bottle

Childproof lock with decoy button

These are old versions of what amount to not much more than childproofing locks. They are ancient Chinese padlocks (actually modern reproductions) that use a "key" that is just a flat piece of metal with a slot in it. When you push the key in, it squeezes together two leaf springs, releasing the latch. No secret—if you've seen one, you've seen them all, and could make a key from a scrap of metal to open any one of them.

ANY LOCK THAT can be opened using only general knowledge of how it works is always going to be very easy to defeat. A really good lock requires some piece of secret information that is unique to that particular lock.

Below is an example of a lock that requires just about the simplest possible form of secret information: where a hidden thing is located. To open this latch, you need a magnetic key. Any magnet will do, but you need to know exactly where to hold it to engage the latch on the other side of the door. That's easy to do when you can see the latch, but harder when it's on the other side of a closed wooden door. Not exactly high security, but good enough to defeat the usual suspects (small children and bear cubs).

But what if you just slowly move the magnetic key all along the edge of the door until it opens? That's a way to figure out the secret information (the location of the latch) without knowing it ahead of time. In other words, it's a way to pick this kind of lock, and it's not very hard!

Old houses in the United States invariably have "mortise" locks, which are opened using a "skeleton key." The keys are not all the same, so you do need some specific secret information about the individual lock. But they are usually the same for all the locks in a given house. Inside the lock there is a half-hearted attempt to make other keys not fit, but really these locks are not meant for security. Their purpose is just to make you think about whether you should really be opening that door, not to stop you from opening it if you are determined.

On the next page we'll see the first type of lock that would actually be a challenge for the average person to get through.

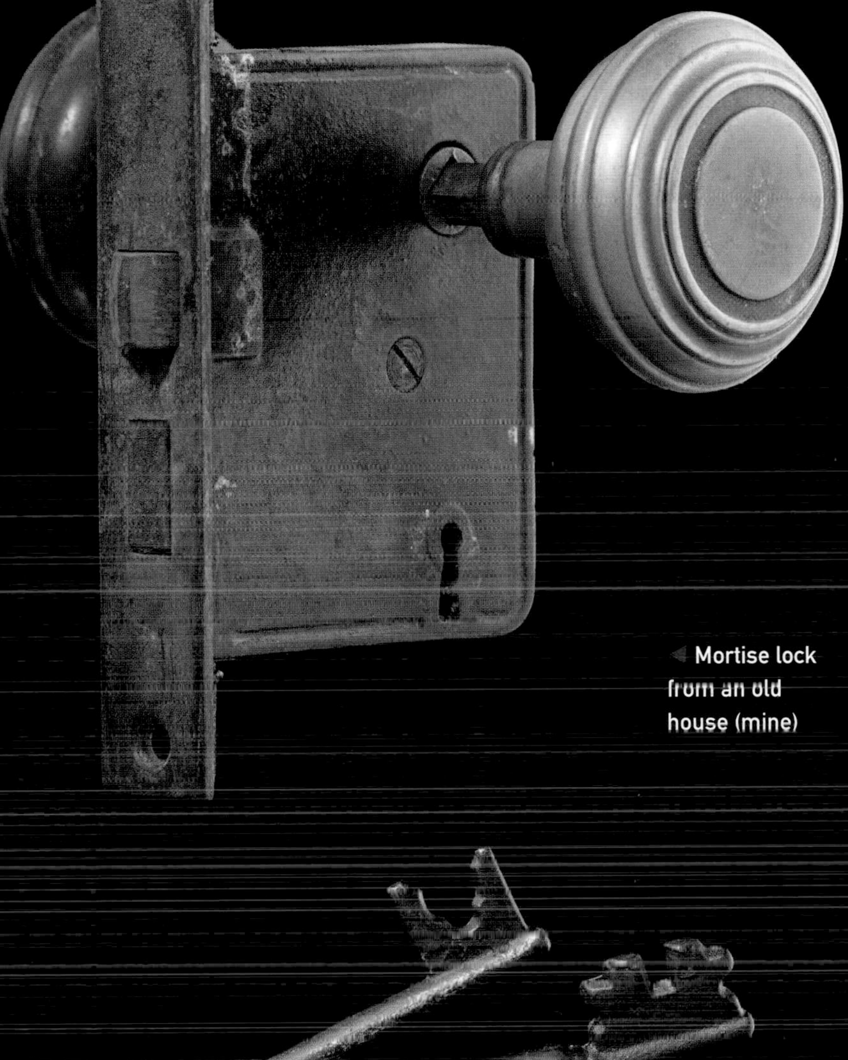

Mortise lock from an old house (mine)

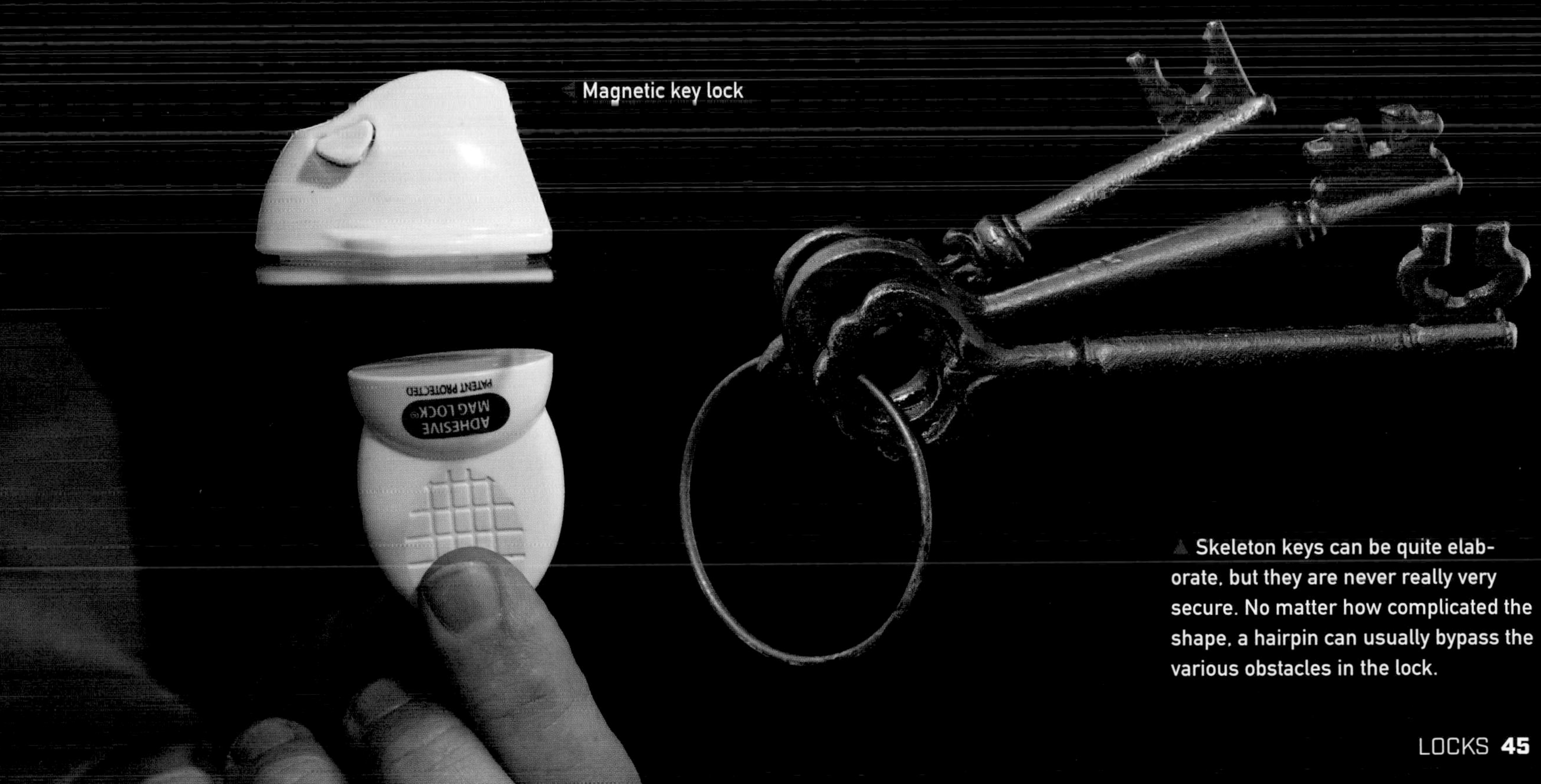

Magnetic key lock

Skeleton keys can be quite elaborate, but they are never really very secure. No matter how complicated the shape, a hairpin can usually bypass the various obstacles in the lock.

Locks with Proper Keys

THESE CRUDE OLD locks are the first examples we've seen that require true unique-for-each-lock secret information to open.

You can see the mechanism clearly in this lock from an old Russian jail cell. When the key is inserted into the lock and turned, it needs to either lift, or not lift, each one of a set of plates inside the lock, so the pin can slide between the gaps in all the plates at the same time, and spring the lock open. If one of the cuts in the key is too low, the plate at the location of that cut won't be lifted high enough. But if the cut is too high for a certain location, the plate will be lifted too high. Either way, the pin won't be able to move and the lock won't open. So the exact shape of the key is the secret information needed to open the lock.

But this lock is still quite easy to get open even without the key. All the plates are within easy reach of the keyhole, and you just have to press them up and wiggle them a bit until the lock pops open. We can do better.

Old Russian jail cell lock

HERE WE COME to the first "real" lock, and it's the most common kind by far: a pin-tumbler lock. If you don't have the key, it is quite difficult to open this lock. Now, when I say "quite difficult," that's relative. A professional locksmith can open just about any pin-tumbler lock in somewhere between seconds and minutes.

Pin-tumbler lock

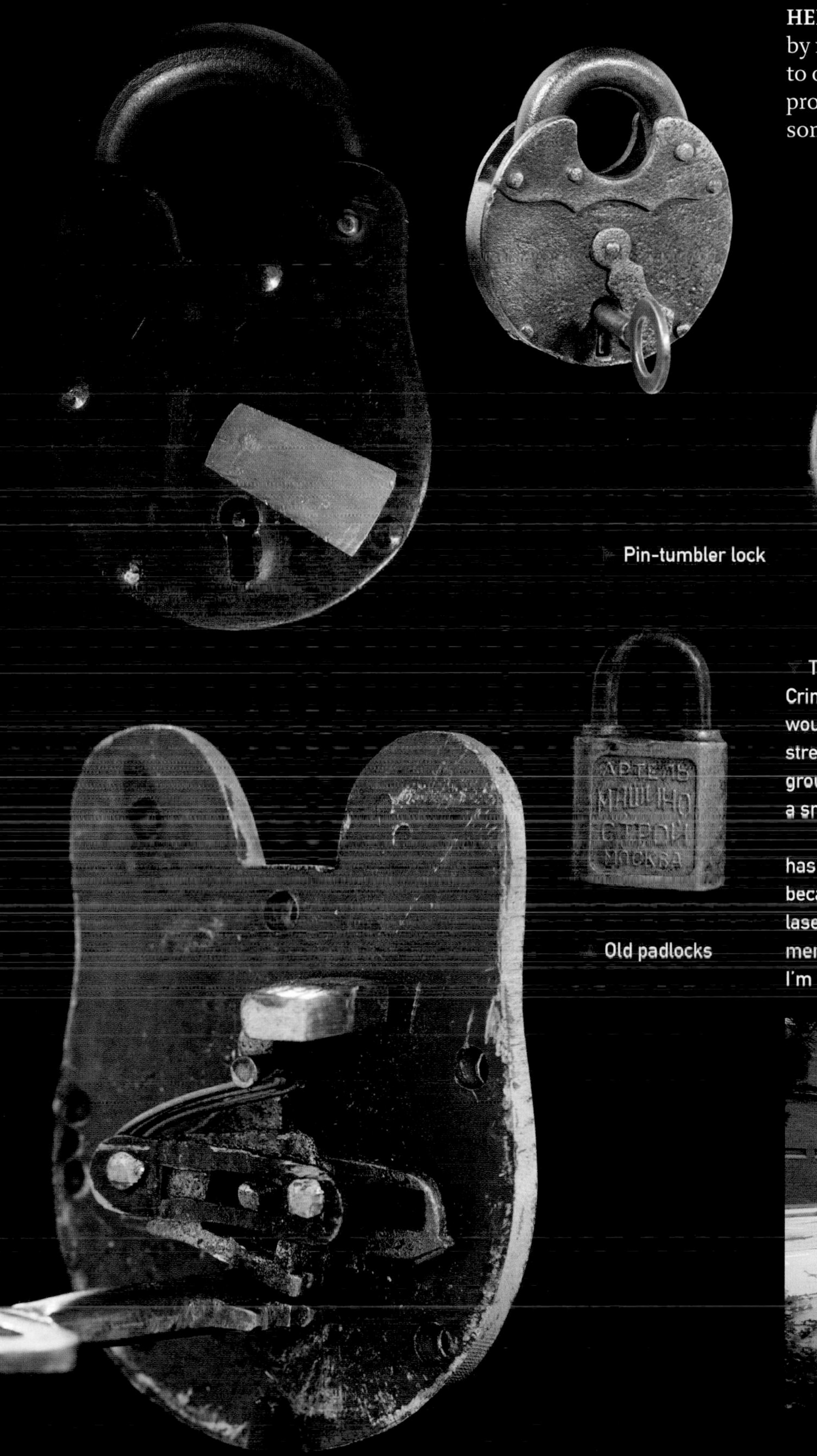

Old padlocks

This more serious pin-tumbler lock is both criminal- and bear-resistant. Criminal-resistant because it's hard enough to pick that it's unlikely a criminal would try (especially since this lock happens to be located literally across the street from the county jail and sheriff's office, which you can see in the background.) And bear-resistant because it's embedded in a thick steel door with only a small window made of inch-thick bulletproof glass.

This is the door to my laser cutter workshop, which is built like a bunker—it has this high-security steel door and a row of bulletproof windows. No, it's not because I'm paranoid, it's because the cheapest space available for housing my laser cutter was a former drive-through bank. It also has water wells in the basement, so as long as we can get in sufficient food supplies, this is definitely where I'm riding out the zombie apocalypse.

Anatomy of a Padlock

PADLOCKS ARE A common place to find pin-tumbler mechanisms. Inside a pin-tumbler lock are pins and a round chamber that turns when you turn the key. You might assume this chamber is called the tumbler, but in fact it's called the barrel of the lock. There is no part of a pin-tumbler lock that is called the tumbler. Don't ask me why. Maybe it's because the pins tumble in the barrel when you turn the key?

Anyway, the pins prevent the barrel from rotating until they are all aligned in exactly the right way by the key. How this works is a bit hard to see even in these visible padlocks, so we're going to look at a model first, then use what we've learned to understand some real locks.

There are a variety of specially made practice locks, including the clear acrylic one on the left, and this real lock that has had enough metal carved away to let you see the pins and barrel inside. These are made to practice lock picking and other locksmithing skills: it's a whole lot easier to pick a lock when you can see exactly what you're doing to the pins. As you get better at lock picking, you learn to do it by feel, and you no longer need to see the pins.

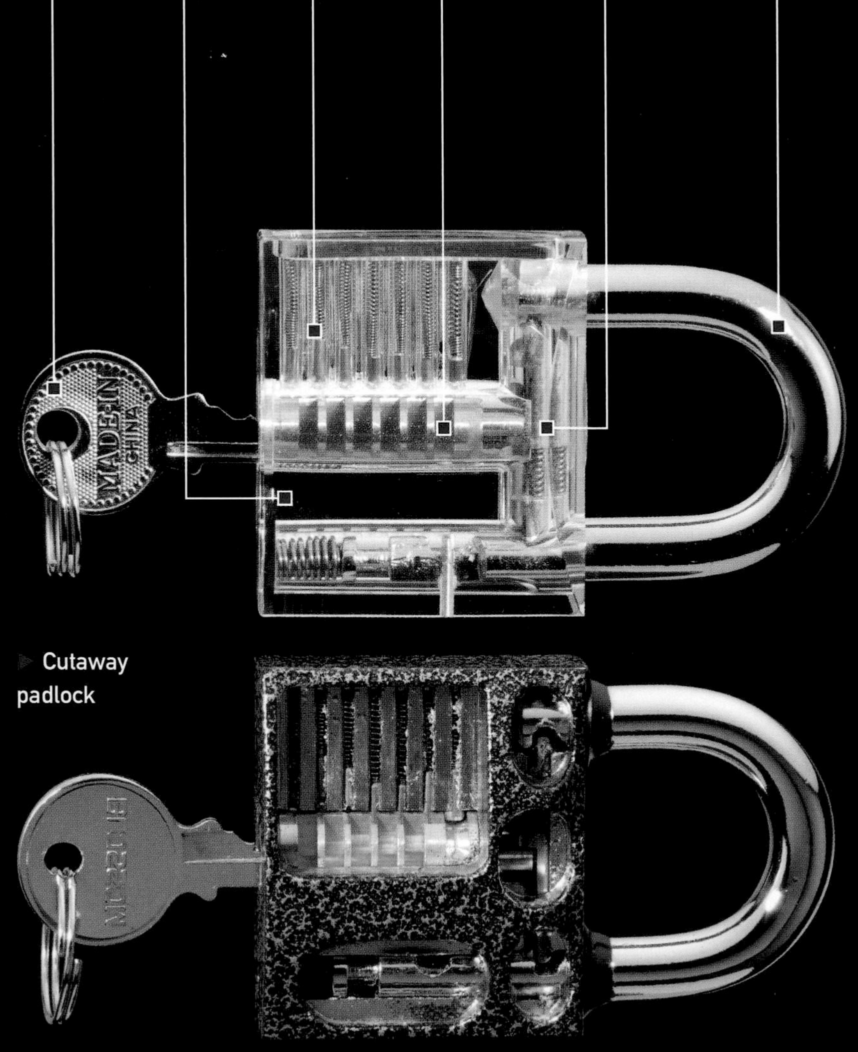

▶ Cutaway padlock

▼ Here's a simplified model of the inside of a pin-tumbler lock. It has five pins, each of which comes in two parts, an upper half (amber) and a lower half (green). A spring pushes each of the pins down from the body of the lock into the barrel below. The pins cross the boundary between the body and barrel, preventing the barrel from moving relative to the body. Notice that, while the amber parts are all the same length, each of the green parts is a unique, different length. When we slide the key in, we'll see why.

1. As you insert the key into the lock, it pushes the pins up against their springs. They ride up and down on the hills and valleys along the top of the key.

2. When the key is pushed in all the way, the length of the green parts exactly match the height of the key cuts, so the tops of all the green parts line up with each other, exactly at the position of the split between the body and the barrel. This is of course no accident.

3. With all the pins lifted up exactly the right amount, the barrel is no longer trapped in place, and it can slide or rotate relative to the body. In a real lock the barrel rotates when you turn the key. Because this is a two-dimensional model of a lock, I've made it so the barrel slides sidewise instead. The principle is the same: the barrel can move only when the key has aligned all the pins.

4. If you put the wrong key in a lock, its hills and valleys won't match the unique lengths of each pin, and the pins will still block the movement of the barrel.

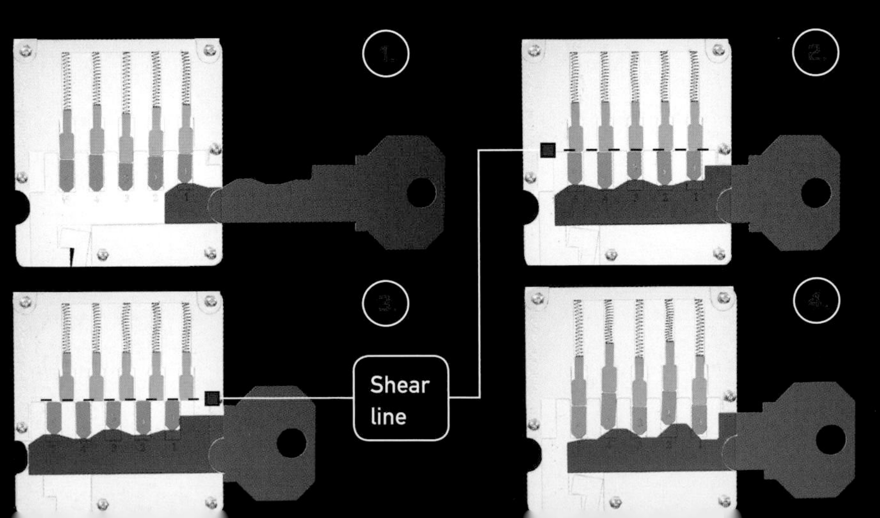

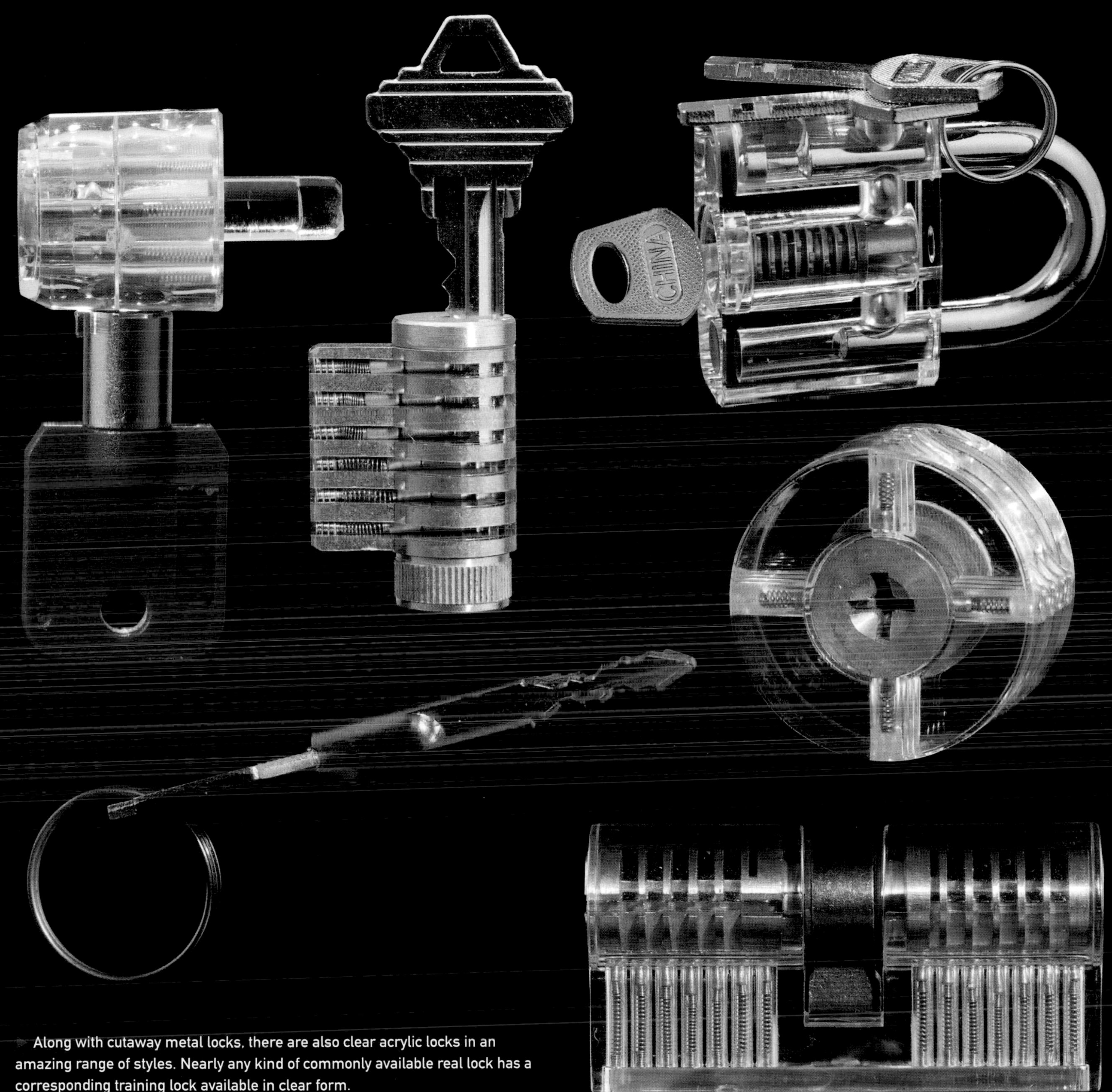

Along with cutaway metal locks, there are also clear acrylic locks in an amazing range of styles. Nearly any kind of commonly available real lock has a corresponding training lock available in clear form.

HOW TO PICK A LOCK

HOW DO YOU open a lock if you don't have the right key? You "pick" it, relying on the fact that no mechanical device is ever perfect. To pick this model lock, you start by applying gentle sideways force to the barrel, which won't move because the amber pins are blocking it. But not all the pins are actually stopping the motion: one of them will always be just a bit thicker, or one of the slots will be a bit narrower, or a bit out of line. That particular pin, called the binding pin, will be the one actually holding the barrel in place. All the other pins are ready to help hold the barrel, but are currently loose in their slots. But which one is the binding pin?

1. To begin picking the lock, you use a lock pick to gently push up on the first pin. As you push it up, one of two things will happen: either nothing if the pin is loose, or a very slight "click" if we got lucky and this is the pin that is preventing movement of the barrel.

2. Most likely the first pin you try will not be the one that is binding, and it will be loose in its slot. When you lower the pick, the pin will fall back into place. Then you need to move on to the next pin.

3. If the first one fails, you keep trying pins until you find one (the third one in this case) where you feel a tiny click and the barrel moves a tiny bit.

4. Because the barrel has shifted, this pin will not fall back down when you lower the pick. Now there will be a *new* binding pin: the second-tightest pin, which might be any one of the others.

5. In a few minutes—or seconds if you're good at it—you can find all of the binding pins, one after another, until you're down to just one pin holding up the tumbler.

6. Lift that pin with the pick and the lock will open. People who make locks try to make this hard to do, but there is no such thing as an unpickable lock (which doesn't stop manufacturers from claiming theirs are).

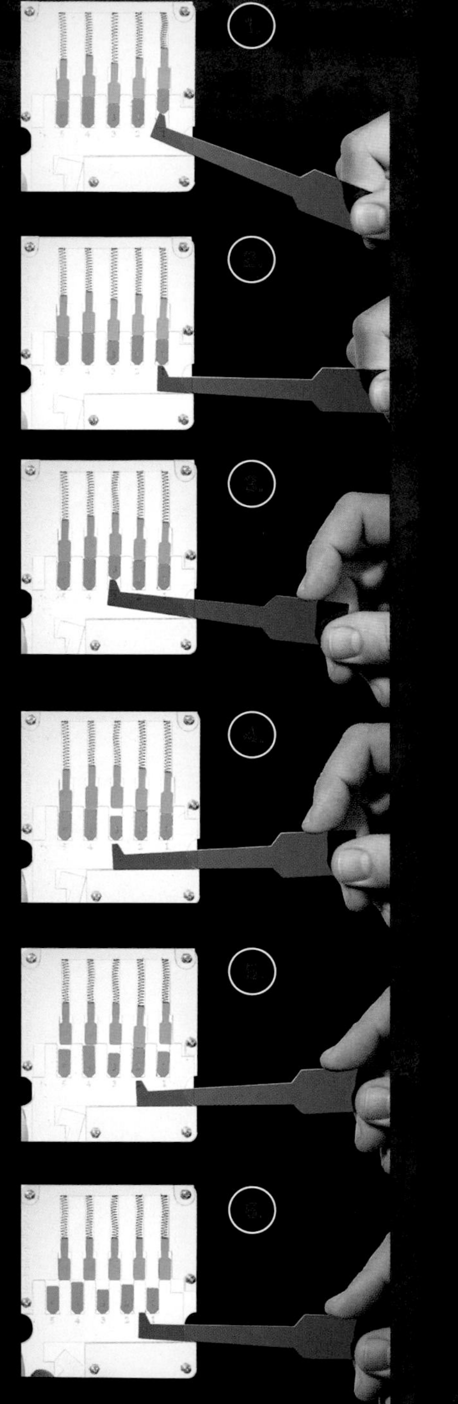

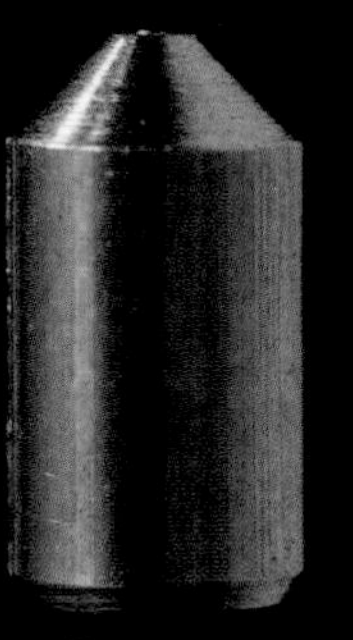

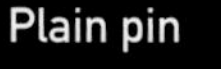

Plain pin

Serrated pin

Serrated pin

Spool pin

Spool pin

Spool pin

Many trick pins are available to make lock picking harder. A "high security" lock may have several different kinds of these in random different positions within the barrel. An experienced locksmith learns to recognize what each type feels like under the pick, and how to defeat each one.

THIS FANCY LOCK is widely advertised as "pick proof" and it certainly looks intimidating. To open it, you have to use its special double key, which has a gear mechanism inside that turns both keys at the same time. If you try to turn either half of the lock separately, it won't open, which would seem to make picking out of the question. YouTube lock-picking star "LockPickingLawyer" does it in two minutes and forty seconds: I will leave you to find the video and see for yourself how it's done.

It is a law of the universe that for every unpickable lock, there is an equal and opposite YouTube video showing how to pick it. Is this a bad thing?

Anyone who has ever locked themselves out of their house or car can confirm that it's actually very *useful* that there are people who can open nearly any lock in a few minutes, while charging you way less than it would cost to replace a broken-down door or smashed window.

I think of it this way: you want a lock that is harder to pick than is worth the time of a criminal who might want to pick it, but not harder to open than other, more destructive ways of getting into the house or car it is protecting.

"Pick-proof" motorcycle lock

PEOPLE SAY THAT if you have a car with a cloth roof, you should never lock the doors. If you do, instead of having a stolen radio, you'll have a stolen radio and a slashed roof—which costs much more to replace than a radio alone. The only lock you'd want on a convertible car is one that's easier to open than it is to cut through cloth, which basically means no lock at all.

The thing to remember about locks and crime is that, contrary to what you see in movies, most criminals are pretty incompetent. They are criminals because they don't have the skills to make an honest living (which is always more fun and satisfying than a life of hiding from the law and scraping by on stolen radios). Learning to pick locks well takes time and intelligence. If you have that, you don't need to be a criminal.

A professional locksmith can pick any ordinary lock in a few minutes, but they—being skilled professionals who can earn a good living opening locks legitimately—don't bother using their skill to steal things. They would rather live an open and honest life, holding their head up high in their community. People who don't feel that way are usually not smart enough to learn to pick locks.

So that's why it's actually fine that the lock on your door can be picked pretty easily. Except in extraordinarily high-security situations, an unpickable lock would just mean that you'd have to replace the whole door if you lost your key.

Processing Equipment

BEYOND THE SIMPLEST "childproof" locks or antique padlocks that open with the same key, all locks are information-processing devices: they require a secret code to open. In the case of a pin-tumbler lock, this code is carried mechanically by the key.

Cut depths are the secret codes to key locks.

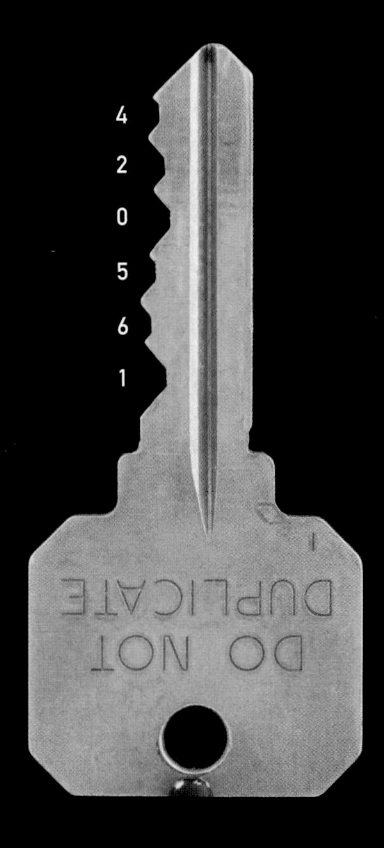

The secret code is the pattern of high and low spots along the length of the key, which is a mechanical way of communicating that information to the lock. For a given brand of lock, there is a table of standard cut depths, numbered typically from zero to ten. So you can define any key by listing the numbers representing the cut depths along the length of the key.

For example, this key has depths 1-6-5-0-2-4. These numbers are the exact equivalent of a PIN (personal identification number) code, like you would use with your debit card or cell phone. The only difference is that the numbers are physically encoded in a metal key, instead of mentally encoded in your head. That's good in some ways, bad in others, but in any case I think it's interesting that locks are an example of pre-electronic-age information-processing equipment.

This is a machine that's used to copy the information in a key. I've removed its covers so you can see the gears inside. A typical key-duplicating machine has a "feeler" that runs along the existing key, connected to a cutting wheel that moves in and out with the feeler, cutting a matching set of grooves in the new key. For a long time, this was the only way you could get copies of keys made, but now that we are living in the information age, people have figured out how to treat keys literally as pure information.

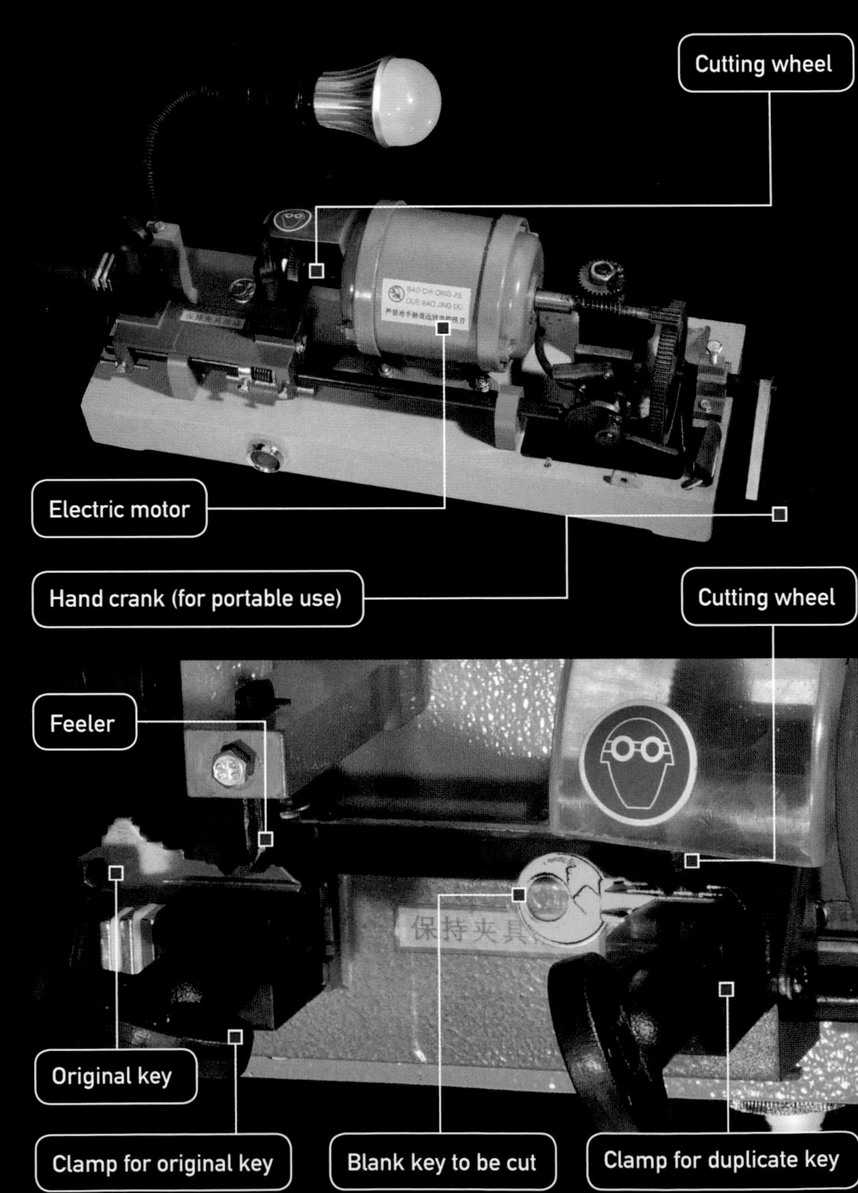

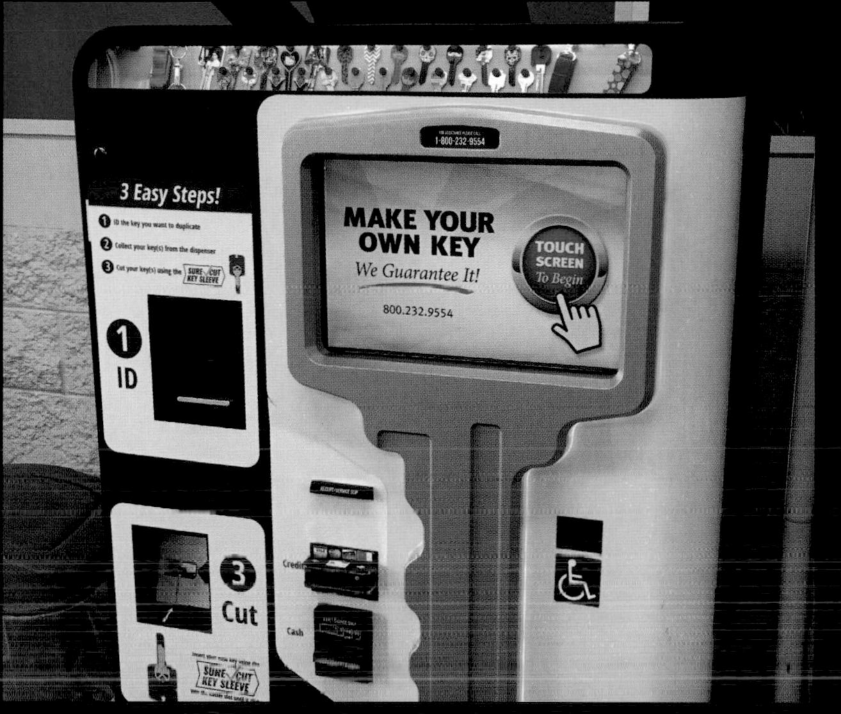

◀ Optical key-duplicating machine

▶ Taking digital reproduction a step further into the "maybe this isn't such a great idea" territory, we find internet-based key-duplicating services. They let you upload a photo of a key, and have a physical copy of that key mailed to the address of your choice. Now, they do promise to be ethical about this: you have to send clear pictures of both sides of the key on a white background. And you have to give them a valid credit card number.

But none of that is too hard to fake with Photoshop and a pre-paid debit card. The front and back images below are actually images of the key on the last page, photoshopped with the cut depths from the paparazzi-style picture above, and the "DO NOT REPLICATE" magically erased. So, OK, you can be a tiny bit worried about someone with a zoom lens using one of these services to break into your house. Maybe plant bushes around the front door?

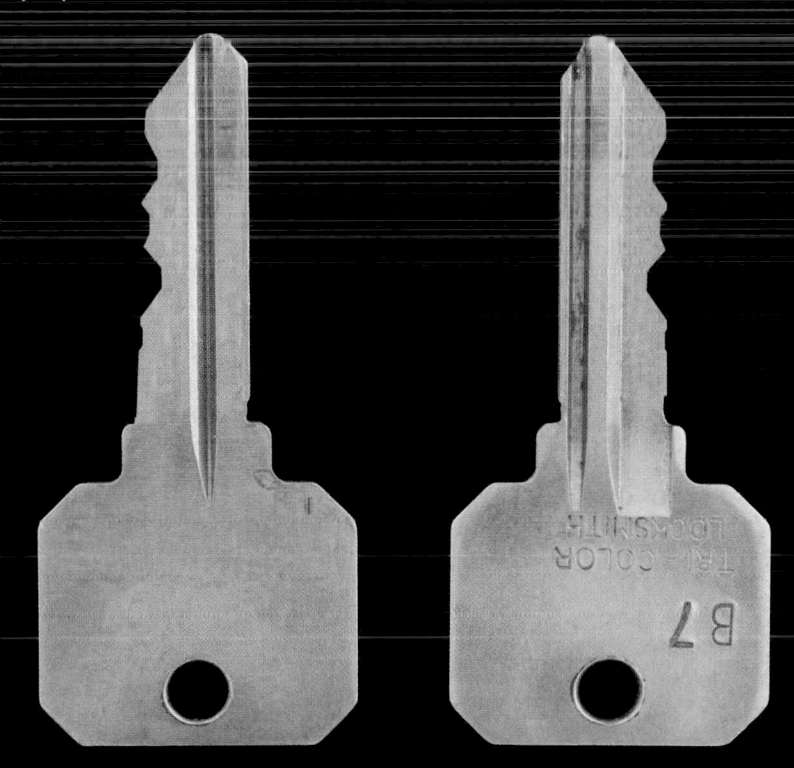

In theory, a photograph of a key contains all the necessary informa-ion to duplicate it. Should you worry about this and always keep your keys hidden? See the prior discussion of criminals being kind of dumb most of the time. Don't worry about it. It's very unlikely that someone s going to try to pull this trick on you.

Copying a key from a photo isn't likely to be done for criminal purposes, but the method is actually used in commonly seen key-duplicating machines: here is one at my local Walmart. Instead of using a mechanical feeler to physically follow the contours of the key, this machine scans it optically—basically takes a picture—and cuts a new key from this information.

A mechanical key-duplicating machine will faithfully re-create the highs and lows of the key you start with—which can be a problem if the original is worn down so that some of the heights are not quite right. If you make a copy of a copy of a copy, eventually the keys will stop working as mechanical errors add up.

But this machine knows the standard cut depths for each type of key it can make, so it can make the new key to exactly the right specifi-cation. As long as the original isn't so worn that it reads as an entirely wrong depth, the new key will be perfect, and you can make as many generations of copies as you like without degradation.

It's exactly the same principle as with sound recordings. If you copy an analog sound recording (LP record, cassette tape, or other such antique technology), the copy is never perfect, and after a few generations of copies it sounds terrible. But if you make digital copies of audio files, every copy is perfect. This key-copying machine takes mechanical keys into the digital age and turns them, temporarily, into pure digital information.

The Flimsy Case of the Master Key

PHOTO-DUPLICATING SERVICES are not the only security flaw built into mechanical key systems. In large commercial buildings with hundreds of doors that each have their own lock, there is usually a "master key" that can open any door. That way the janitors, maintenance people, or owners of the building can get into any room without having to carry literally hundreds of different keys.

If you lose the master key, that's a *big* problem. Anyone who finds it is able to open every single door. Even if you recover the key, you can't be sure it didn't get copied, so you may end up having to rekey every single lock in the whole building.

An even bigger problem is that each individual lock contains the information needed to create a new master key. Each and every person living in the building has everything they need, within their own door, to open any door in the building!

To understand why, let's look at how master-keyed locks work.

▼ Locks with master keys have their pins split into three parts (shown here in amber, blue, and green) instead of the usual two. That means there are two different points on each pin that will allow the barrel to slide or rotate, which in turn means that for each pin, two different keys can both align that pin at a height that will allow the lock to open.

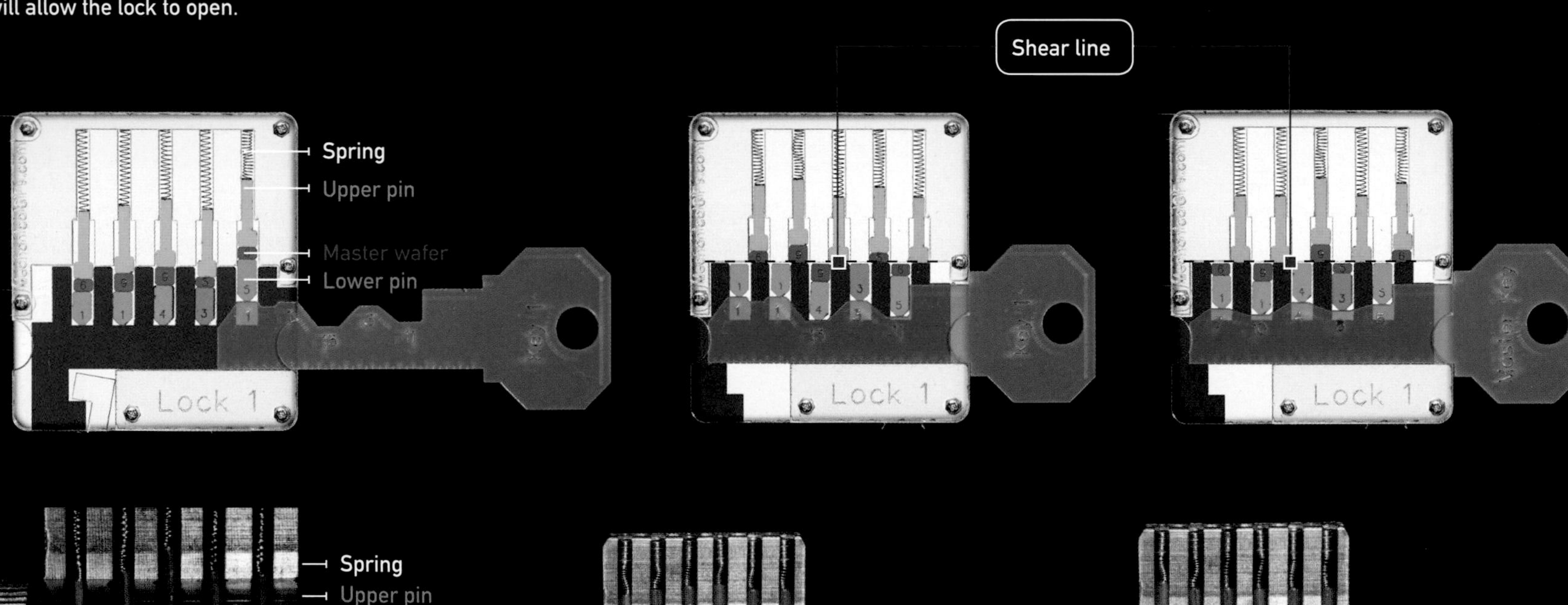

▲ With this key inserted, the lock opens with one set of splitting points all aligned.

▲ This completely different key also opens the same lock, by aligning the pins at their *other* splitting

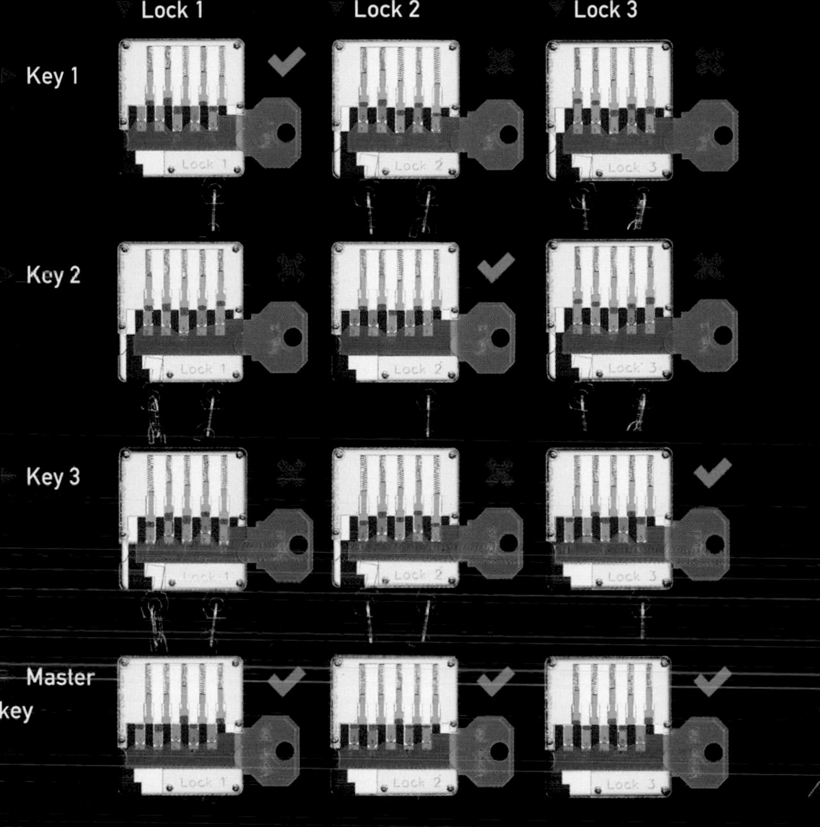

Here are three different locks. Each of them can be opened by its own individual key, and those individual keys won't open any of the other locks. As we saw on the last page, each lock can also be opened by a second, different key. By carefully choosing the heights of each of the two components to the pins, I have arranged it so that this second key is the *same* for all three locks. That's the master key: it is one key that can open all three locks.

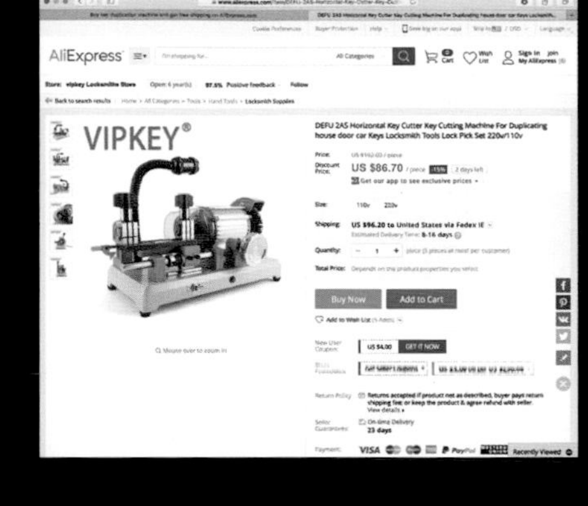

These days you can buy a key-duplicating machine for less than $100, no questions asked. It does not magically refuse to work on "Do Not Duplicate" keys. Combine this with easy, no-questions-asked access to lock picks, electric lock-picking tools, key blanks, instructional videos, and so on, and there's really no meaningful security left in keyed locks (beyond the very important fact mentioned earlier that criminals are nearly all pretty dumb).

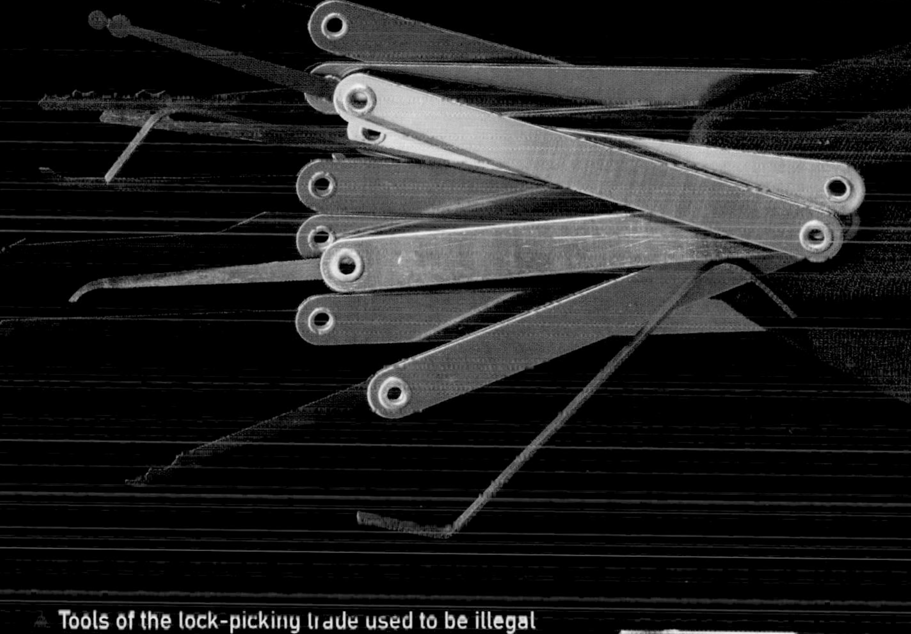

Tools of the lock-picking trade used to be illegal to sell to anyone without a locksmith's license. Good luck enforcing that in the world of eBay and direct-from-China web marketplaces.

Here's the security problem: if you have any one of these three locks, and the individual key that opens it—as you would if you lived in the building— then you can figure out what the other opening position is for each pin. Those alternate positions for each pin, taken together, equal the master key. (There are several ways to do this in practice: you can take the lock apart and look at the pins, or you can use trial and error to cut a test key, filing it down a bit at a time in each position. As always, there are YouTube videos with the details.)

How is it possible that people lived for decades in buildings where basically any one of their neighbors could get into their apartment by watching a few YouTube videos? I think it's largely because YouTube didn't exist for most of those decades. In the past, locksmithing was a secretive profession. The tools were hard (or even illegal) to buy, and the knowledge of how keys and master keys work was known only to those who had been accepted into the profession. Marking a key as "Do Not Duplicate" actually meant something, because the only people who owned key-duplicating machines were reputable locksmiths who respected those markings.

Lots of variations on the standard pin-tumbler lock have been invented over the last hundred years. Some of them have multiple sets of pins facing in different directions. Others have the pins arranged in a ring. Some of them are pretty darn hard to pick, but *all* of them have been picked, with tools ranging from a fancy electric tool to the back side of a disposable ballpoint pen (which is used to pick this ring kind, called a tubular lock).

THE MYSTERY OF THE VINCARD LOCK

THIS IS A REAL anomaly, the only lock of its kind that I am aware of. I only know about it because locks of this type were used until recently by a charming chain of private-pool-in-every-room hotels. It looks like a key card lock, as you see on pretty much every modern hotel door, except that it's completely mechanical.

Modern key card locks are computer-controlled electronic devices whose cards encode secret information either on magnetic stripes (like pre-chip credit cards) or in near-field radio responder chips. But in this VinCard lock, the secret information is instead a unique pattern of punched-or-not-punched holes in a two-dimensional grid. When you insert the card, these holes engage with a set of pins in the body of the lock. The lock will only open if there are holes in all the right places, and no holes where there are not supposed to be holes.

I would love to see inside this lock, but I don't dare take it apart. It took me months of hunting and pretty-please asking the management of the hotel chain if they could possibly locate one of their old locks. They finally managed to dig up this single unit. I have no idea what's happened to the rest of the probably fairly small number ever made.

The elusive VinCard lock

This is the closest thing I could find to the VinCard lock. It's a tiny little travel lock, meant to be used to lock up a suitcase during flight. It has a much-simplified key card with just a few holes punched in it. (It also has a key slot that allows the TSA inspectors to open the lock and check for bombs. Which means this lock can be opened by anyone who has spent the $3 to get a universal TSA key. Which means it is effective only against people who really have absolutely no intention of stealing anything from your suitcase.)

Could this lock be picked? Of course—it's a lock, it can be picked. But simply being rare gives it an advantage. Even a criminal who's somehow managed to learn to pick a regular lock isn't likely to have any idea how to deal with this one. The proof of how rare it is? There is, to my great astonishment, *no YouTube video on how to pick a VinCard mechanical hotel door lock.*

That's called "security by obscurity" and it's generally acknowledged to be a rather weak form of security. It works only for things that no one cares very much about. The minute someone halfway competent wants to figure out a way in, they won't have too much trouble. That's not a problem for a charming chain of pool hotels, but the same way of thinking—no one knows how this works, they'll never figure it out—has led to countless thousands of internet security breaches, with the personal information of basically everyone on the planet having been stolen several times over by now.

THERE ARE SOME types of locks that are genuinely more difficult to pick than others. This very chunky padlock is quite difficult to cut apart or break (just because it's made of such thick metal). It's also quite difficult to pick, because it uses a fundamentally different system: instead of a pin-tumbler inside, it has a "disk detainer" mechanism. However, even this lock can be picked in a few minutes by an experienced lock picker using a special tool.

To make locks more secure, we need to start by making them more *abstract*. We need to separate the secret information needed to open the lock from any physical object (key). In other words, the key needs to become pure information. This is not a new idea: pure-information keys have existed for several thousand years, with the first examples found in Greece and Rome. Ismail al-Jazari describes one in his famous book of 1206.

But the modern version was not perfected until just over a hundred years ago. The best of them are *almost* impossible to pick.

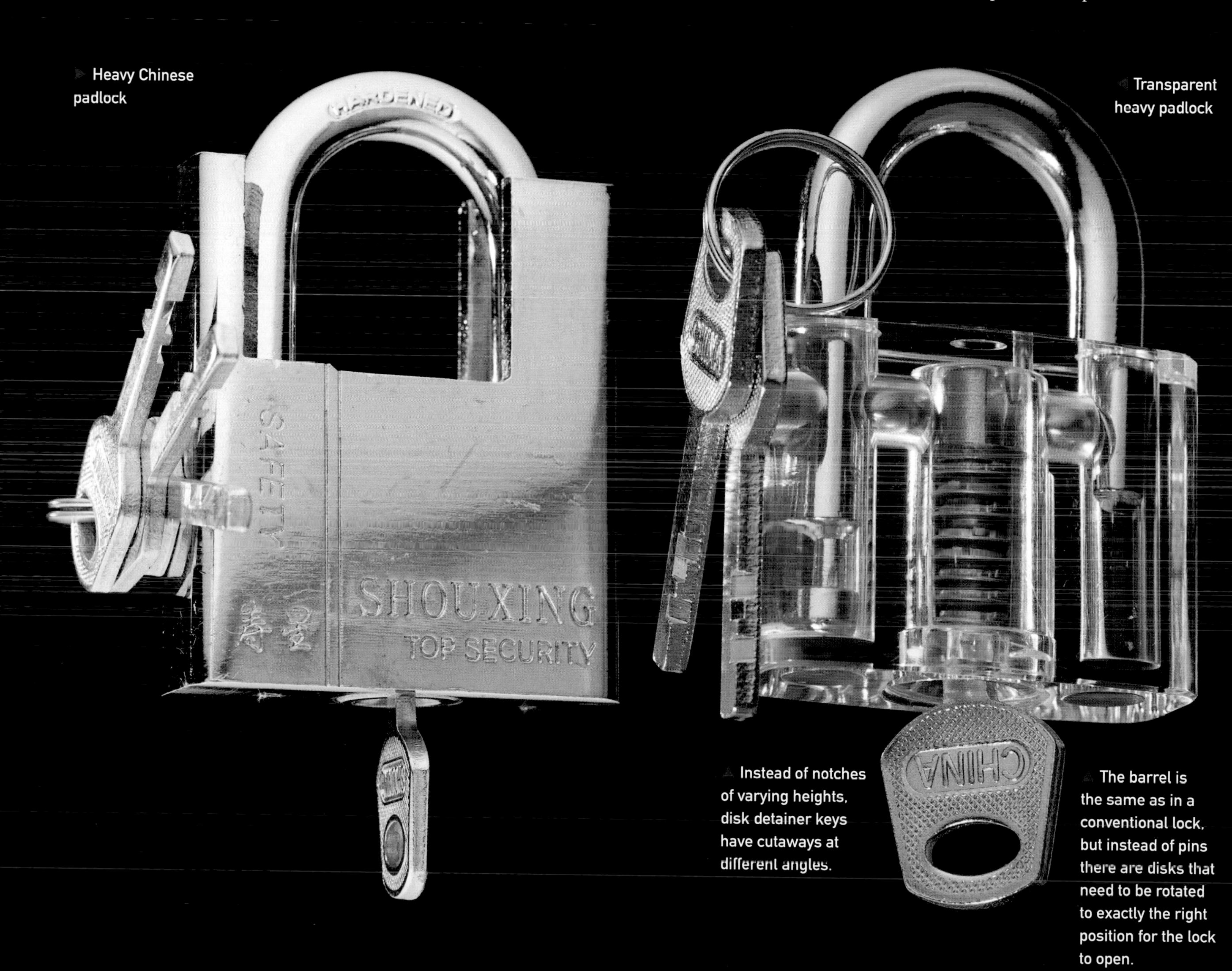

Heavy Chinese padlock

Transparent heavy padlock

Instead of notches of varying heights, disk detainer keys have cutaways at different angles.

The barrel is the same as in a conventional lock, but instead of pins there are disks that need to be rotated to exactly the right position for the lock to open.

Combination Locks

THE CLASSIC PURE-information lock is of course the *combination lock*. The "key" is a sequence of numbers: if you know those numbers, you can open the lock, and if you don't, you can't (at least not easily). Because there is no requirement to make a physical key, you can keep the combination completely secret just by never writing it down anywhere. There's nothing to lose, except your mind.

▼ The oldest combination locks used a set of rings next to each other, with numbers or letters around the rim. (These are Chinese characters because this is a reproduction ancient Chinese lock.)

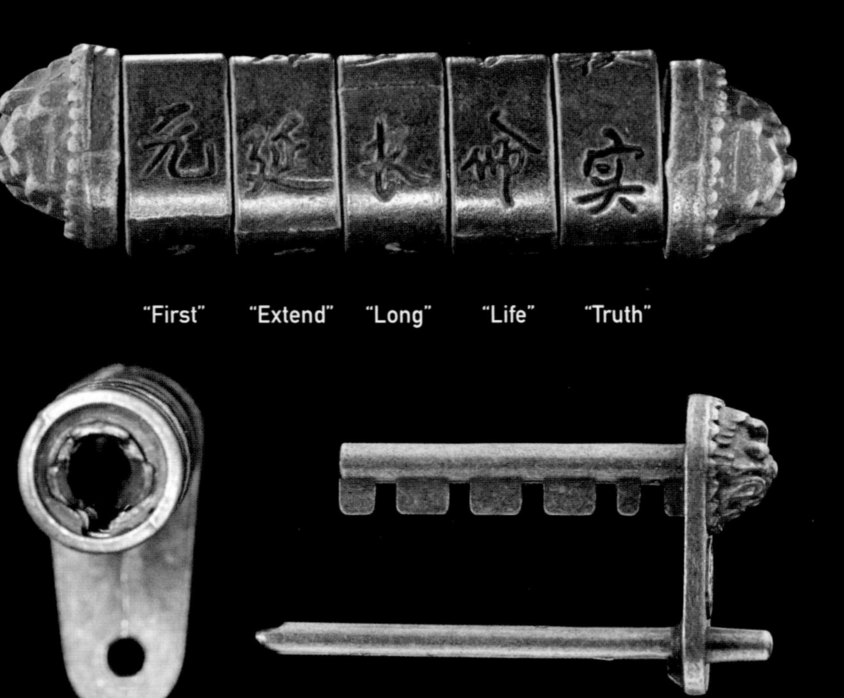

Under one of the symbols on each ring, invisible from the outside, is a notch cut out of the ring. This is called the "gate." When all the gates are lined up with each other, the lock opens. The gates line up, of course, only when the disks have been rotated to line up the correct set of numbers—the combination.

The simplest of these locks are easy to pick. There's always some way to feel for the notches using a thin piece of metal slipped between the disks. Or, with a really basic one like this, you can just put some pressure on the release mechanism and feel how the disks turn, or don't turn. Just as when picking a pin-tumbler lock, one disk will always be the one currently holding back movement of the part that is supposed to come out: that disk will be harder to move, until it suddenly clicks into place.

Modern padlocks of this style work in exactly the same way, but are harder to pick for two reasons. First, they are mechanically more precise; that makes it harder to feel any difference from one tumbler to the next, or to fit a metal feeler into the gap around the tumblers.

What really slows down picking is the "false gates." When you are trying to pick the lock, you constantly feel that you've found the right position for one of the tumblers, but actually you've just found one of the false gates, which allows the shackle to move a tiny bit, but won't actually release it. The difference between a false gate and the real one can be very hard to tell.

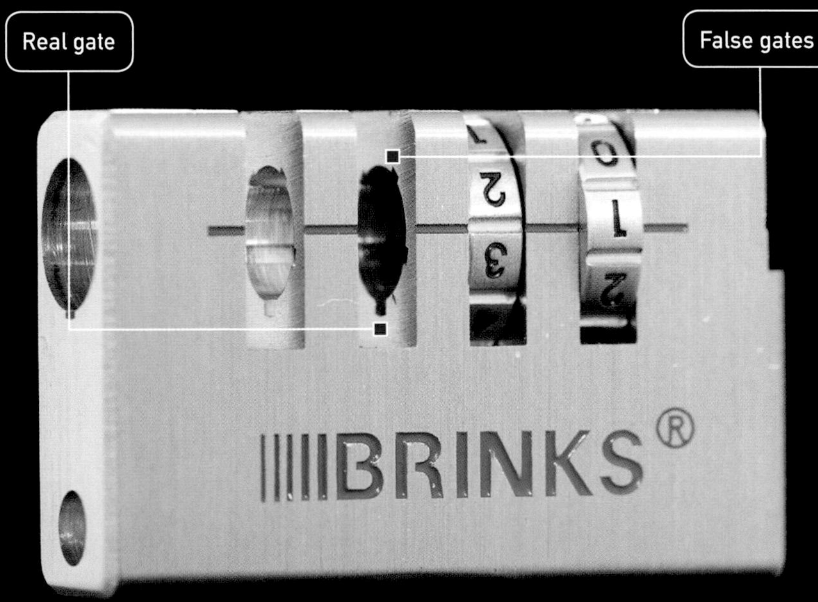

WITH FOUR TUMBLERS of ten letters each, there are 10 x 10 x 10 x 10 = 10,000 possible four-letter words this lock can spell. Of course many of them are not actual words, but a simple bit of computer programming narrows it down to this list of 952 words that occur in a standard dictionary. (This is in and of itself a security flaw: it's a reasonable guess that people will pick actual words for their combination, which means there are many fewer combinations you would need to check to pick the lock.)

Some hint of how they came up with the letters may be found in this fact: if you write a program to search for what setting of the wheels will spell the *most different words at the same time*, you get exactly one combination that spells seven different words all around the lock: DUAL, LOOK, FAST, BIKE, SLED, PENS, HYMN. (The remaining three, in case you're curious, are MHLM, TNRP, and WRTY.)

They obviously picked these letters so they could set the lock to show these words on display in the store. They probably sent the lock home with an intern to see if they could come up with any really embarrassing words. When the lazy intern came back with nothing worse than, say, "DANK," they decided it was good to go. This demonstrates once again that computer programming is a useful tool in product design, because if I were them, I would not have shipped this lock. (And no, I can't call out which words would have been a no-go for me, because then they wouldn't let us sell this book in schools. But they're in the list if you care to look for them.)

SLED SLAM SLAP SLAY SLAT SLOP SLOT SLOE SEED SEES SEEN SEEM SEEP SEEK SEND SENS SENT SELL SERE SETS SETT SEAS SEAN SEAM SEAL SEAT SHED SHES SHAD SHAM SHAY SHAT SHOD SHOP SHOT SHOE SNAP SNOT SUED SUES SUET SUNS SUNK SUMS SUMP SULK SURD SURE SUSS SUSE SONS SONY SOME SOLD SOLS SOLE SORT SORE SOTS SOAP SOAK SOON SOOT SAND SANS SANK SANE SAME SALK SALT SALE SARS SATE SASS SASK SASE SAKS SAKE SINS SINK SINE SIMS SILL SILK SILT SIRS SIRE SITS SITE SIAN SIAM PLAN PLAY PLAT PLOD PLOP PLOY PLOT PEED PEES PEEN PEEP PEEL PEEK PEND PENS PENN PENT PELT PELE PERM PERL PERK PERT PETS PETE PEAS PEAL PEAK PEAT PEON PEST PEKE PYLE PYRE PHAT PRES PREP PREY PRAM PRAY PRAT PROD PROS PRON PROM PROP PUNS PUNY PUNK PUNT PUMP PULP PULL PULE PURL PURE PUTS PUTT PUSS PUKE POEM POET POND PONY PONE POMS POMP POLS POLY POLL POLK POLE PORN PORK PORT PORE POTS POOS POOP POOL POSS POSY POST POSE POKY POKE PANS PANT PANE PALS PALM PALL PALE PARS PARK PART PARE PATS PATE PASS PAST PIED PIES PINS PINY PINK PINT PINE PIMP PILL PILE PITS PITY PITT PISS PIKE HEED HEEP HEEL HENS HEMS HEMP HEME HELD HELM HELP HELL HERD HERS HERE HEAD HEAP HEAL HEAT HESS HYMN HUED HUES HUEY HUNS HUNK HUNT HUMS HUMP HUME HULL HULK HURL HURT HUTS HUSK HOED HOES HONS HONK HONE HOMY HOME HOLD HOLS HOLY HOLT HOLE HORN HOTS HOOD HOOP HOOK HOOT HOSP HOST HOSE HOKE HAND HANS HANK HAMS HALS HALL HALT HALE HARD HARM HARP HARK HART HARE HATS HATE HAAS HASP HAST HAKE HIED HIES HIND HINT HIMS HILL HILT HIRE HITS HISS HIST HIKE MLLE MEED MEEK MEET MEND MEME MELD MELT MERE METE MEAD MEAS MEAN MEAL MEAT MESS MYST MUMS MULL MULE MURK MUTT MUTE MUSS MUSK MUST MUSE MOET MONS MONK MONT MOMS MOLD MOLL MOLT MOLE MORN MORT MORE MOTS MOTT MOTE MOAN MOAT MOOD MOOS MOON MOOT MOSS MOST MANS MANN MANY MANE MAMS MALL MALT MALE MARS MARY MARL MARK MART MARE MATS MATT MATE MASS MASK MAST MAKE MIEN MIND MINN MINK MINT MINE MIME MILD MILS MILL MILK MILT MILE MIRY MIRE MITT MITE MISS MIST MIKE TEED TEES TEEN TEEM TEND TENS TENN TENT TEMP TELL TERN TERM TEAS TEAM TEAL TEAK TEAT TESS TESL TEST TYRE TYKE THEN THEM THEY THEE THAD THAN THAT TREY TREK TREE TRAD TRAN TRAM TRAP TRAY TROD TRON TROY TROT TUES TUNS TUNE TUMS TULL TURD TURN TURK TUTS TUSK TOED TOES TONS TONY TONE TOMS TOME TOLD TOLL TOLE TORS TORN TORY TORT TORE TOTS TOTE TOAD TOOL TOOK TOOT TOSS TOKE TANS TANK TAMS TAMP TAME TALL TALK TALE TARS TARN TARP TART TARE TATS TATE TASS TASK TAKE TIED TIES TINS TINY TINT TINE TIME TILL TILT TILE TIRE TITS TIKE WEED WEES WEEN WEEP WEEK WEND WENS WENT WELD WELL WELT WERE WETS WEAN WEAL WEAK WEST WYNN WHEN WHEY WHET WHEE WHAM WHAT WHOM WHOP WREN WRAP WUSS WOES WONK WONT WOLD WORD WORN WORM WORK WORT WORE WOAD WOOD WOOS WOOL WOST WOKS WOKE WAND WANK WANT WANE WALD WALL WALK WALT WALE WARD WARS WARN WARM WARP WARY WART WARE WATS WATT WASP WAST WAKE WIND WINS WINY WINK WINE WIMP WILD WILY WILL WILT WILE WIRY WIRE WITS WITT WISP WIST WISE DEED DEEM DEEP DENS DENY DENT DELL DEAD DEAN DEAL DEON DESK DYED DYES DYNE DYAD DYKE DRAM DRAY DRAT DROP DUES DUEL DUET DUNS DUNN DUNK DUNE DUMP DULY DULL DUTY DUAL DUOS DUSK DUST DUSE DUKE DOES DONS DONN DONE DOME DOLL DOLT DOLE DORM DORY DORK DOTS DOTE DOOM DOSS DOST DOSE DANK DANE DAMS DAMN DAMP DAME DALE DARN DARK DART DARE DATE DIED DIES DIEM DIET DINS DINK DINT DINE DIMS DIME DILL DIRK DIRT DIRE DIAS DIAM DIAL DION DISS DISK DIST DIKE LEES LEEK LEND LENS LENT LETS LEAD LEAS LEAN LEAP LEAK LEOS LEON LESS LEST LYNN LYME LYLY LYLE LYRE LYON LUNE LUMP LULL LURK LURE LUTE LUST LUKE LONE LOME LOLL LORD LORN LORE LOTS LOTT LOAD LOAN LOAM LOOS LOON LOOM LOOP LOOK LOOT LOSS LOST LOSE LAND LANK LANE LAMS LAMP LAME LARD LARS LARK LATS LATE LAOS LASS LAST LASE LAKE LIED LIES LIEN LIND LINK LINT LINE LIMN LIMP LIMY LIME LILY LILT LIRE LITE LION LISP LIST LIKE FLED FLEE FLAN FLAM FLAP FLAY FLAK FLAT FLOP FLOE FEED FEES FEEL FEET FEND FENS FELL FELT FERN FETE FEAT FESS FEST FRED FREY FRET FREE FRAN FRAY FRAT FROM FUEL FUND FUNK FUMS FUMY FUME FULL FURS FURN FURY FURL FUSS FUSE FOES FOND FONT FOLD FOLL FOLK FORD FORM FORK FORT FORE FOAM FOAL FOOD FOOL FOOT FANS FAME FALL FARM FART FARE FATS FATE FAST FAKE FIND FINS FINN FINK FINE FILM FILL FILE FIRS FIRM FIRE FITS FIAT FISK FIST BLED BLTS BLAT BLOT BEES BEEN BEEP BEET BEND BENT BELL BELT BERN BERM BERK BERT BETS BEAD BEAN BEAM BEAK BEAT BESS BEST BYES BYRD BYRE BYTE BRED BRET BRAD BRAS BRAN BRAY BRAT BRAE BROS BUNS BUNK BUNT BUMS BUMP BULL BULK BURS BURN BURP BURY BURL BURK BURT BUTS BUTT BUOY BUSS BUSY BUSK BUST BOND BONN BONY BONK BONE BOLD BOLL BOLT BOLE BORN BORK BORE BOTS BOAS BOAT BOOS BOON BOOM BOOK BOOT BOSS BOSE BAND BANS BANK BANE BALD BALM BALL BALK BALE BARD BARS BARN BARK BART BARE BATS BATE BAAS BAAL BASS BASK BAST BASE BAKE BIND BINS BILL BILK BILE BIRD BITS BITE BIAS BIOS BIOL BIKE

Dial Combination Locks

TUMBLER COMBINATION LOCKS, especially the silly ones, are inherently low-security. A well-made dial combination lock, on the other hand, can be extraordinarily hard to pick. Generations of school children have struggled with these locks: three turns to the left past 37, then two turns to the right past 82, oh darn, wait...start over. Teacher, can you help me? This sums up the first week of middle school throughout the United States.

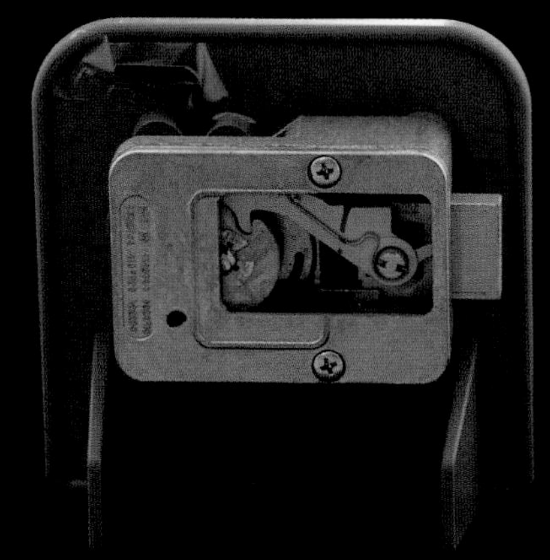

These are demonstration models of fairly high-end combination locks. Both are of a design originally made by the Sargent and Greenleaf company, but now made in essentially identical form by many different manufacturers (the patent expired decades ago). These salesman's samples are real locks exactly like the ones used on many safes, but the cases have been partially removed or cut away to show the workings inside. Although they are built much more precisely and robustly than a cheap school locker lock, the basic idea inside is largely the same.

MY STUDIO IS in a former bank building and one of the best perks is this honest-to-goodness bank vault located just off the main room. It has a serious door: hundreds of pounds of steel and concrete, with a big dial, and a lever you have to dramatically lift up to open the door whenever visitors come. When the door is closed, steel pegs slide out in all directions to lock it and hold it in place on all four sides.

What's at the heart of all this high security? A combination lock mechanism *exactly the same* as the display model shown to the left. Why such a puny lock on a such a big door? Because the strength of the door doesn't come from the combination mechanism itself being strong: it comes from everything else about the door being strong, and from the way the combination lock mechanism is isolated from the strong parts of the door.

On the back of the door we see a set of steel bars connected to the pegs that move out into the surrounding door frame when the door is locked. The handle on the front of the door is connected to these bars and pegs through a series of linkages. A deadbolt connected to the combination lock mechanism can slide into place to block these linkages, preventing the pegs from being withdrawn.

The combination mechanism itself doesn't have to be strong, because everything is set up so if you try to force the door open, something inside will give way before any force can be transmitted to the pegs holding the door closed.

The author's studio vault door

Vault door with back cover removed

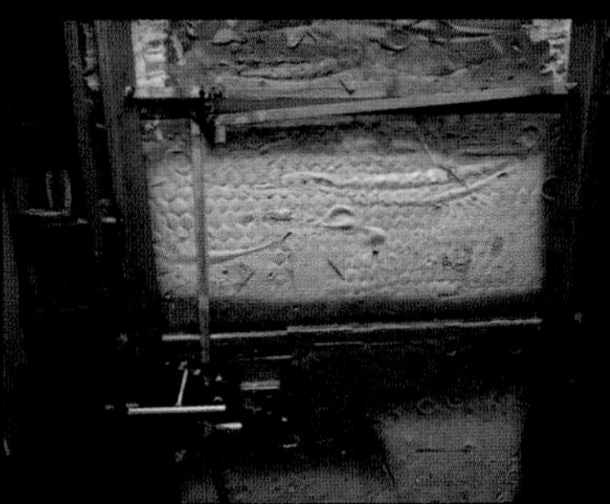

Exposed combination lock

If you try the obvious move of pulling up *really hard* on the handle, this nut will just slip. It's much weaker than the deadbolt holding it in place.

If you cut or smash off the dial on the outside and then drive a rod through the hole behind it, you can pretty easily knock the entire combination mechanism off the back side of the door. Which does you absolutely no good, because that just releases this "relocker," which slams the deadbolt into place and makes the door impossible to open.

This "fusible link" is there to frustrate anyone who tries to cut through the door with a torch. As soon as the back side of the door heats up, the low-melting-point solder holding the two halves of this link together will melt, allowing them to separate, again slamming the deadbolt into place. At that point the only way in is with a cutting torch and a jackhammer to smash out the fireproof concrete liner.

I love my little vault, but upstairs in the same building is the mother of all vault doors. This is a serious piece of steel! Two feet thick, 20 tons (40,000 pounds, 18,000 kg), with pegs the size of your leg. But inside is a combination lock mechanism that is really not much different from the old display sample we saw earlier.

The mechanism is more complicated than the one on the last page, but that's mostly because it includes redundancy and a time-lock feature, where the vault can be made impossible for *anyone* to open overnight. The basic idea is the same: the locking mechanism itself is relatively small and weak, but it's isolated from any attempt to apply force from the outside.

(Time locks are used to guard against one of the methods of getting into a vault: kidnapping the bank manager and holding them until they tell you the combination. With a time lock, the manager is no more able to open the lock than anyone else.)

There are two identical combination lock mechanisms (one is shown here with the back cover removed). Both are set to the same combination, and the safe opens if either one of them is unlocked. Why this duplication? So that if either of the locks fails in some way, you can still get into the safe. You can see there are four code wheels, meaning this is a four-number combination lock.

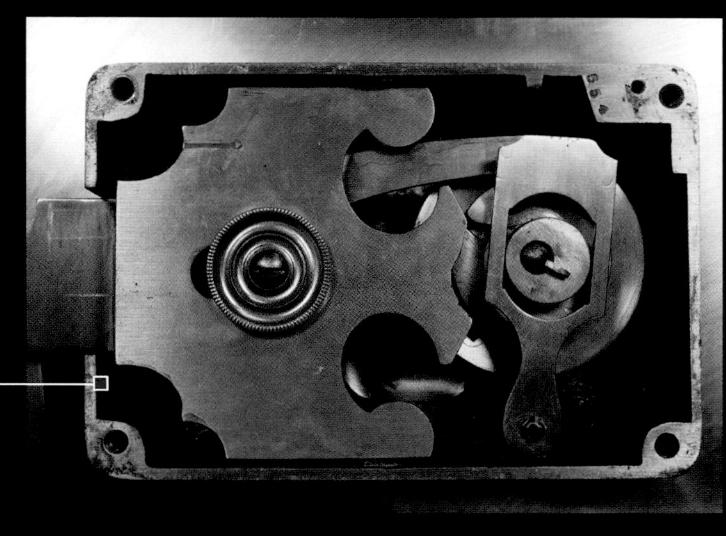

The time-lock mechanism is even more redundant than the combination locks: it has no fewer than FOUR separate clock units! To set the time lock, you use a square-hole key to crank all four of the clocks up to however many hours you want the safe to be unopenable (up to a maximum of 120 hours, or five days). You have to set all four clocks, and the safe will be openable again if any one of them gets all the way back to zero. In other words, three out of four clocks could fail and you could *still* open the safe. The redundancy in the clocks and combination locks is there for a simple reason: if you ever *couldn't* open the door, you would really be in a serious pickle. If the combination locks failed, you *might* be able to pick them, but quite possibly not: they are of a very secure design. If the time lock fails, you're completely out of luck. The best option then might be jackhammering through the thick concrete walls of the vault, as cutting into the door could do far more expensive damage.

Here's the funny thing about the huge, incredibly impressive and impenetrable bank vault door on the previous page: the actual vault behind it is tiny (above). The door was mostly just for show in the lobby of the bank, to make people think what a great bank this must be to have such a fancy vault door. Apparently it didn't work, since the bank closed down. The real vault is neither this show-off vault nor the dumpy vault in my studio (which was to keep spare cash for the drive-through windows).

This is the *real* vault, located in the basement away from the public areas of the bank. The door is strong, but not fancy. And behind it is a *huge* room, lined with reinforced, fireproof concrete. This is where the great bulk of the bank's valuables were kept.

The Anatomy of a Combination Lock

NOW THAT WE'VE seen some places where quality combination locks can be found, let's take a look at how they work inside, using this transparent model. Is there actually a good reason why you always have to turn one way three times, then the other way twice, then back again once? Yes: this confusing dance of turns is fundamental to how these things work.

The heart of every dial combination lock is a set of *tumblers*, or code disks. Each one has two important features: a notch where the outside edge is cut away, and pins that stick out on both sides of the disk. The number of disks in the lock determines how many numbers there are in the combination. So a typical lock with a three-number combination will have three disks (or two disks plus a notch on the dial itself that determines the third number). The positions of the pins relative to the notches determine the numbers in the combination.

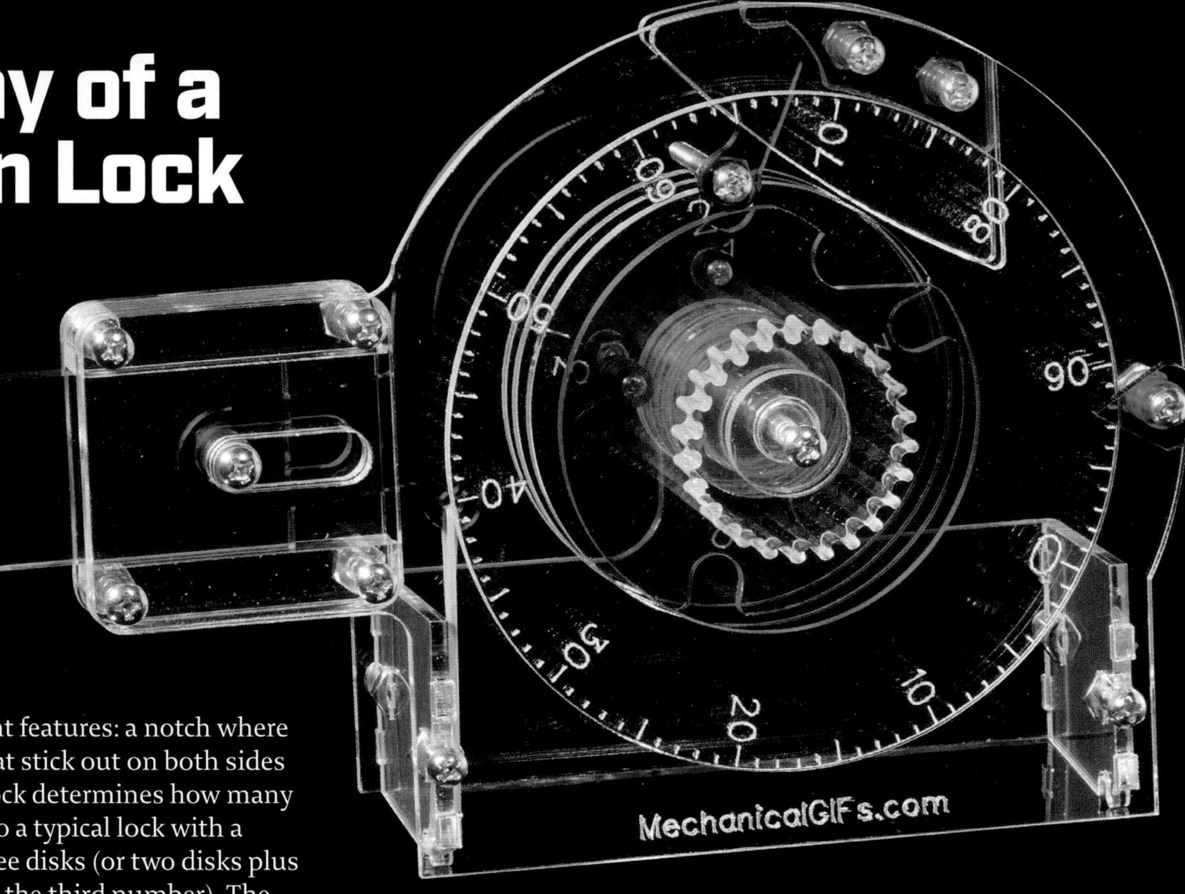

A look inside a combination lock

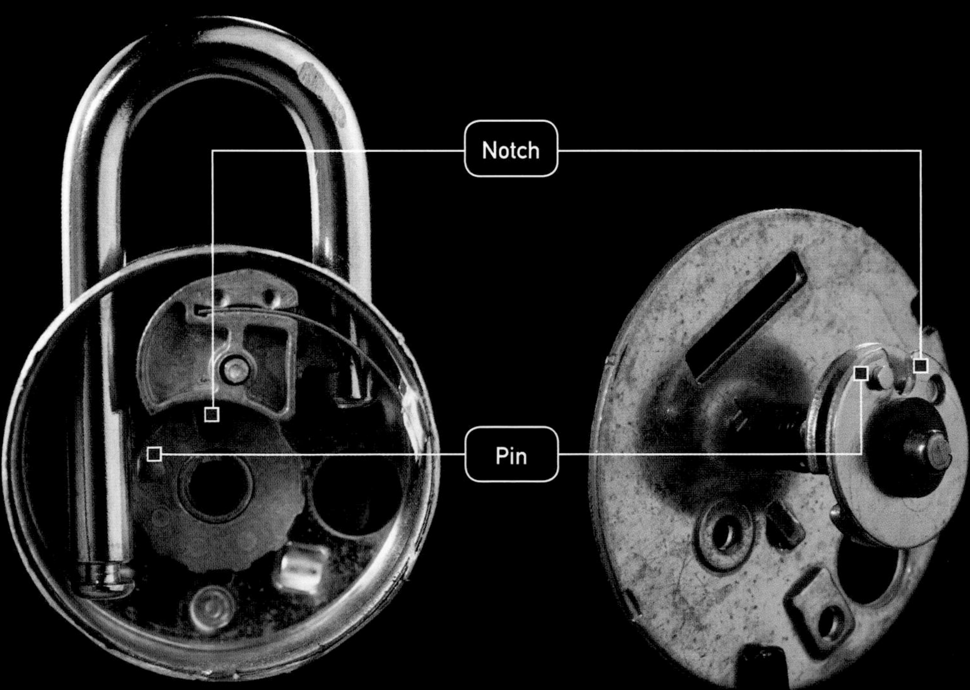

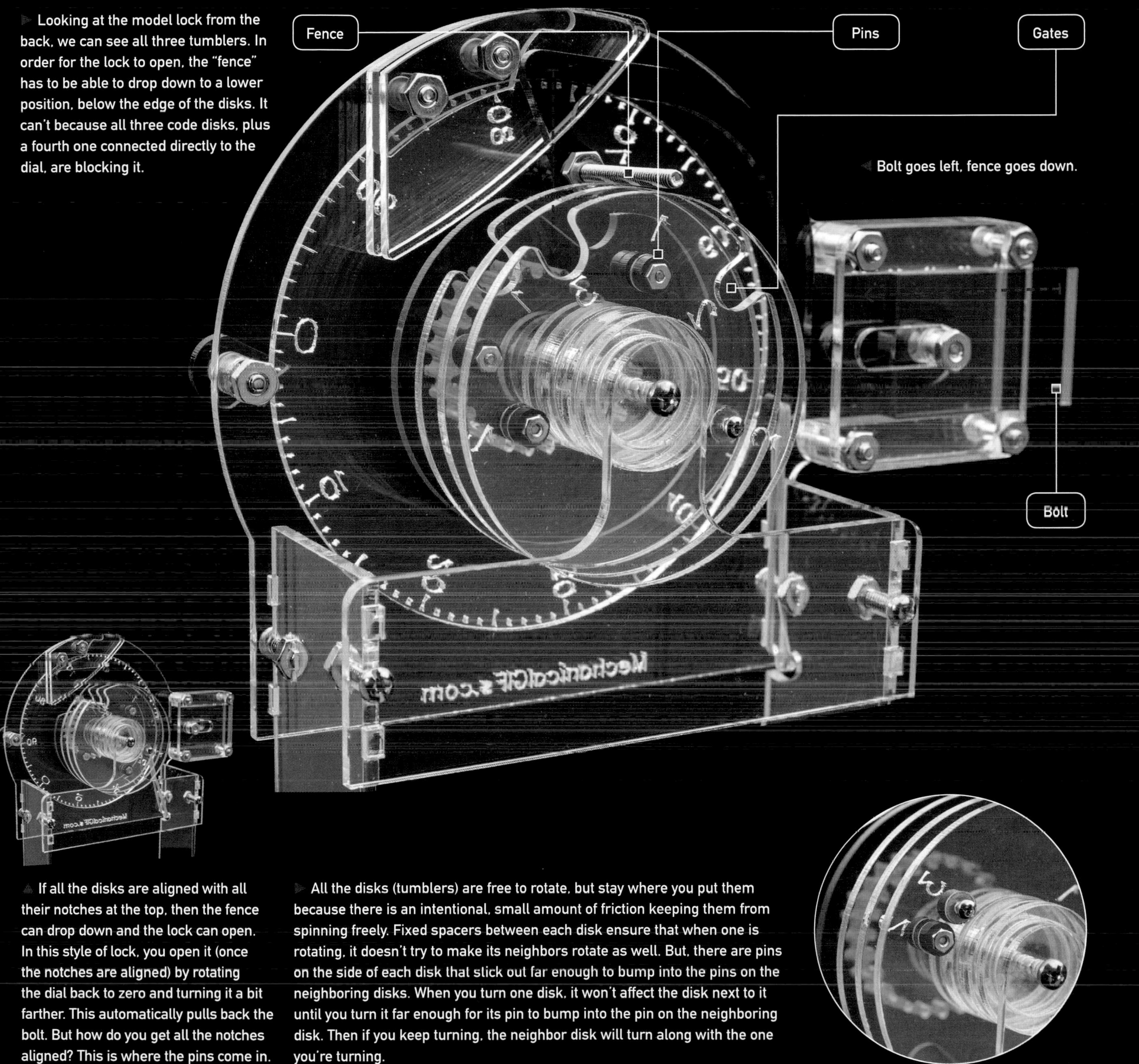

Looking at the model lock from the back, we can see all three tumblers. In order for the lock to open, the "fence" has to be able to drop down to a lower position, below the edge of the disks. It can't because all three code disks, plus a fourth one connected directly to the dial, are blocking it.

Bolt goes left, fence goes down.

If all the disks are aligned with all their notches at the top, then the fence can drop down and the lock can open. In this style of lock, you open it (once the notches are aligned) by rotating the dial back to zero and turning it a bit farther. This automatically pulls back the bolt. But how do you get all the notches aligned? This is where the pins come in.

All the disks (tumblers) are free to rotate, but stay where you put them because there is an intentional, small amount of friction keeping them from spinning freely. Fixed spacers between each disk ensure that when one is rotating, it doesn't try to make its neighbors rotate as well. But, there are pins on the side of each disk that stick out far enough to bump into the pins on the neighboring disks. When you turn one disk, it won't affect the disk next to it until you turn it far enough for its pin to bump into the pin on the neighboring disk. Then if you keep turning, the neighbor disk will turn along with the one you're turning.

▼ **1.** The dial (seen here from the back side, looking from inside the lock) is what you turn to enter the combination. It has a pin of its own, which bumps into the pin on the first code disk, which pushes the pin on the second disk, which pushes the pin on the final, third disk.

▼ **2.** Therefore, in order to get the cutout on the edge of the third disk rotated to where it needs to be under the fence, you have to turn, in this case, *three* times around to the left, then stop at the first number in the combination (50, as you can see from the position of the dial under the pointer on the far left).

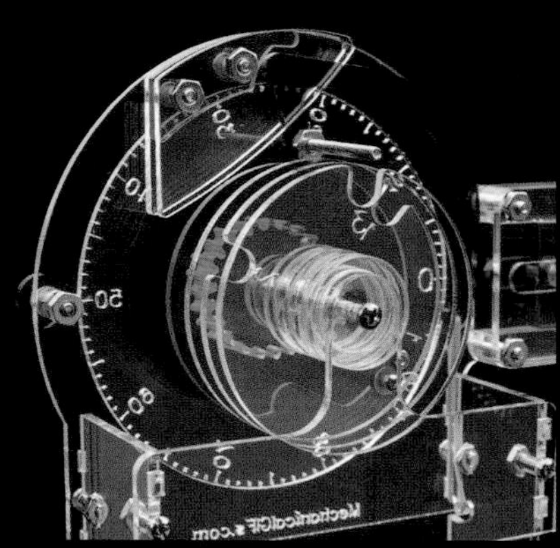

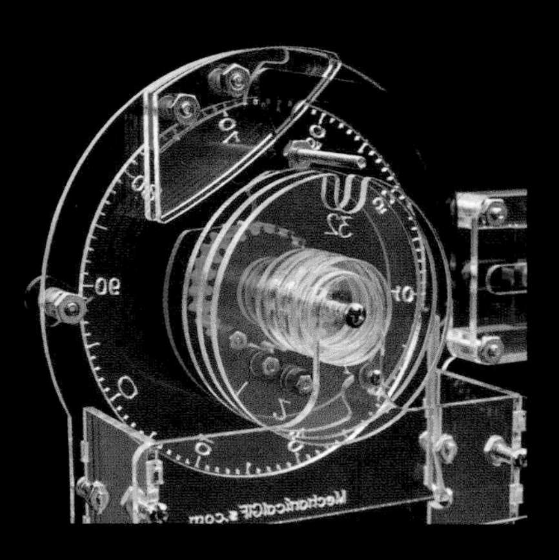

▲ **3.** Next we turn the other direction, to the right. For the first nearly complete turn of the dial, nothing happens: the pin on the dial moves away from the pin on the first disk and just turns around not touching anything. But after almost one full turn, it's back around, now on the *other side* of the pin on the first disk. If you keep turning farther, the first disk and then the second disk will start turning.

▲ **4.** After *two* full turns, the second disk will start turning. The cutout will be aligned when you stop at the second number in the combination (90). If you go even a bit too far, you can't just turn the dial backward: if you do, the pins will just separate and the second disk will not move. You would have to go backward two whole turns to get the second disk to move backward to where it needs to be. Since you can't (in a real lock) see what you're doing, you really just need to start over from the beginning.

Picking a Combo Lock

▼ **5.** After getting the second disk aligned, you switch direction again, but this time you only have to turn the dial *one* full turn to pick up the first disk. Stop at the third and final number in the combination (40) and all three cutouts (gates) will be aligned properly underneath the fence (the bar on top, which is going to drop down into the gates). In a simpler type of combination lock you would now be finished: you would try the handle, or pull up on the shackle of a padlock, which would push the fence down, allowing the lock to open. But this is a model of a fancy safe lock, which has one more trick. Notice that there is one special disk, the orange one, which is connected directly to the dial and has a specially shaped cutout.

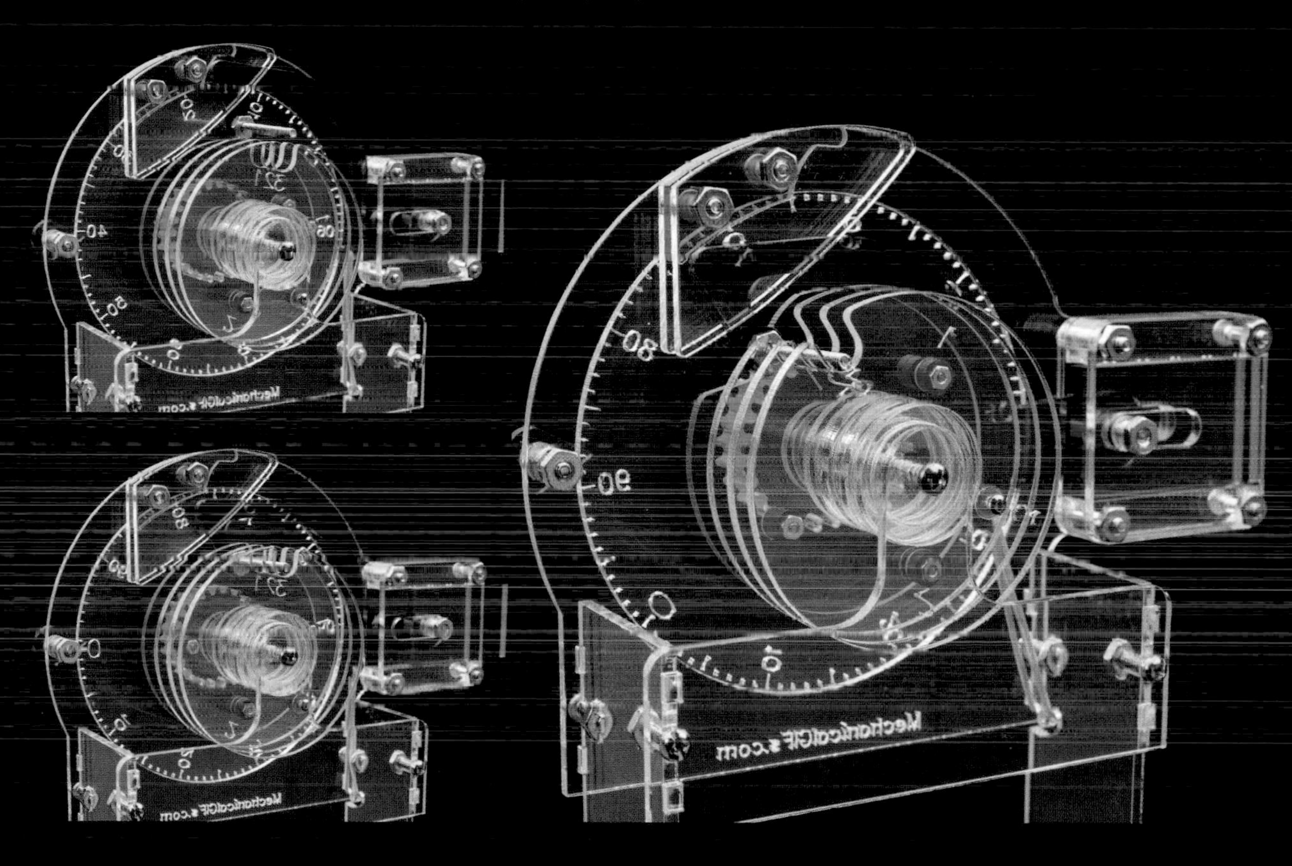

▲ **6.** If you turn the dial back to zero, the fence is finally able to drop down into all four cutouts (three gates on the code disks, and the fourth, special gate on the orange disk).

▲ **7.** Turn the dial a little bit farther and the reason for the special shape of the cutout on the orange disk is revealed: it grabs the fence and pulls the bolt back, automatically opening the lock. This is both a clever way of using the dial itself as the way the lock is opened, and an important security feature.

(To purchase a build-your-own model of this transparent combination lock visit mechanicalgifs.com.)

IN A SIMPLE combination lock, it's possible to put pressure on the mechanism by pulling on the shackle much like we used a tool to apply pressure to the barrel in the pin-tumbler lock. This pressure pushes the fence down onto the code disks, so that when we rotate them, we can feel subtle clicks as the gates go by underneath the fence.

In this fancier model, the orange disk holds the fence up and away from the gates: there is no way to feel them going by as the disks rotate, because there is no contact between the fence and the gates (except for the one point around each full turn of the dial when the gap in the orange disk "tests" the gates). These locks are very, very difficult to pick. A few people can do it in a matter of minutes, but most locksmiths end up drilling a hole next to the dial and using an inspection camera to watch the gates go by as the dial is turned.

THE BEST WORST SAFE

WHERE I LIVE we have a restaurant called Cracker Barrel and I love that place. It's like an artificial time capsule. In other words, it's not actually preserved from the past, it's newly made to look like it's old (it is, in fact, a modern chain restaurant with hundreds of locations). The main feature of every Cracker Barrel is a gift shop almost as big as the rest of the restaurant, filled with similarly new-old items. Candy bars made to look like they did in the 1950s, "classic wooden toys" from a "simpler time," down-home country-style farm clothes, and overly cute sayings painted on "old" barn siding (made in China, no doubt).

It's all very fake, but gosh darn it, every once in a while it strikes home, as was the case when I found this beautiful thing. I had one of these when I was a kid! Yet here is a brand new one, exactly the same in every detail as the one I remember. (Repeat that experience enough times with medium-old people and you have a real business.)

It has a really crappy two-wheel combination lock mechanism (trivial to pick by feel), but when you open the door there's a really loud RING-RING-RING alarm, like a bicycle bell gone mad, that runs for what seems like an hour. The idea, of course, is that you can't open this thing without alerting the whole house, including your sibling who probably doesn't want you opening their safe.

Electronic Locks

COMBINATION LOCKS ARE one step in the direction of pure information: instead of a physical key, the key is a set of numbers you have to remember. But they are still mechanical devices with many limitations. To go beyond those limitations we need to move to locks that go even further in the direction of pure information: electronic locks, and ultimately software locks.

The two locks pictured on this and the following page both have numerical keypads, and you open both of them by punching in a numerical code. But inside they are completely different, and they have their own set of advantages and disadvantages.

It may surprise you to know that the mechanical one you see here is on my studio door, and I have five more of them on other doors at the studio and my house. I have zero electronic keypad locks.

Why did I choose mechanical locks? They ***just work***. Individual battery-operated keypad locks are irritating, unreliable bits of technology, and when they fail, you're locked out of your house. These mechanical locks are robust, reliable, and likely to last for decades. I don't need the hassle of replacing broken electronic locks every few years, and I don't need any of their fancy features.

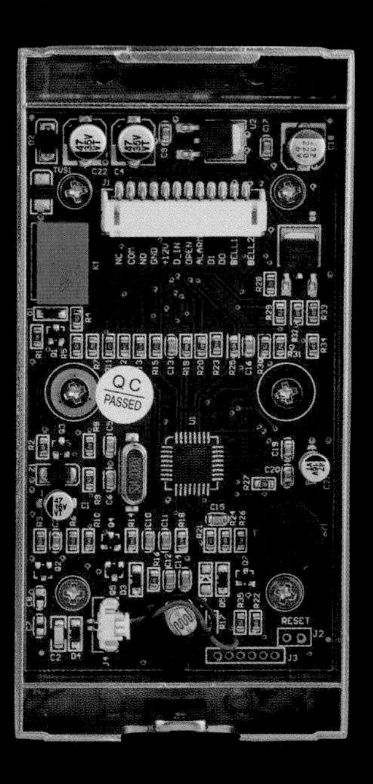

This is an electronic keypad lock. It can be programmed to accept codes of any length. It can be given more than one code, which means that creating a "master key" is possible by programming many locks to have both a unique and a shared code. Temporary codes can be assigned so you can give someone access for only a limited amount of time, or only at certain times of the day. Codes can be changed remotely, and every use of a code is logged, so you know who opened the door and when. What's not to like about this lock?

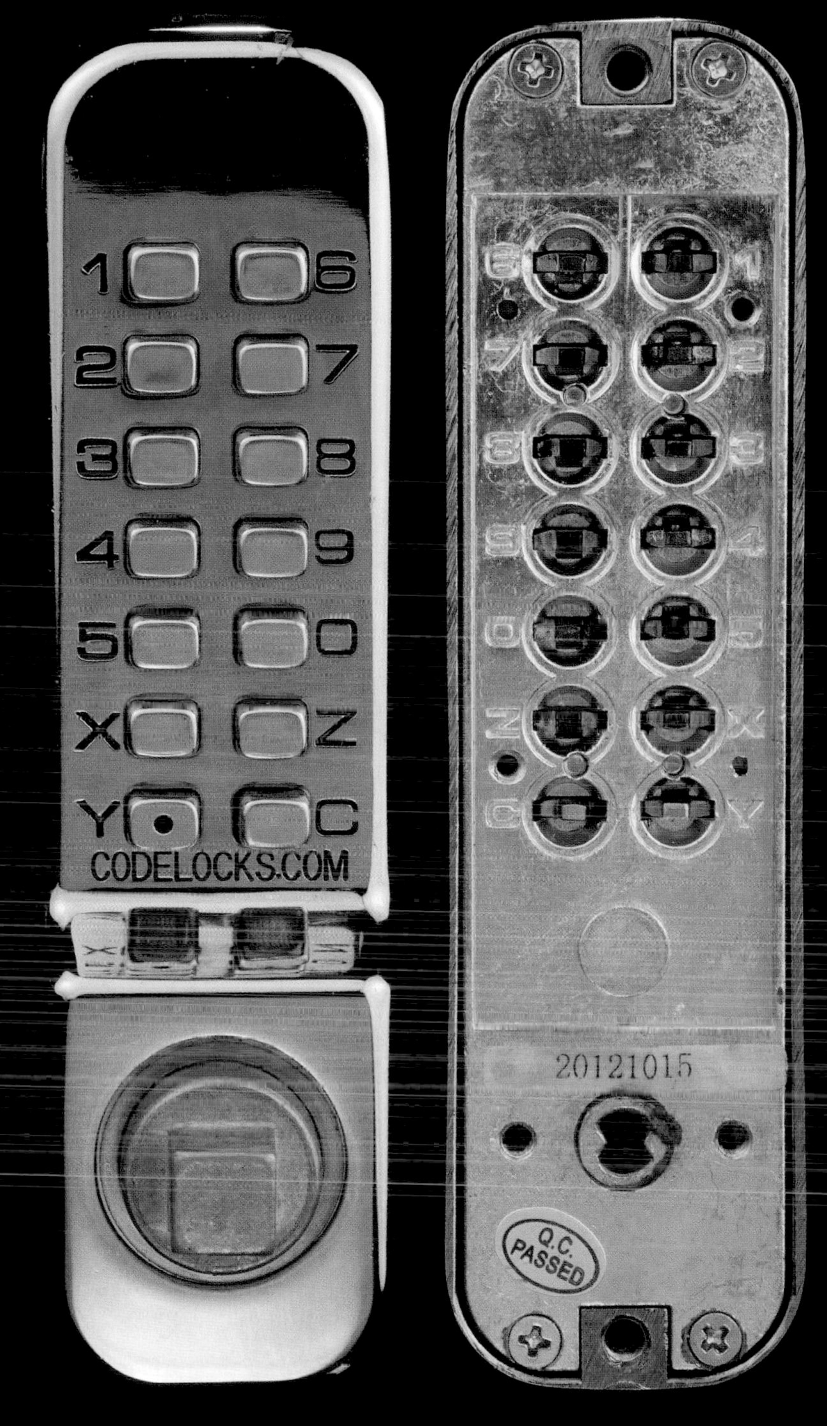

▲ This is a purely mechanical lock. It's very clever inside, but has many limitations compared to the electronic version. The numerical code is actually just a list of which digits you must push to open the lock: it doesn't matter what order you push them in, or whether you push them more than once. (In other words, for each digit you either have to push it or not push it before trying the handle.) That means there are many fewer possible codes, making the lock easier to pick. Changing the code required disassembling the whole thing and swapping out small notched plugs. Of course you can't have more than one code, or any of the other fancy features of the electronic version.

I prefer mechanical locks for my own doors, but in other situations the advantages of electronic locks are undeniable. Running a hotel with old-school key locks is nothing but trouble. With centrally controlled locks on every door, the front desk can create keys that work for just the days you've paid for, other keys that let housekeepers in while recording who entered what room when, and so on.

Before electronic locks, the standard system was for hotels to use pin-tumbler key locks with one unique key per room, and a master key that could open them all. Many hotels required guests to turn in their key at the front desk every time they left the property. To encourage this, the keys were attached to heavy, awkwardly shaped key fobs that would be impossible to forget in a pocket.

These days the practice persists only in quaint "boutique" hotels, where the practice itself is quaint but also annoying. It forces you to interact with a human being every time you come and go, even if it's 2 AM and the front desk person is also the owner and they are asleep. The last time I experienced this was surprisingly recently: I was giving a talk in a small city in Italy in 2018 and for some reason the organizers put me up in a tiny eight-room hotel that was run by nuns. It was, in fact, *in* the 800-year-old nunnery (convent) where said nuns lived. They made it clear that no one was getting back into their room after 10 PM (and no impure thoughts allowed).

Biometric Locks

LONG CONFINED TO science fiction movies, "biometric" locks are now pretty common. This one opens when you press your finger on a glass sensor, which captures an image of your fingerprint and compares it to a list of authorized fingerprints. But this lock goes *way* beyond the call of duty by supporting a total of **five** different ways of opening it. (Which of course means there are also five different approaches for hacking into it, along with the several other methods of getting through a door that don't involve attacking the lock.)

Combination locks are more secure than key locks in part because there is no physical object that incorporates the secret information (the code) needed to open the lock. This lock in some ways goes backward, because the RFID (radio frequency identification) tag that can open it is a physical object that incorporates the secret information (in other words, it's a physical key just like the metal key of a mechanical lock). However, clever encryption technology means that, if the system is designed properly, it is impossible to extract and copy this secret information from the RFID key, even if you have unlimited access to it. In the next section we'll talk about how it's possible to have a key that can be used, but not copied.

I have some skeptical things to say about this lock, mainly to do with its incredible overkill of features. It's not even the worst offender in overkill—there are models that add face recognition to the mix—but all in all, I have to say I'm impressed. It's a very solidly built piece of semi-security. And it feels good in your hand, which is really the most important thing about a device you will touch and use several times every day, and which you trust to keep you safe from bad people and bears alike.

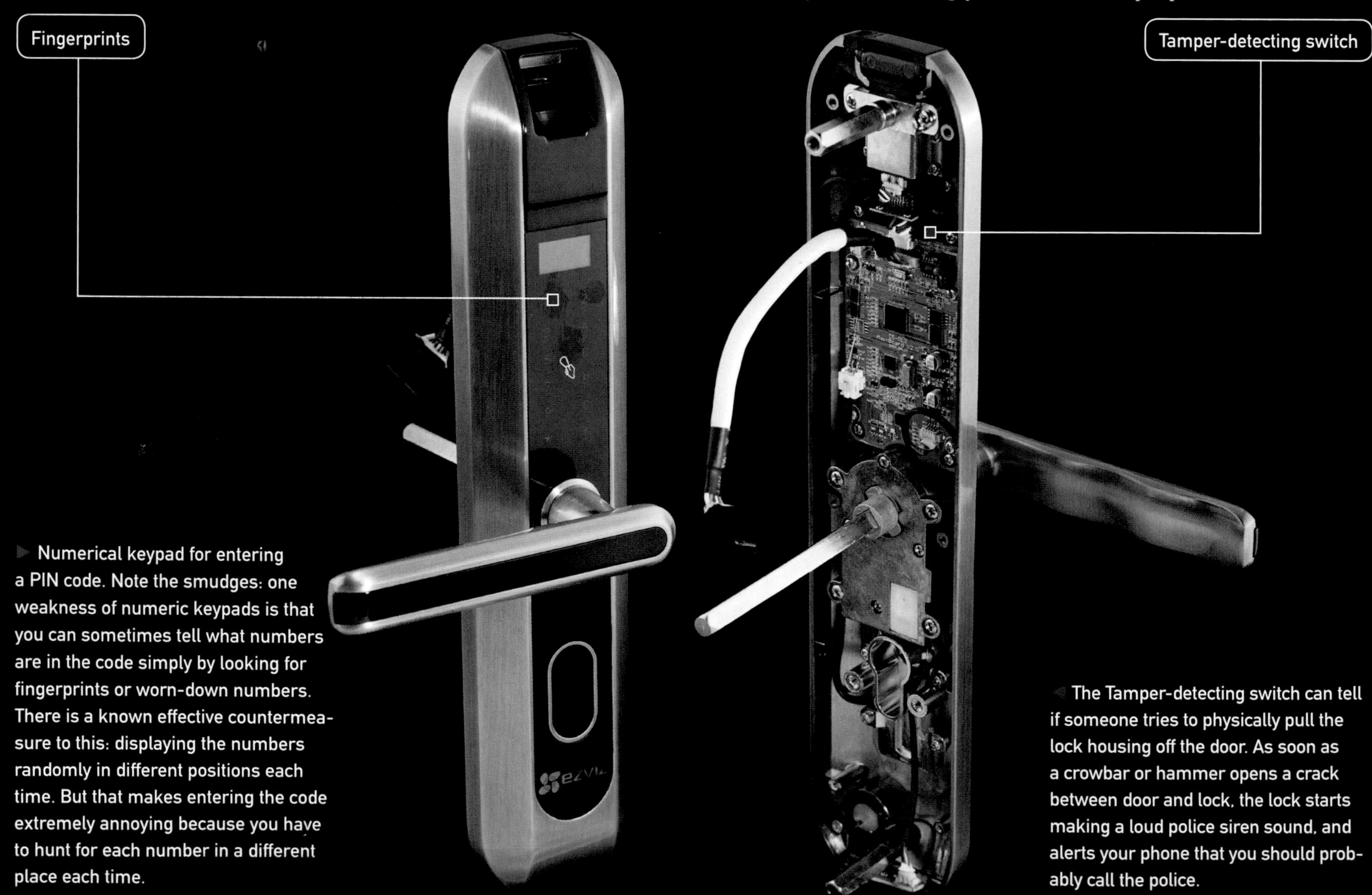

Numerical keypad for entering a PIN code. Note the smudges: one weakness of numeric keypads is that you can sometimes tell what numbers are in the code simply by looking for fingerprints or worn-down numbers. There is a known effective countermeasure to this: displaying the numbers randomly in different positions each time. But that makes entering the code extremely annoying because you have to hunt for each number in a different place each time.

The Tamper-detecting switch can tell if someone tries to physically pull the lock housing off the door. As soon as a crowbar or hammer opens a crack between door and lock, the lock starts making a loud police siren sound, and alerts your phone that you should probably call the police.

This lock does not mess around with the physical side of security. This bracket and another like it on the bottom move out of the way from the lock body when the deadbolts are engaged. They are meant to be attached to rods that reach all the way to the top and bottom of the door, engaging vertical deadbolts at both ends to make it harder to kick the door in.

Eight AA batteries power this contraption—until they run out, at which point only the mechanical key or the emergency USB power source will still work.

If the batteries are dead and you don't have the physical key, you can *still* get into this lock, by connecting a USB phone-charging cable to the socket at the very bottom of the lock. The designers were bound and determined that this lock is *going to open* come hell or high water—except not high water, because the circuitry is definitely not waterproof.

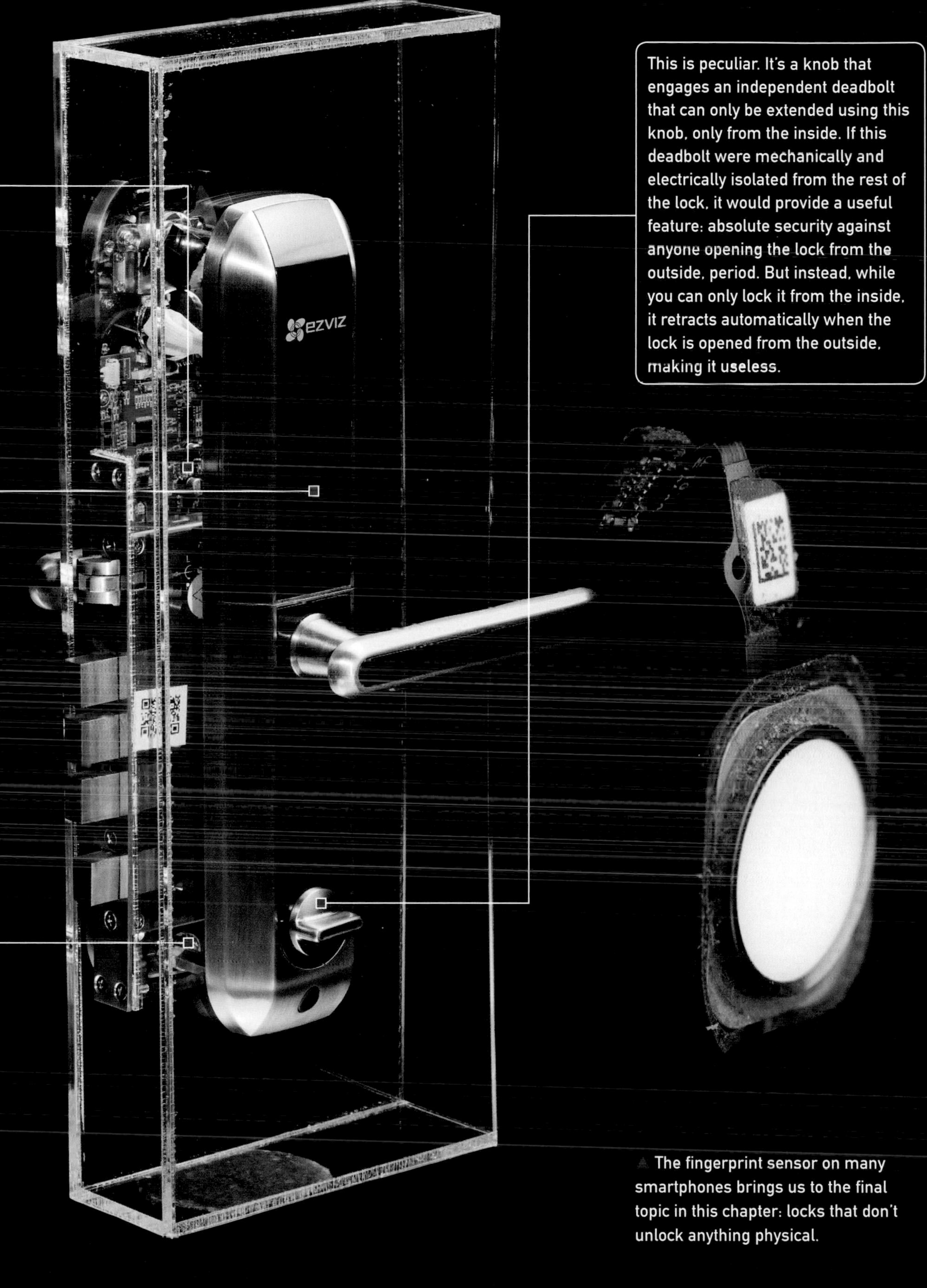

This is peculiar. It's a knob that engages an independent deadbolt that can only be extended using this knob, only from the inside. If this deadbolt were mechanically and electrically isolated from the rest of the lock, it would provide a useful feature: absolute security against anyone opening the lock from the outside, period. But instead, while you can only lock it from the inside, it retracts automatically when the lock is opened from the outside, making it useless.

▲ The fingerprint sensor on many smartphones brings us to the final topic in this chapter: locks that don't unlock anything physical.

Locks of the Most Abstract Kind

WE DISCUSSED HOW combination locks are more abstract than key locks, but they're still physical objects that attempt to secure access to a literal door. The concept of locks has in modern times been extended into entirely abstract, electronic situations where there is no lock or door. And yet, many of the same styles of lock exist in this virtual form as well.

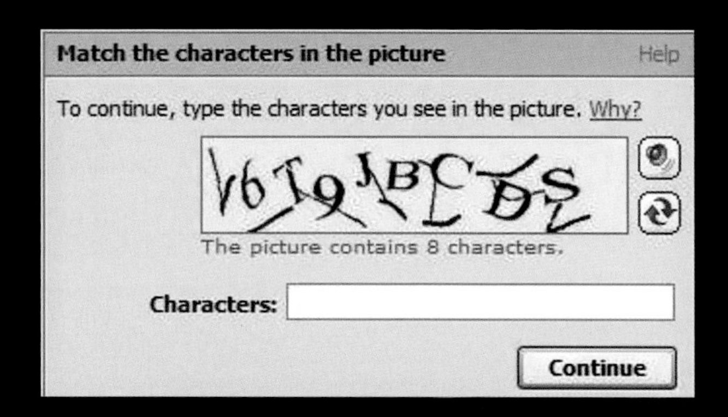

For example, the exact virtual-world equivalent of the childproof bottle is known as a CAPTCHA, which stands for Completely Automated Public Turing test to tell Computers and Humans Apart. "Turing test" is a reference to the mathematician and early computer theorist Alan Turing, who realized—far before most people had even conceived of the issue—that as computers became more and more powerful, it would become more and more difficult to tell whether you were talking to a human or to a computer. He proposed that, if the day ever came when there was a computer you could not tell from a human, no matter how long you talked to it, then that computer would *be human* in any meaningful sense of the word.

In his day the Turing test was a hypothetical open-ended conversation between a human tester and an artificial intelligence trying to prove itself worthy of the name human. In the modern incarnation of CAPTCHAs it is instead a computer testing *us* to see if *we* are human enough. How degrading is that?

Two-factor authentication keys are the rough equivalent of the physical key in a pin-tumbler lock. Many kinds of online accounts can be set up so you can only log in to them if you are in physical possession of one of these devices that has been associated with the account. Each one contains a unique code, just like the code held by the pattern of high and low spots on a metal key.

A smartphone unlock code is an example of the virtual equivalent of a combination lock. Your unlock code corresponds to the numerical code that opens a combination lock. It doesn't open anything physical, but it opens up a world of time-wasting games and social media. A typical six-digit unlock code on a cell phone is, to start with, about as secure as a three-number mechanical combination lock code (where each number has two digits). In both cases there are a million possible codes. But the electronic version can pull a trick not available to a dumb mechanical device.

Have you ever put in the wrong code too many times and had your phone lock you out for a while? Or maybe your annoying brother did it to you on purpose? The length of time you get locked out for is based on how many wrong guesses you've tried, and it's calculated to make it nearly impossible to get into someone's phone by trial and error unless you get really lucky, or they picked a really dumb code.

Login passwords are another example of a key that unlocks a virtual door, whether it's to your laptop, your Facebook account, or whatever. Instead of just a few numbers, a typical password can be a dozen letters or more long, each of which can be one of around seventy easily typed letters, numbers, and symbols. That makes it *much* harder to guess one, or find it by exhaustive search through all the possibilities.

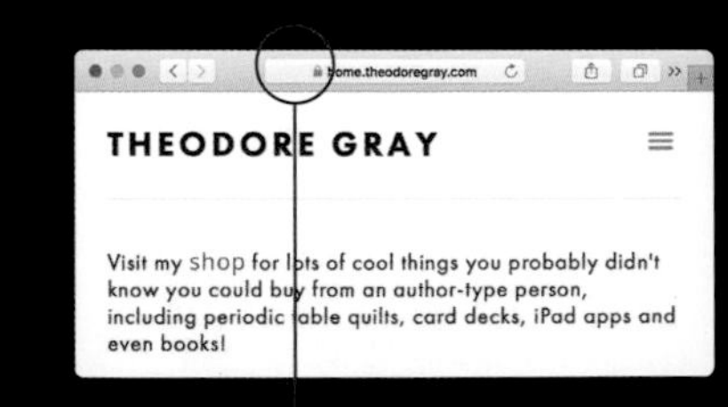

This little padlock icon is a sign that your web browser is making use of perhaps the most surprising, unexpected, and powerful discovery in practical mathematics in the second half of the twentieth century: public key cryptography.

YOU'VE PROBABLY READ about hackers stealing passwords from, well, just about every website that's ever stored a password. But unless the operator of the website is really incompetent—which unfortunately many of them are—it's not entirely trivial to get usable passwords from the files the hackers can steal off a server. That's because the file that stores the passwords for all the users on a system does not actually contain any of the passwords. In fact, it does not contain any information from which it is possible to recover the passwords.

Wait, how does that work? If there's no way to get the passwords out of the passwords file, how does the website decide whether you entered the correct password? Surely it has to compare the one you entered to the one it has stored? Actually, no, it doesn't. There is a more clever way of doing it.

Instead of storing your password, the file stores a *hash* of your password. A hash is a seemingly random string of digits that is derived from your password through an impossible-to-reverse series of mathematical operations called a hash function. Basically your password is scrambled beyond any hope of recognition or recovery. No reasonable amount of computing power can un-hash the hash to get back the password. But, and this is crucial, the scrambling operation is completely *deterministic*, meaning completely predictable. If you give the hash function the same password, it will always come up with the same hash value.

```
Bob 0822ddcd7bfb0d56afd3a57b00ae52d8
Alice f7c91bfa7e547dbe685c364b8b0e2cb0
Eve fd15f772f164ab08e9a4c4014ac12559
Carol 125e4f72c2c1272ed6ca2959fea20644
```

When you try to log in, the system takes the password you entered and applies the hash function to it, then compares the seemingly random string it gets to the seemingly random string it has stored. If they match, then it knows you must have entered the right password, even though it does not have that password stored in a recoverable form.

So when hackers steal a file full of passwords, all they get is a pile of hash values that cannot be used to log in to anything. To turn those hash values into usable passwords, they have to try to guess millions of passwords, hash them all using the correct hashing function, and compare those hash values to what's in the file.

If you've used a simple password, like a word in the dictionary, they are going to find a match pretty quickly. If you used something difficult with lots of odd characters in it, the hackers will have to work much harder, trying many more combinations, before they happen on one that works. That's why websites are always trying to get you to use passwords that you'll never be able to remember.

In this era of universal hacking and rampant stealing of personal data, many people choose to password-protect the data on their computers. It's tempting to make the analogy that this is like locking your disk up in a safe, but this is actually a very misleading way of thinking about it. If you lock your disk up in a safe, all someone has to do is cut a hole in the side of the safe and pull the disk out. They don't need the code to the lock.

But if you "encrypt" the contents of the disk—which is what happens when you add password protection to it—the computer doesn't just put a lock around the outside of the data, it completely scrambles the contents of the disk, mixing it up with the text of the password in a way that makes it impossible to read without knowing the password. Even if hackers steal all the data off the disk, it will be unintelligible to them unless they can guess the password, or steal that too.

Modern cryptography (the technology of scrambling data) is so good that, to the best of anyone's knowledge, *no one* can break a properly encrypted disk, as long as the password is long and difficult enough—not the hackers, not the police, not even the national security agencies of any country.

So, while they may not be real things, software locks are more secure than any physical lock has ever been.

Public Secrets

SECRET CODES HAVE been used to encode messages for thousands of years. The beautiful thing about an encoded message (that is, a message that has been mixed up and scrambled together with a secret password) is that it can be sent out in the open—broadcast over the radio if you like—and the enemy won't be able to do anything with it.

Both sides need to know the secret password so the receiver can decode what the sender has encoded. If that password ever leaks to the enemy, it's game over: any intercepted communications can then be read by the enemy, and you have to establish a new, different secret password.

In times of war setting up a new password was difficult and often dangerous: a trusted courier had to hand-carry the password from sender to receiver, without getting caught and without ever losing sight of the code, lest it be stolen by the enemy. If the receiver was a spy behind enemy lines, or a forward outpost in a battle zone, this could be a huge problem.

But what else could you do? Obviously both sides need to know the same secret password, and equally obviously, they can't just shout the password over the radio!

This was all quite obvious, until 1969 when it became no longer true. In that year a method was invented, though not published until several years later, whereby both sides could shout their passwords over the radio, and yet still no one could make any sense of the intercepted communication between them.

THE MATHEMATICS OF PUBLIC KEY CRYPTOGRAPHY

A SECRET COMMUNICATION system that lets you broadcast the password to the whole world, yet the message stays secret? Wrap your head around that one! Let's see how it's done. (Warning: it involves numbers so big you could wrap them around your head several times.)

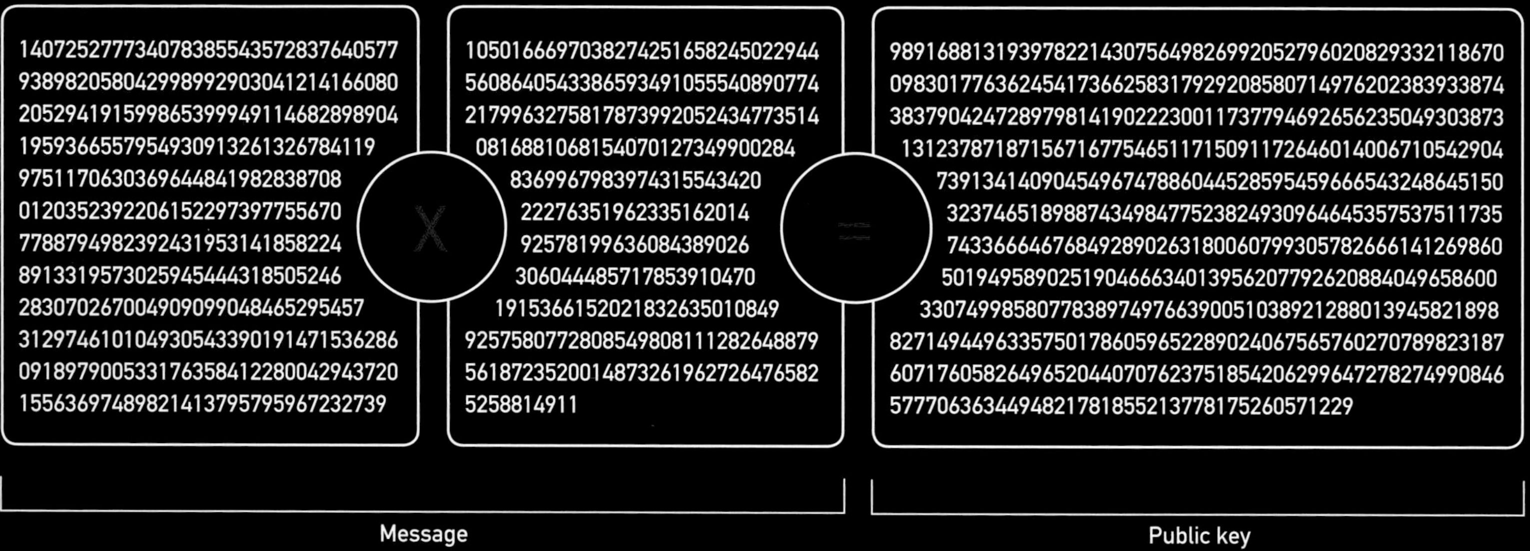

1407252777340783855435728376405779389820580429989929030412141660802052941915998653999491146828989041959366557954930913261326784119975117063036964484198283870801203523922061522973977556707788794982392431953141858224891331957302594544431850524628307026700490909904846529545731297461010493054339019147153628609189790053317635841228004294372015563697489821413795795967232739

X

1050166697038274251658245022944560864054338659349105554089077421799632758178739920524347735140816881068154070127349900284836996798397431554342022276351962335162014925781996360843890263060444857178539104701915366152021832635010849925758077280854980811128264887956187235200148732619627264765825258814911

=

989168813193978221430756498269920527960208293321186700983017763624541736625831792920858071497620238393387438379042472897981419022230011737794692656235049303873131237871871567167754651171509117264601400671054290473913414090454967478860445285954596665432486451503237465189887434984775238249309646453575375117357433666467684928902631800607993057826661412698605019495890251904666340139562077926208840496586003307499858077838974976639005103892128801394582189882714944963357501786059652289024067565760270789823187607176058264965204407076237518542062996472782749908465777063634494821781855213778175260571229

Message

Public key

Message

Public key

Encrypted message

Encrypted message

Private key

Message

A prime number is a number that cannot by divided by any other number. For example, 10 is not prime, because you can divide it by 2 and 5. But 11 is prime, because it can't be divided by anything other than 1 (which divides any number so it doesn't count) and 11 (which is itself, so it doesn't count either).

Figuring out what other numbers divide a given number is easy enough when the number is fairly small. For example, my computer will instantly tell me that 23458204234234234323409349 = 409 x 21821 x 103130479 x 25486474879 (and that each of those four numbers is prime, so they can't be broken down into any smaller factors).

But if you have a number with many hundreds of digits, like the ones above, it gets to be very difficult indeed, especially if the number is made up of only two other very large prime numbers. In my example, a 600-digit number is the result of multiplying together two 300-digit numbers, both of which are prime. No computer in existence, and no computer we have yet imagined, would be able to figure out the two 300-digit prime numbers if it had only the 600-digit number to start with. (If you're reading this years after it was written, you might have to increase the size of the numbers for this statement to remain true. But no matter how fast computers get, there will always be numbers long enough that they can't realistically be factored.)

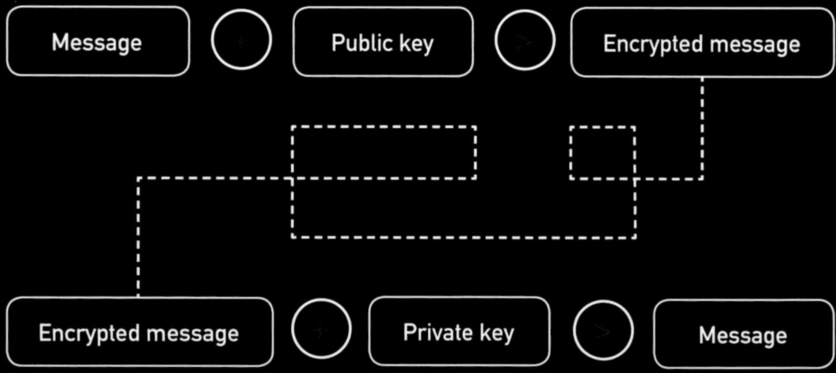

The brilliant idea behind what is called public key cryptography is that you encode a message using the 600-digit multiplied-together number as the secret code, but do it in such a way that the message can only be decoded by someone who *also* knows the two 300-digit factors.

To set up a bi-directional, secure, secret communication channel, each side picks two roughly 300-digit prime numbers (which is easy to do). These two numbers are their "private key" and they keep it secret.

Define

$$n \equiv p\,q$$

for p and q primes. Also define a private key d and a public key e such that

$$d\,e \equiv 1 \pmod{\phi(n)}$$

$$(e, \phi(n)) = 1,$$

where $\phi(n)$ is the totient function, (a, b) denotes the greatest common divisor (so $(a, b) = 1$ means that a and b are relatively prime), and $a \equiv b \pmod{m}$ is a congruence. Let the message be converted to a number M. The sender then makes n and e public and sends

$$E = M^e \pmod{n}.$$

To decode, the receiver (who knows d) computes

$$E^d \equiv (M^e)^d \equiv M^{e\,d} \equiv M^{N\,\phi(n)+1} \equiv M \pmod{n},$$

since N is an integer. In order to crack the code, d must be found. But this requires factorization of n since

$$\phi(n) = (p-1)(q-1).$$

Both p and q should be picked so that $p \pm 1$ and $q \pm 1$ are divisible by large primes, since otherwise the Pollard p − 1 factorization method or Williams p + 1 factorization method potentially factor n easily. It is also desirable to have $\phi(\phi(p\,q))$ large and divisible by large primes.

It is possible to break the cryptosystem by repeated encryption if a unit of $\mathbb{Z}/\phi(n)\mathbb{Z}$ has small field order (Simmons and Norris 1977, Meijer 1996), where $\mathbb{Z}/s\mathbb{Z}$ is the ring of integers between 0 and $s-1$ under addition and multiplication (mod s). Meijer (1996) shows that "almost" every encryption

This pair of numbers does not need to be communicated to anyone, just stored safely at home.

Then they multiply those two numbers together to get a roughly 600-digit number, which they can shout from the rooftops. That is their "public key." Everyone can know that number, and use it to encode messages. Those messages can only be decoded if you know the two 300-digit numbers, so only the person who published the public key—the one who knows the private key—can read the messages.

Both sides exchange their public keys out in the open: it doesn't much matter if the enemy sees them or not, because both sides also closely guard their private keys at home. Now both sides can send secret messages to each other that only the other one can decode. And they never have to exchange *anything* through a secure channel. No one has to brave crossing enemy lines to deliver a locked suitcase containing a single sheet of paper with a secret password on it. Whole sections of standard-issue spy movies have become obsolete.

The math is a bit more complicated than I've described, but the key "hard part" that makes the system secure is the difficulty I've described of factoring very large numbers. Additional clever variations of the scheme can be used to, for example, verify that a given public key actually came from the person you think it did, or that a given document was in fact created by that person (this is called a digital signature).

This form of public key cryptography underlies the entire world of modern communication, from web browsers to secure bank communications to its ultimate application, blockchain ledgers and public distributed currencies like Bitcoin and Newton.

We have now arrived at the end of the chapter on locks with pretty much the most abstract possible form of lock: a cryptographic system based on the branch of mathematics called number theory, which is widely regarded, even by mathematicians, as the branch most likely to give you a headache.

CLOCKS

CLOCKS WERE THE first really *good* machines. They were developed earlier, had more complex mechanisms, and were more widely valued than any other intricate mechanical device. As leading representatives of mechanical things, clocks stayed neck and neck with steam engines for centuries, until the invention of muscle cars and the iPhone.

Early Clocks

THE FIRST CLOCKS were not the clever devices they would eventually become, but they did work. A stick in the ground will tell time as long as the sun is shining and you know which way is north. As the sun moves through the sky, the shadow of the stick shows what is now known as "local solar time." (It used to be known as just "time" until new forms of time were invented, including time zones, daylight saving time, and Coordinated Universal Time.)

Add tick marks calibrated for the local angles of the sun at different times and you have what's called a sundial. It's still basically a stick in the ground, but now it's a fancy calibrated brass stick in the ground. Sundials do genuinely work. Once properly placed, they will tell accurate time for as long as the sun shines and the earth turns. (Actually not. See page 132.)

Of course, the problem with a stick in the ground is that it stops working the minute a cloud passes overhead. And at night you've really got a problem.

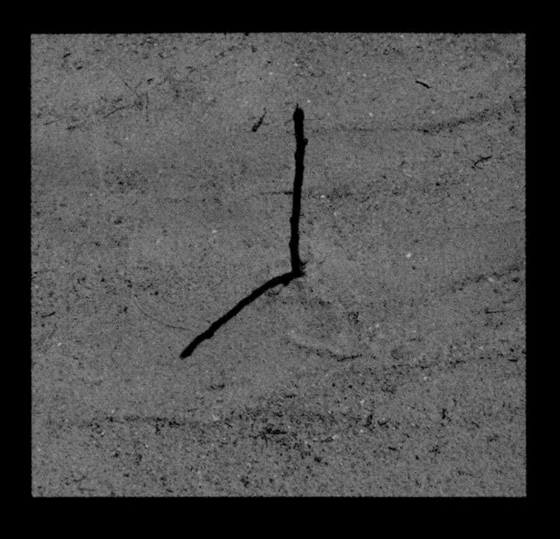

▲ A stick in the ground casts a shadow that tells you the time of day.

◀ A sundial is just a calibrated stick in the ground.

▼ Unfortunately, a stick in the ground is foiled by clouds or night.

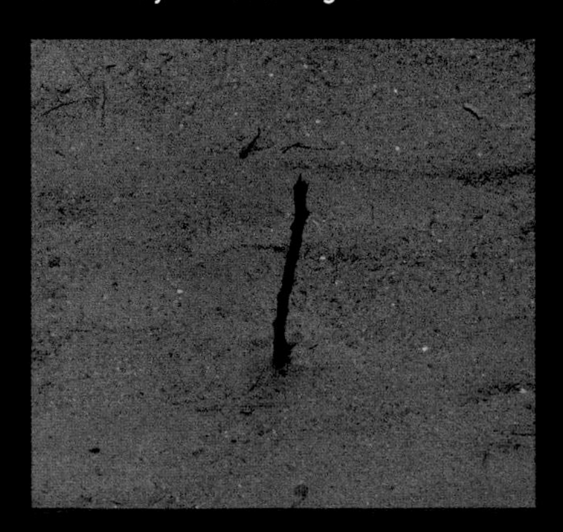

Hourglasses

SAND FALLING AT a predictable rate through a hole is another classic pre-mechanical form of clock. Most "hourglasses" don't actually run for a whole hour. Nor do three-minute egg timers run for exactly three minutes.

A random collection of hourglasses

Hourglasses meant to be used on sailing vessels were hung from a hook, which kept them level (and thus running at a consistent rate) as the ship tilted back and forth. A sliding mount let you flip them over to restart them when they ran out.

By changing the shape of the hourglass and adding calibrated marks, you can read off intermediate times as the sand slowly fills up.

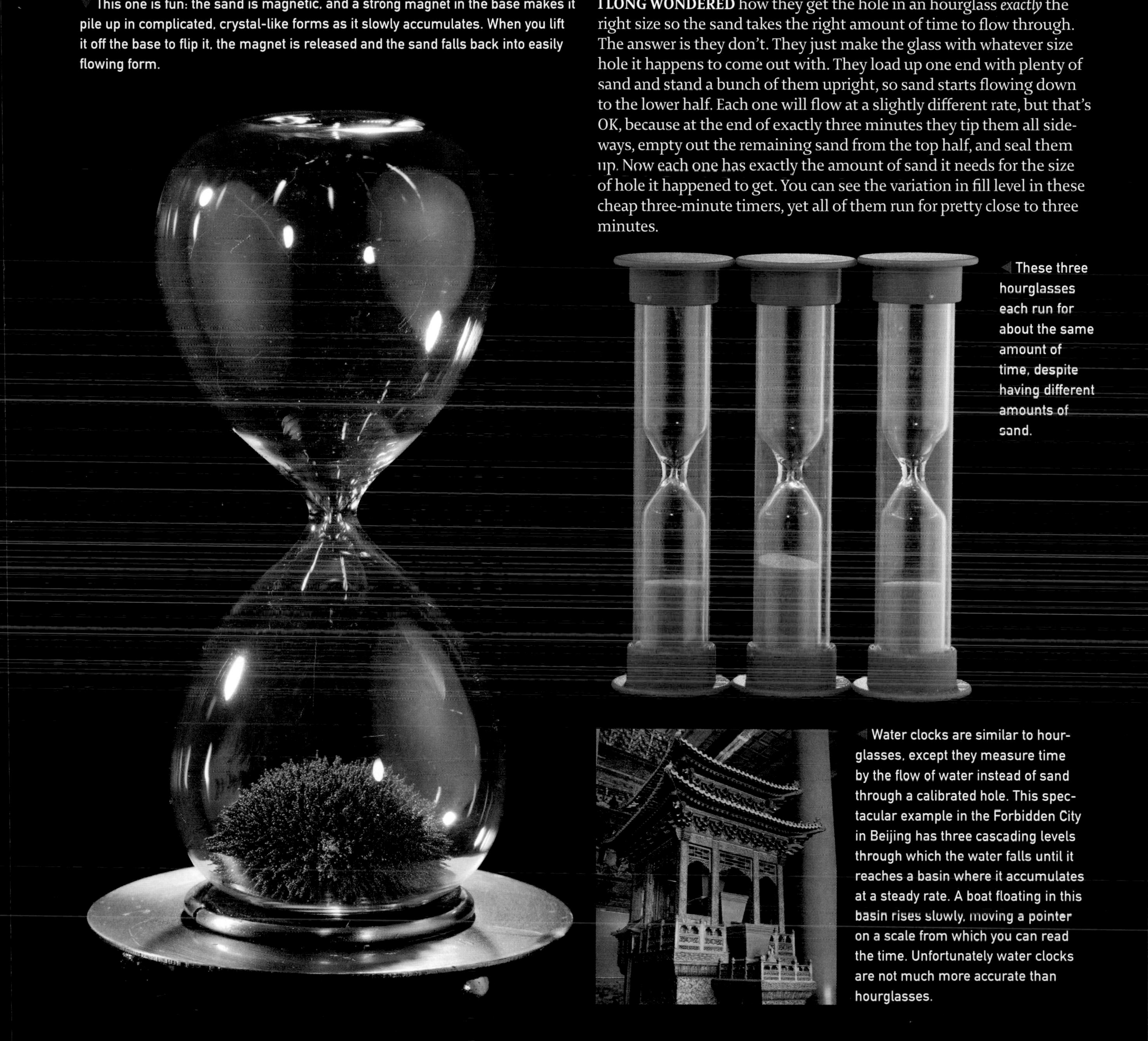

This one is fun: the sand is magnetic, and a strong magnet in the base makes it pile up in complicated, crystal-like forms as it slowly accumulates. When you lift it off the base to flip it, the magnet is released and the sand falls back into easily flowing form.

I LONG WONDERED how they get the hole in an hourglass *exactly* the right size so the sand takes the right amount of time to flow through. The answer is they don't. They just make the glass with whatever size hole it happens to come out with. They load up one end with plenty of sand and stand a bunch of them upright, so sand starts flowing down to the lower half. Each one will flow at a slightly different rate, but that's OK, because at the end of exactly three minutes they tip them all sideways, empty out the remaining sand from the top half, and seal them up. Now each one has exactly the amount of sand it needs for the size of hole it happened to get. You can see the variation in fill level in these cheap three-minute timers, yet all of them run for pretty close to three minutes.

These three hourglasses each run for about the same amount of time, despite having different amounts of sand.

Water clocks are similar to hourglasses, except they measure time by the flow of water instead of sand through a calibrated hole. This spectacular example in the Forbidden City in Beijing has three cascading levels through which the water falls until it reaches a basin where it accumulates at a steady rate. A boat floating in this basin rises slowly, moving a pointer on a scale from which you can read the time. Unfortunately water clocks are not much more accurate than hourglasses.

Candle Clocks. Yes, Candles.

THE RATE AT which a candle burns down is another way of telling time, approximately. As long as your candles are all made the same way, from the same wax, with the same wicks, they should burn at pretty close to the same rate. So with a calibrated candle holder maybe you can tell what hour it is, roughly.

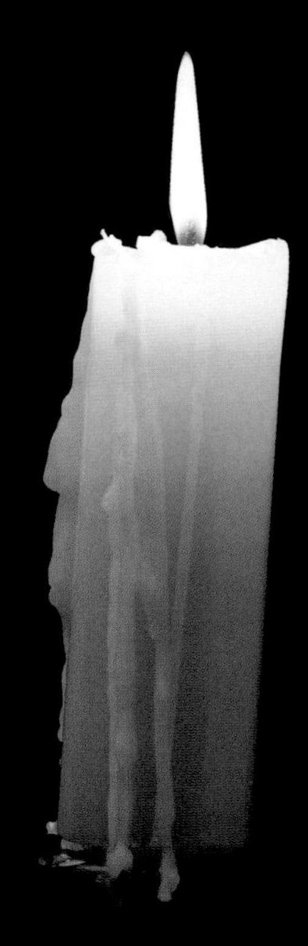

The position of the shadow shows hours passed since the candle was lit.

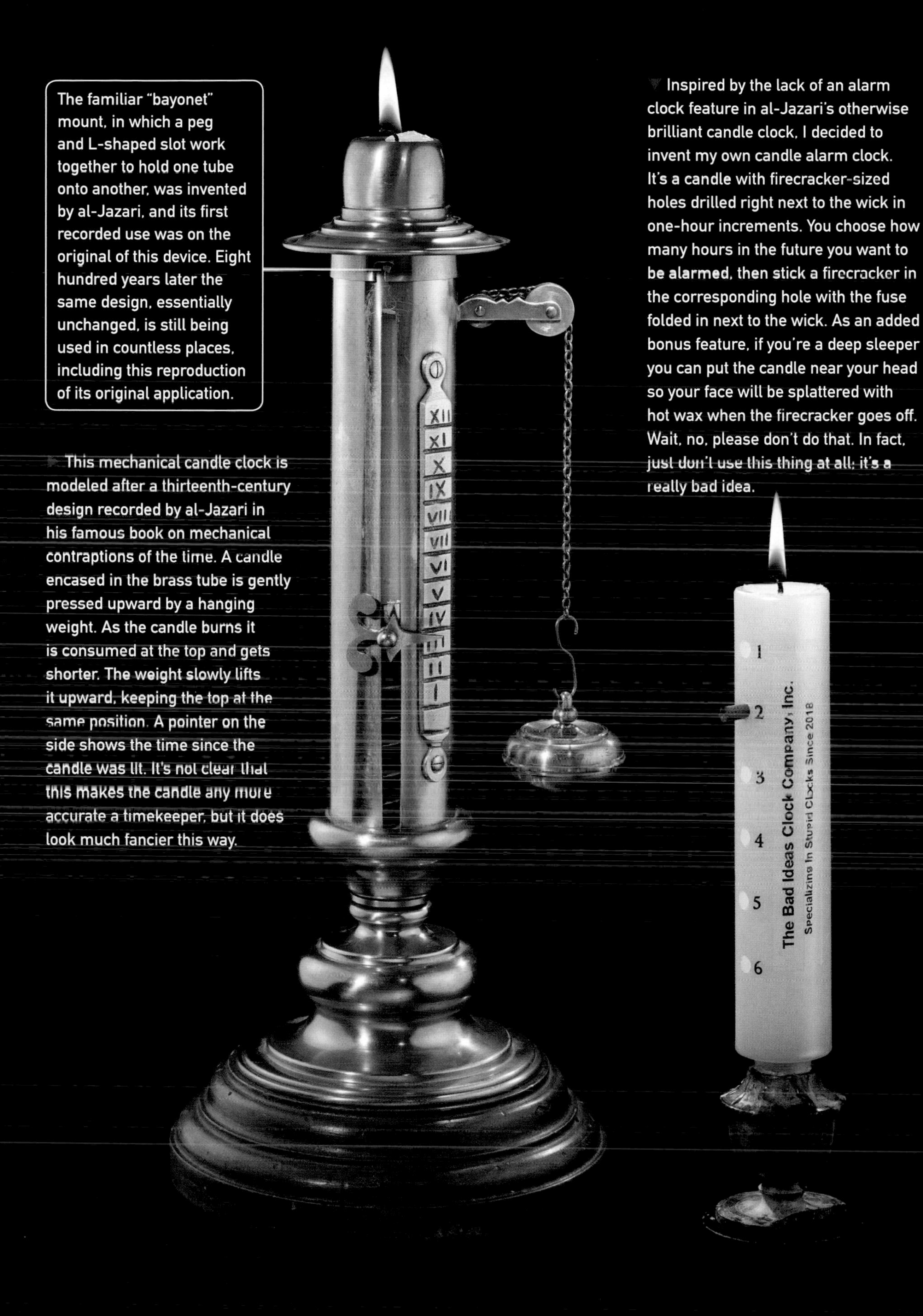

The familiar "bayonet" mount, in which a peg and L-shaped slot work together to hold one tube onto another, was invented by al-Jazari, and its first recorded use was on the original of this device. Eight hundred years later the same design, essentially unchanged, is still being used in countless places, including this reproduction of its original application.

▶ This mechanical candle clock is modeled after a thirteenth-century design recorded by al-Jazari in his famous book on mechanical contraptions of the time. A candle encased in the brass tube is gently pressed upward by a hanging weight. As the candle burns it is consumed at the top and gets shorter. The weight slowly lifts it upward, keeping the top at the same position. A pointer on the side shows the time since the candle was lit. It's not clear that this makes the candle any more accurate a timekeeper, but it does look much fancier this way.

▼ Inspired by the lack of an alarm clock feature in al-Jazari's otherwise brilliant candle clock, I decided to invent my own candle alarm clock. It's a candle with firecracker-sized holes drilled right next to the wick in one-hour increments. You choose how many hours in the future you want to be alarmed, then stick a firecracker in the corresponding hole with the fuse folded in next to the wick. As an added bonus feature, if you're a deep sleeper you can put the candle near your head so your face will be splattered with hot wax when the firecracker goes off. Wait, no, please don't do that. In fact, just don't use this thing at all: it's a really bad idea.

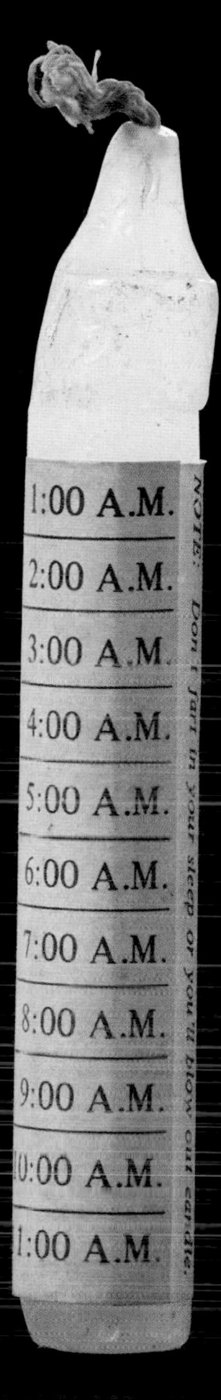

▲ You might think I have a monopoly on really bad ideas for candle alarm clocks, but you'd be wrong. This commercially made bad idea has instructions on the other side (which I have not shown in order to protect the delicate sensibilities of my readers) that say you should insert it into a certain part of your body, up to the desired hour mark, and then light it before going to bed facedown. I cannot overstate what a terrible idea that would be, and somehow I don't think the people who sell these expect anyone to actually use it.

The Greenwich Observatory

SUNDIALS MAY SEEM crude (and that candle alarm clock certainly is), but until 1955 the most accurate time in the world was kept by what amounted to a series of glorified sundials. The first ones used to keep official time were installed in 1675 in this room in the famous Greenwich Observatory in London. Instead of a stick casting a shadow, it used a "quadrant" to measure the angle of the shadow cast by the sun on a calibrated scale, a bit like a protractor.

Every day the observatory staff reset their best mechanical clock to the observed time of noon. (For practical reasons they actually measured a time on either side of noon when the sun was at equal elevations. The time exactly between those two is the observed time of

The best mechanical clocks of the era could only keep time accurate to about five seconds per day, but by resetting these clocks from the sun every day, they were able to maintain a universal time standard that was never more than five seconds off.

A clock face on the outer wall of the observatory, still running today, communicates Greenwich Mean Time to the world. A giant red ball on the roof, visible from the heart of London, drops at exactly noon each day, allowing anyone in the city to re-synchronize their clock with the standard. These traditions are carried on for sentimental, and I suppose tourist-related, reasons. But on page 124 we will encounter a wristwatch that uses a remarkably similar system, up-

The main problem with measuring time from the position of the sun is that the sun is really big and has fuzzy edges, making it hard to say *exactly* where it is in the sky. So, starting in 1721, and continuing all the way until 1955, time was determined by measuring the exact moment that certain stars crossed directly overhead. Because stars are so far away, they appear as almost infinitely small points of light whose positions can be measured with tremendous precision. The Troughton 10-foot Transit Instrument is basically a sundial for stars. Instead of the shadow of the sun, it observes the light of the stars.

Sundials, candles, and transit instruments can keep extremely good time, but they aren't really clocks as defined in this chapter. I mean, they have no moving parts—what sort of a clock is that? And yet, even in ancient times there existed highly complex devices not unlike modern mechanical clocks. (See the sidebar on the Antikythera mechanism.)

So, now that we've beat around the bush long enough with clocks that have no clockworks and clockworks that are not clocks, let's take a look at the first really good clockwork clocks.

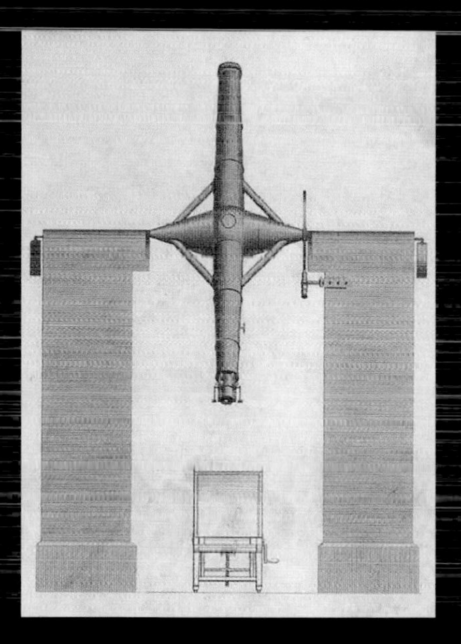

An illustration of the Troughton 10-foot Transit Instrument

This ball atop the Greenwich Observatory drops at exactly noon every day.

The Greenwich Observatory has been famous for a very long time, in large part because it defines the origin of the lines of longitude (degrees east and west on the surface of the earth). I'm pretty sure they didn't have a gift shop in 1675, but today they do, and in it you can buy this hat to remember the day you stood on the prime meridian (the line that defines zero longitude).

ANTIKYTHERA RECONSTRUCTION

THIS IS PERHAPS the single most remarkable ancient artifact ever discovered. There are hundreds of massive pyramids and thousands of astonishingly beautiful ancient sculptures, but this thing is unique and nothing like it has ever been found from the ancient world. It was clearly the product of one or more mechanical designers so far ahead of their time that their work must have seemed magical to those around them. Its equal in fine machining would not be seen again for 1,500 years.

This is the Antikythera mechanism. In 1901 it was found in a shipwreck where it had lain for at least two thousand years. (This picture is a computer reconstruction based on careful x-ray analysis of the original mess found at the bottom of the ocean, which was almost hopelessly corroded and encrusted.)

Technically, it's not a clock: it doesn't tell time. It's more of a mechanical calendar and astronomical prediction machine. Using dozens of connected gears, the dials and pointers on its face show the movements of the sun and the moon, predict eclipses, and spell out the dates of the ancient Olympic games—for many decades beyond the time it was made.

The Antikythera mechanism may not be a clock, but it is filled with *clockwork*, a word used to describe the sorts of gears, springs, wheels, levers, and dials that are characteristic of mechanical clocks.

Pendulum Clocks

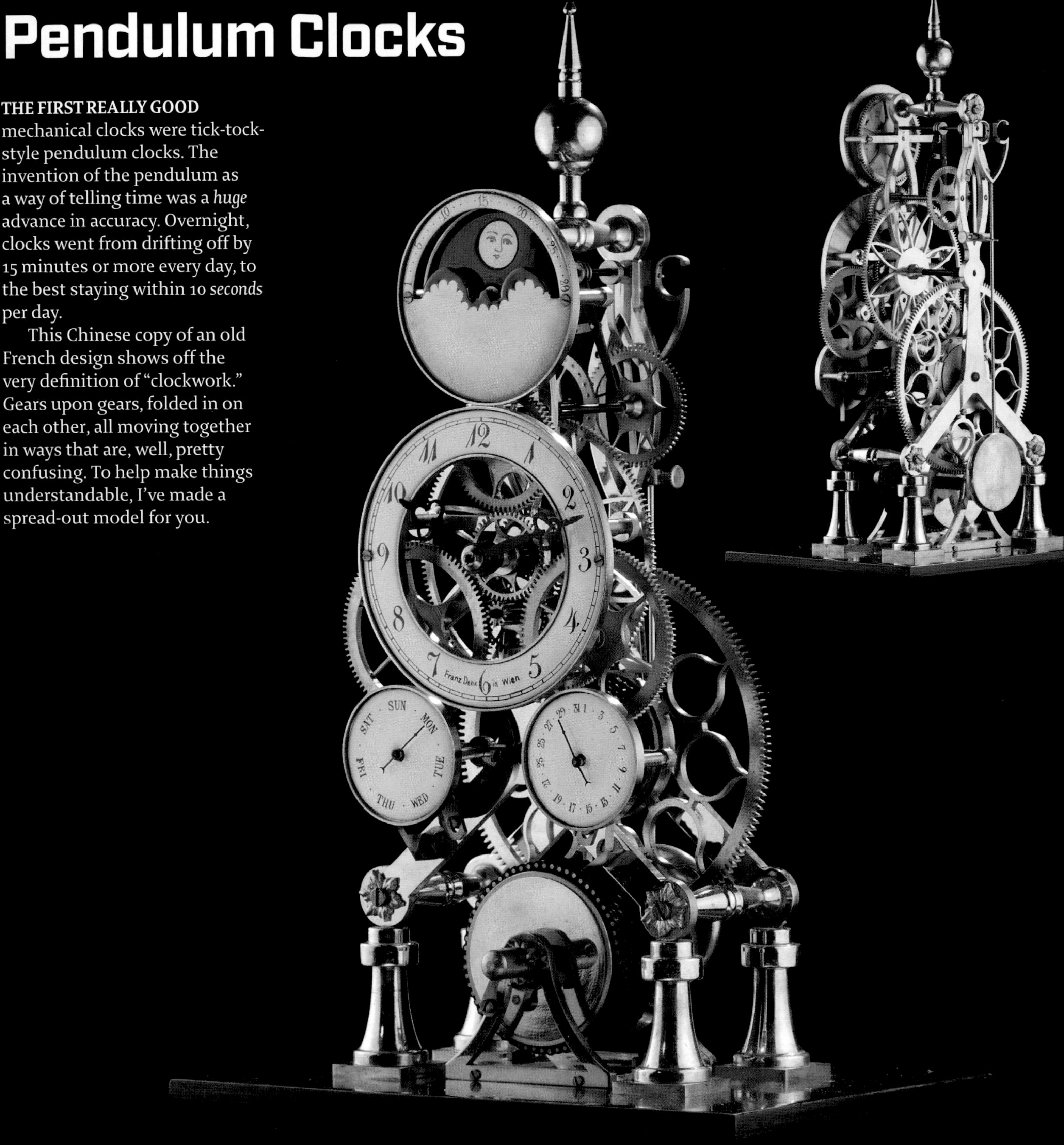

THE FIRST REALLY GOOD mechanical clocks were tick-tock-style pendulum clocks. The invention of the pendulum as a way of telling time was a *huge* advance in accuracy. Overnight, clocks went from drifting off by 15 minutes or more every day, to the best staying within 10 *seconds* per day.

This Chinese copy of an old French design shows off the very definition of "clockwork." Gears upon gears, folded in on each other, all moving together in ways that are, well, pretty confusing. To help make things understandable, I've made a spread-out model for you.

FOR THE BENEFIT of those of you born in the present millennium, a primer on clocks of the old school. This is the universal symbol of time, the clock face. There are three "hands," which all turn around the same central point. Around the outside there are two sets of marks: one set of 60 small tick marks for seconds and minutes, and one set of 12 large tick marks for hours. This clock is showing 10:14:35 (14 minutes and 35 seconds after 10 o'clock).

The three hands on a clock are all turning around the same point, but they are turning at wildly different speeds. It takes 60 times longer for the minute hand to go around than the second hand, and the hour hand is another 12 times slower. For every complete turn of the hour hand, the second hand has to go around 720 times. This huge difference in speed is created by a mess of gears on the back of the clock.

The large tick marks with numbers (Roman numerals in this case) mark the hours.

The small tick marks, often not numbered, mark the minutes and seconds. You're just supposed to know that each hour mark corresponds to an interval of 5 minutes (for the minute hand) or 5 seconds (for the second hand). Each quarter of the circle is 15 minutes or 15 seconds, respectively.

The "hour hand" moves forward one large tick mark per hour, and takes 12 hours to go around one full turn. This hour hand is pointing about one-quarter of the way from the 10 mark to the 11 mark. That's because we are 14 minutes (and 35 seconds) past 10 o'clock.

The "minute hand" moves forward one small tick mark each minute, and takes one hour (60 minutes) to go around one full turn. This minute hand is pointing a little over halfway from the 14 minute mark to the 15 minute mark. That's because we are 35 seconds—about halfway—between those two minutes.

The "second hand" moves forward (in the "clockwise" direction) by one small tick mark each second, and takes one minute (60 seconds) to go around one full turn. This second hand is pointing at the 35 second mark.

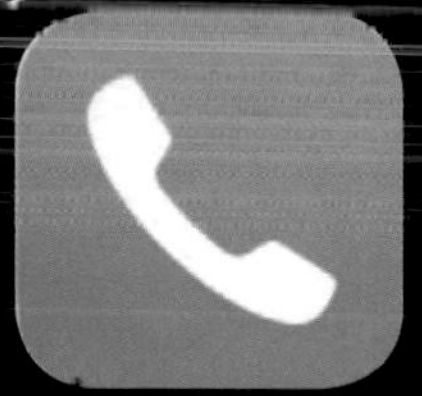

▲ There is something amusing in the fact that the telephone app on today's iPhones has an icon that looks like a phone no one has used in twenty years, and the clock app looks like a clock most people rarely see anymore. (It's also amusing that a separate telephone app is necessary on a device that is, by name, already a phone. But, the truth is, making phone calls is one of the iPhone's least commonly used functions.)

THIS IS A model of a pendulum clock in which the gears have been spread out on a line so each one can be seen and appreciated separately. Instead of one clock face with three hands, there are three separate clock faces: one for the seconds, one for the minutes, and one for the hours. We'll look at each of the parts of this clock separately over the next few pages.

This gear chain is what keeps the three hands of the clock turning at exactly the right speeds relative to each other. As long as the escapement is doing its job ticking off seconds at the correct rate, the rest of the clock will follow along perfectly. The beautiful thing about gears is that they can't slip: you could turn the second hand a million times and, unless something broke, the hour hand would still be in exactly the right place. (A million times might sound like forever, but it's less than two years of normal operation: a million minutes is 694 days.)

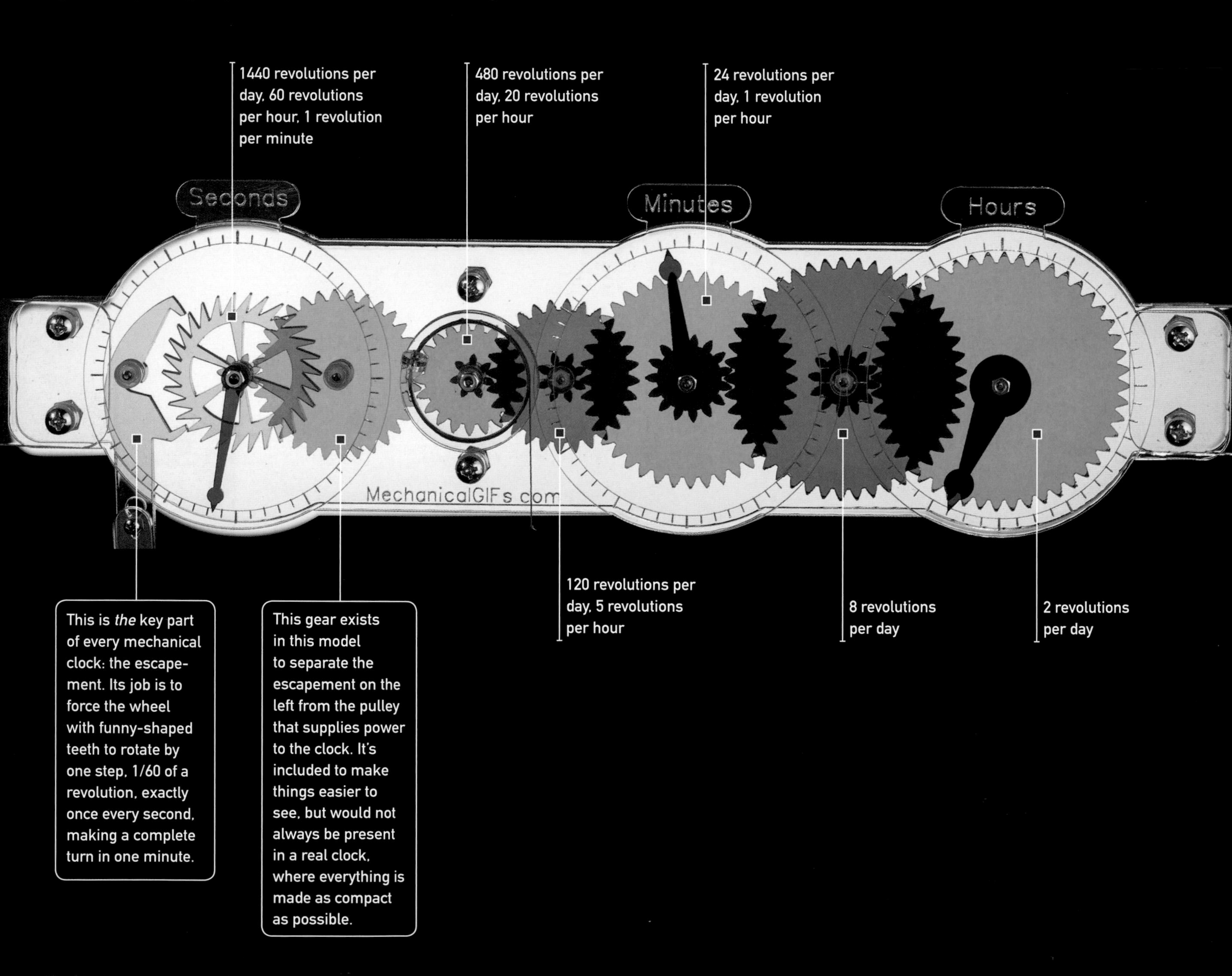

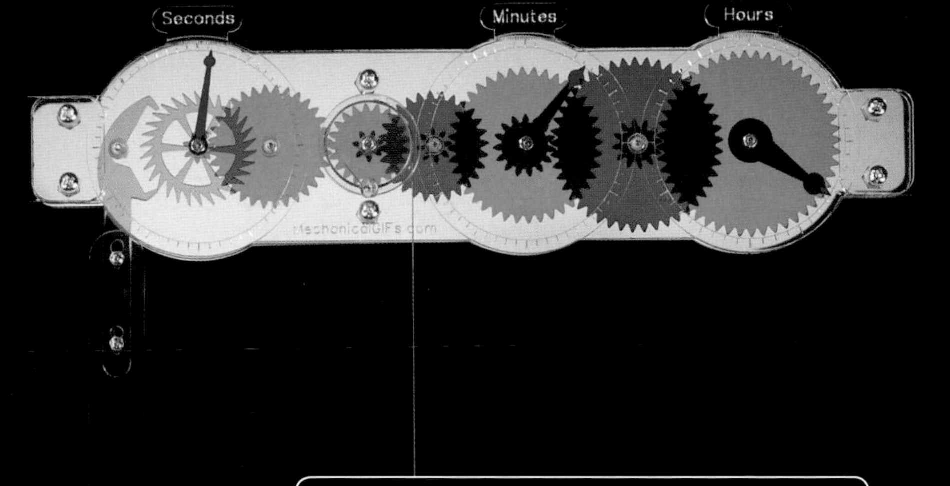

This small gear has 10 teeth and the large gear has 30 teeth (3 times as many). When the small gear turns around once, 10 teeth will have passed by the point of contact. The big gear will also have moved forward by 10 teeth, but because it's bigger and has 3 times as many teeth, it will have gone around only 1/3 of a full turn. Only after three complete turns of the small gear will the big gear have gone around once. We say there is a 3-to-1 gear ratio between these two.

This sliding joint lets you adjust the length of the pendulum to fine-tune the speed of the clock. In a real clock adjustments of a small fraction of a millimeter can throw the clock off by several minutes a week.

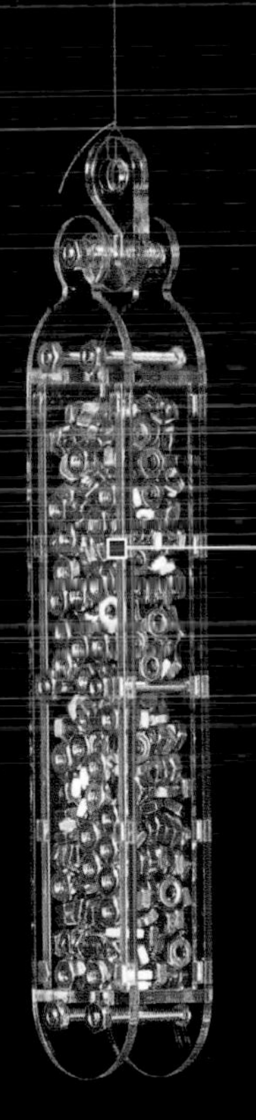

This hanging weight is what supplies energy to the clock. Each time the second hand ticks one notch, the pulley rotates a small amount, lowering the weight a bit and thus releasing a bit of potential energy. When the weight reaches the floor (typically after a few days to a week in a real clock) the clock needs to be rewound by lifting the weight back up to the top. In real clocks these weights are typically hollow brass cylinders filled with lead pellets, solid lead, or cast iron. I filled mine with #6 hex nuts because I have a lot of #6 hex nuts.

The gear ratio between the second hand and the minute hand in a clock needs to be 60 to 1 (the minute hand turns 60 times slower than the second hand). If you give the second hand a gear with 6 teeth (which is about the smallest number of teeth you can have on a gear), the gear on the minute hand would need to have 6 x 60 = 360 teeth. That's a bit nutty: a gear with that many teeth needs to be huge! And there's another problem with this arrangement: when two gears are meshed with each other, they turn in opposite directions. So if the second hand in this model is going forward (clockwise) the minute hand will be going backward (counter-clockwise). There is a much better way: a compound gear chain.

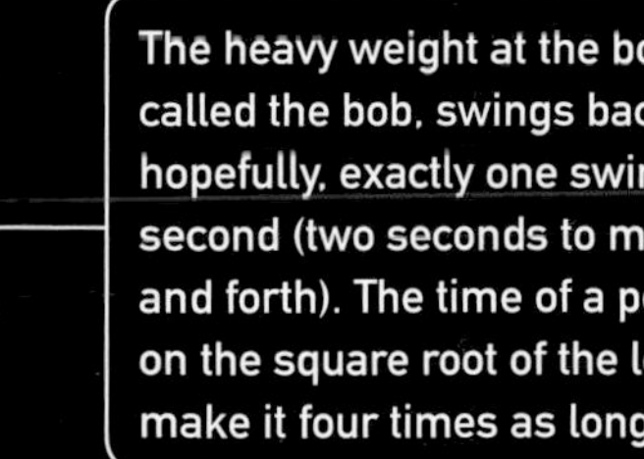

The heavy weight at the bottom of the pendulum, called the bob, swings back and forth at a rate of, hopefully, exactly one swing past the center per second (two seconds to make a complete cycle back and forth). The time of a pendulum's swing depends on the square root of the length of the pendulum: make it four times as long and it swings half as fast.

COMPOUND GEAR CHAINS

THE KEY IDEA in a compound gear chain is this: a pair of gears, one small and one large, fastened to each other on a common axle. Because they are rigidly connected, they will always make the same number of revolutions when they are turning. But because they have different numbers of teeth, other gears meshing with them will see a different number of teeth go by per revolution.

Using a two-stage gear chain from seconds to minutes (as in the example below) is a pretty common design in clocks, but it's not the physically smallest set of gears that can achieve this ratio. Going one step further to a three-stage gear chain can get you an even more compact design. This is what I chose to use for the complete model clock on the previous page, and on the next page is a simplified excerpt, showing just the gears that connect the second and minute hands. The first gear (connected to the second hand) has 7 teeth. It meshes with another 7-tooth gear whose only purpose is to reverse the direction of rotation (because otherwise the minute hand would turn the wrong way). This second small gear meshes with a gear that has 21 teeth: that creates a 3-to-1 reduction in speed (21/7 = 3). The 21-tooth gear is connected on the same shaft to another gear with 7 teeth, which in turn is driving a gear with 28 teeth. That creates a 4-to-1 reduction (28/7 = 4). The 28-tooth gear is on a shaft with a gear that has 8 teeth, which in turn meshes with a gear with 40 teeth. That creates a 5-to-1 reduction (40/8 = 5). All together, these three reduction ratios equal a 60-to-1 ratio (3 x 4 x 5 = 60).

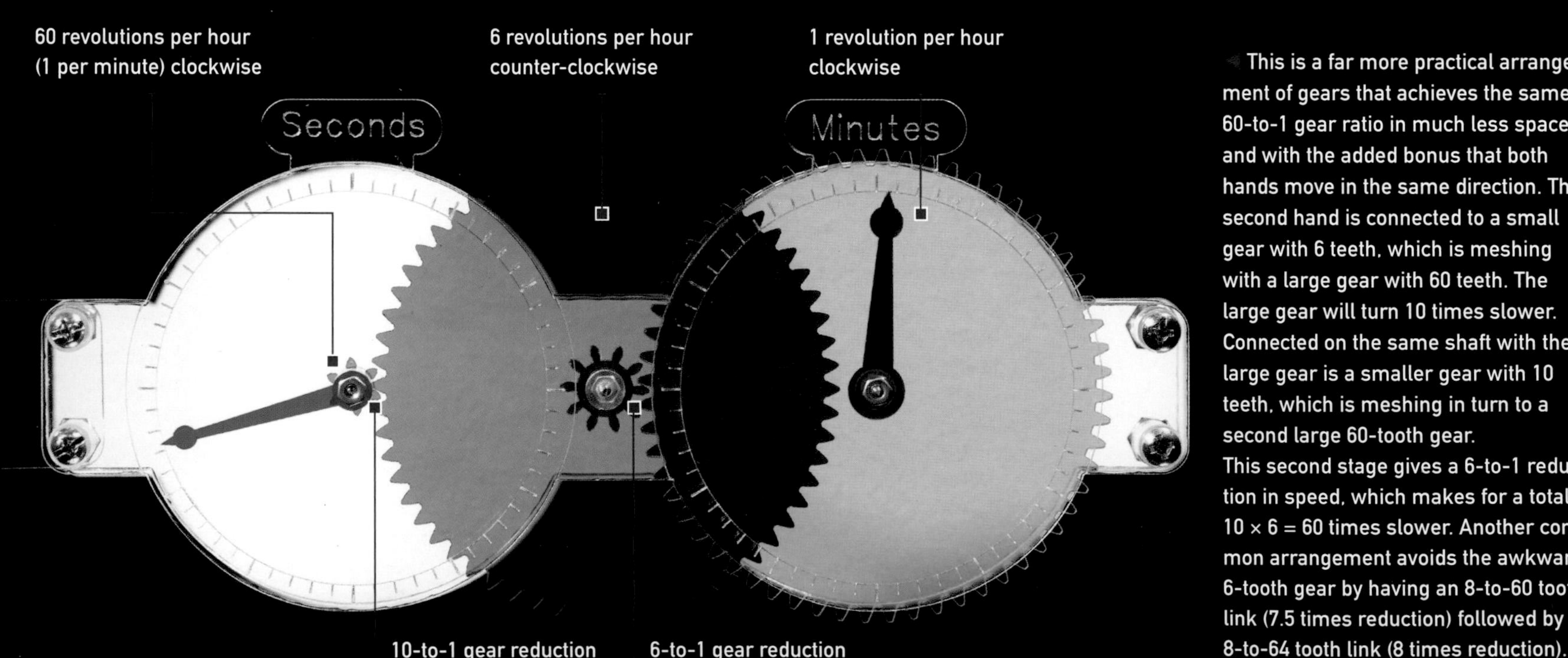

This is a far more practical arrangement of gears that achieves the same 60-to-1 gear ratio in much less space, and with the added bonus that both hands move in the same direction. The second hand is connected to a small gear with 6 teeth, which is meshing with a large gear with 60 teeth. The large gear will turn 10 times slower. Connected on the same shaft with the large gear is a smaller gear with 10 teeth, which is meshing in turn to a second large 60-tooth gear.
This second stage gives a 6-to-1 reduction in speed, which makes for a total of 10 × 6 = 60 times slower. Another common arrangement avoids the awkward 6-tooth gear by having an 8-to-60 tooth link (7.5 times reduction) followed by an 8-to-64 tooth link (8 times reduction). This works because 7.5 x 8 = 60.

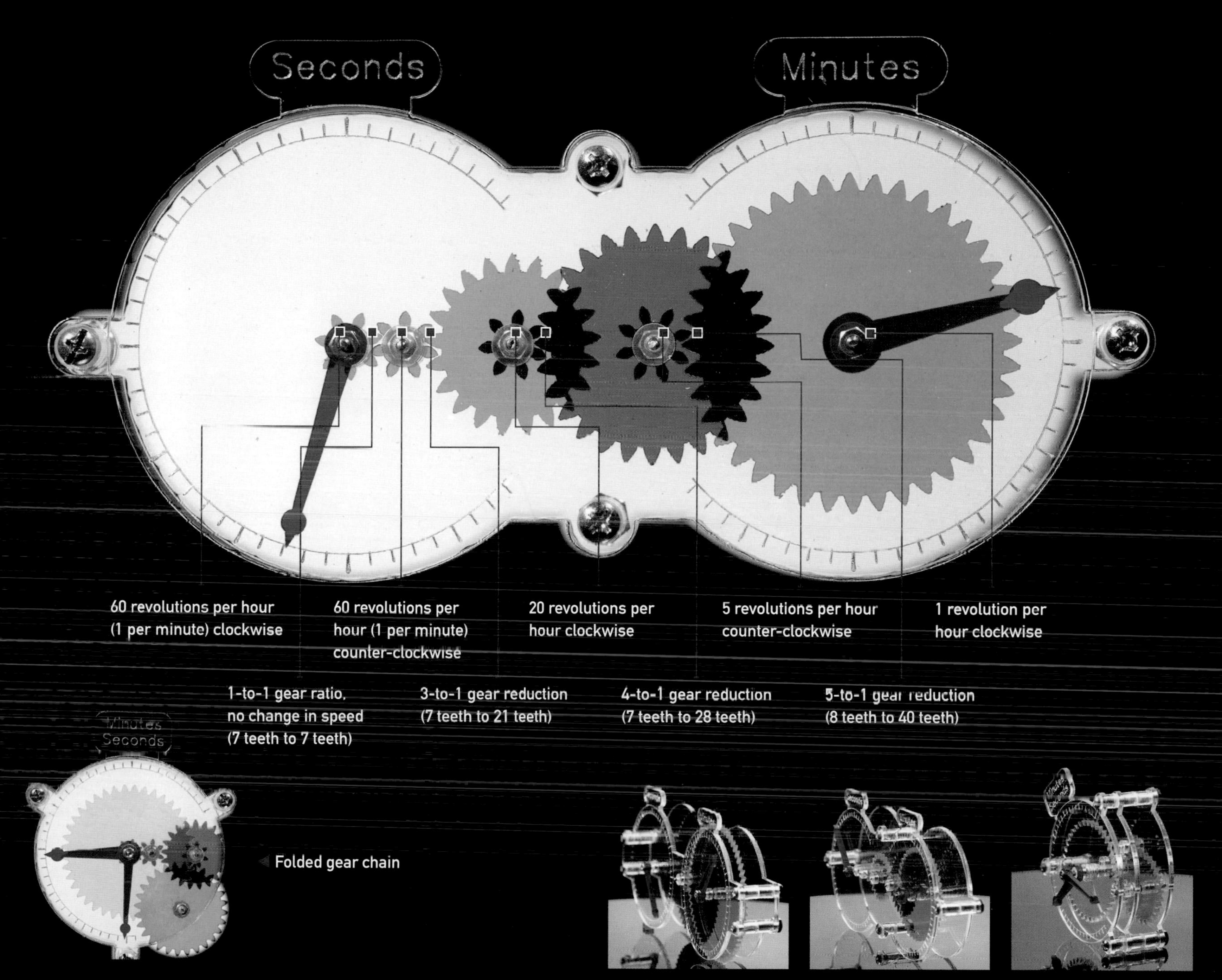

IN A REAL CLOCK, the 60-to-1 gear chain has to be folded in on itself to put the second and minute hands in the same place. The second hand is connected to a solid shaft that goes through a hollow shaft connected to the minute hand, allowing them to turn independently around the same center. (The hour hand, not included in this model, adds a third hollow shaft around the minute hand's hollow shaft.) This model contains exactly the same gear chain as the one above, but it's harder to understand because the gears are all bunched together. It only gets worse when you add the hour hand.

The price you pay for folding up a gear chain, aside from making it confusing to look at, is that you need to add more layers. Any time the axle of one gear falls within the area of another gear, their bearings need to be in separate layers of the mechanism. In the complete clock model on the previous page I spent a lot of time figuring out a set of gear ratios that not only multiply together to the correct overall ratios, but also have each gear a bit bigger than the previous one, so the whole thing could fit in one layer. That's possible in my model only because it's spread out: there's no way to do it in a clock where all the hands are together on one center.

ESCAPEMENTS

THE ESCAPEMENT IN a clock has two functions: let the toothed escapement wheel turn exactly one notch each time the pendulum swings back or forth (the timekeeping function), and give the pendulum a tiny push on each swing to keep it going (without which the pendulum would eventually stop).

At first, escapements combined these two functions into single zones on the "pallets" (the two angled bits at the ends of the "anchor" that touch the teeth of the escapement wheel). Combining the functions meant that the shapes of the pallets had to be a compromise between what was best for timekeeping and what was best for pendulum-pushing. In 1675 the design shown here, the "deadbeat escapement," was invented. Forty years later it became popular, and today, three hundred years later, it is used on most pendulum clocks.

The exact design of the escapement is important because clocks are all about accuracy, and accuracy at the level required for a good clock comes only by rigorously eliminating every possible source of error, at every point in the mechanism. If you want your clock to be accurate to, say, one second per month, that is an accuracy of one part in 2,592,000 (the number of seconds in a month). For a mechanical device, that is a *very* high degree of accuracy!

Mechanical clocks this accurate do exist: the best in the world, a modern re-creation of John Harrison's "Clock B" at the Greenwich Observatory in London, is good to about one second in six months, or one part in 15,552,000. Reaching that extreme level of accuracy requires taking the escapement to another level, as shown on the next page.

Friction is the enemy of durability and accuracy. Notice that there is a point in the movement of the escapement on this page when the paddle is sliding against a tooth of the escapement wheel. The wheel must maintain a certain amount of torque so it can push the pendulum at the center of its movement, so there must necessarily be a certain amount of force pressing the tooth and paddle together while they are sliding. This is bad news.

While it might seem impossible, an escapement can be built in which no part ever slides against another while under any load. The result is a *movement without friction.*

▶ With the pendulum at one end of its swing, a tooth of the escapement wheel is touching the outside curved surface of the right-hand paddle of the escapement. This surface is a segment of a circle with its center at the pivot point of the pendulum. As the pendulum swings, this surface does not change its distance from the pivot point, so the escapement wheel remains stationary.

◀ For a brief interval at the center of the pendulum's swing, the tooth of the escapement wheel is pushing against this short, angled surface. At this point the pendulum is given a push by the torque of the wheel.

▶ At the other end of the pendulum's swing the other palette is engaging with another tooth of the escapement wheel, again on a circular surface that slides without causing the escapement wheel to turn in either direction.

▶ On its way back in the other direction, the pendulum is again given a little push at the center point of its movement, this time from the thrust plane on the other palette.

JOHN HARRISON INVENTED this elegant and remarkable "grasshopper" escapement to use on his Sea Clocks (see sidebar). It operates without friction: at no point is there one part sliding or even rolling against another, except when there is no force pressing them together.

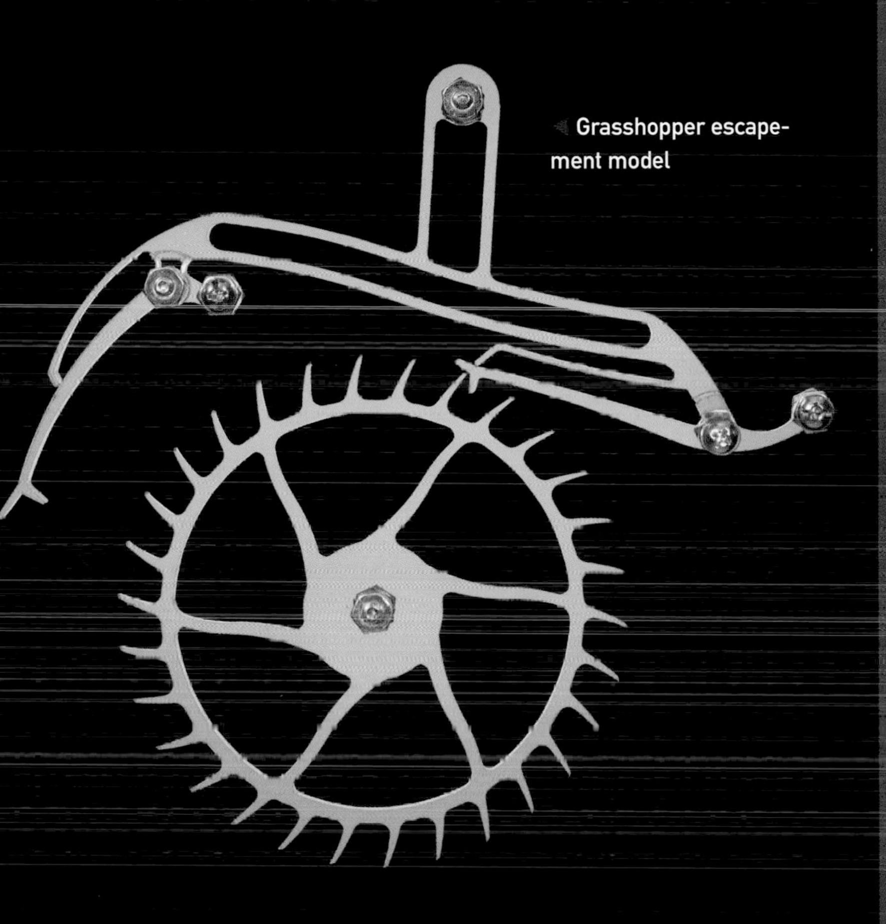

Grasshopper escapement model

The two arms work alternately to release each other. One arm is held in place by light pressure from one of the teeth on the escapement wheel, until the other arm moves into position and rotates the escapement wheel slightly backward, releasing the first arm without either of them having slid against the teeth they are engaging with. Then the cycle repeats with the two arms trading roles. Remarkably, even the bearings that the two arms rotate on never turn when under load: the arms are in motion only when there is no pressure on those bearings.

The simple deadbeat escapement is clever, but in retrospect pretty obvious. The grasshopper is on a whole other level. It takes some study to convince yourself that it's even possible for an escapement to operate without friction. To have conceived of this from scratch is a testament to just what a brilliant designer Harrison was.

THE HARRISON H1 CLOCK

NAVIGATING SAILBOATS across the ocean with any degree of predictability requires being able to measure your position on the earth to within a few miles. The British fleet was losing too many ships for lack of a reliable way to do this, so in 1714 the British government set a prize for anyone who could solve the problem.

It was well known at the time that you can determine your location anywhere on earth if you have an accurate clock. For every minute your clock is off, your measured position will be off by about 16 miles (25 km). If you lived in the 1700s and could make a clock accurate to within about a minute over the course of a two-to-three-month ocean voyage, problem solved and you win enough money to retire in comfort.

Ordinary pendulum clocks of the day were accurate enough, except for one small problem: they absolutely did not work on ships. The smallest movement completely throws off a regular pendulum's swing.

The person who solved the problem, through decades of painstaking work and brilliant engineering insights, was John Harrison. His quest earned him the admiration of many, and toward the end of his life, a fair bit of prize money.

Harrison's first idea was a pair of dumbbell-shaped bars swinging in opposite directions and pushed back and forth by very fine coiled springs instead of gravity. Because the balls were perfectly balanced about their centers of rotation, movement of the clock did not impart any torque on the rods, and thus did not, in theory, change the speed of their movement.

This clock, the Harrison "H1" sea clock, is in the museum at the Greenwich Observatory in London. As beautiful as it is, it was not good enough to win the prize. To do that, Harrison had to abandon large moving weights to create an entirely new style of clock, whose descendants are the mechanical wristwatches of today.

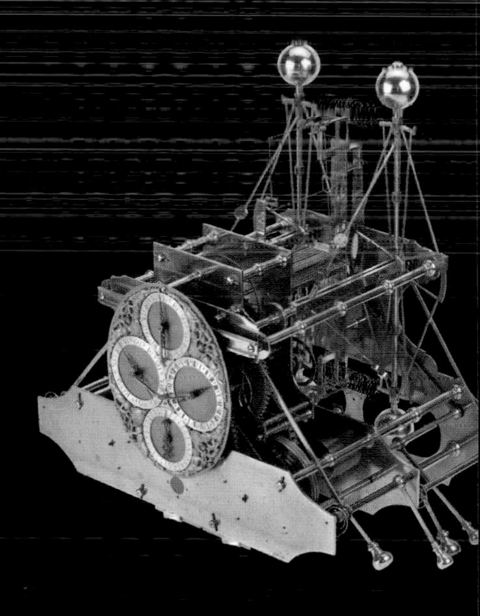

John Harrison's great sea clock

CLOCKS ARE A mess of details. To make the transparent model clock shown on page 88, I had to create laser cutting patterns for each of the parts. Most of them were pretty easy. For example, there are online tools to create gear shapes automatically and download them as files the laser cutter can use directly. I just needed to decide how many teeth I wanted on each gear. Most of the other parts are simple shapes I could easily draw using my favorite CAD (Computer Aided Design) program.

But the escapement was a nightmare.

I started by tracing pictures of escapements, but they didn't work very well, and every modification I tried to get them to work better just seemed to make things worse. I didn't understand the counter-intuitive constraints that caused every escapement I made to either allow the wheel to spin freely between ticks (bad) or lock it up completely (worse).

When trying to solve an engineering problem, there are two approaches. You can cast about aimlessly without really understanding what you're doing until you find something that works. Or you can analyze the situation mathematically and calculate the correct solution—always the more reliable technique, if you can do it.

It might sound like I'm saying that using math and calculation is always the better plan, but there is value in casting about as well. You can learn a lot about the parameters of the situation by trying a lot of things that don't work. These failures help illuminate the boundaries of what isn't possible, and teach you what the issues are going to be that need proper understanding. (And sometimes, for easy problems, messing around is all you need.)

Escapements, apparently, are not in the easy category. Nothing worked for me until I finally went back and read up on the theory and correct geometry of the deadbeat escapement. Using this information, plus a bit of messing around to compensate for the laser cut width (guided this time by the knowledge of how the shapes needed to relate to each other), I was able to get, in about three tries, an escapement that worked better than any of my previous attempts. Another half-dozen tries later it was working quite nicely. And I understood much better how each modification of the shape would affect its movements.

I feel a bit bad that I wasn't able to design a good escapement on my own. But I'm consoled by the fact that it took nearly 400 years to advance from the first mechanical escapement (the verge escapement of 1275) to the escapement that finally allowed for really accurate clocks (the anchor escapement of 1657). It then took a further 18 years for the deadbeat escapement design I'm using to be developed. The fact that I couldn't do it in an afternoon, even with a laser cutter, isn't really too surprising.

A sad pile of the author's failed escapements

Now that we've studied the working parts of a clock, let's look again at the beautiful clock from a few pages ago and see if it makes sense this time.

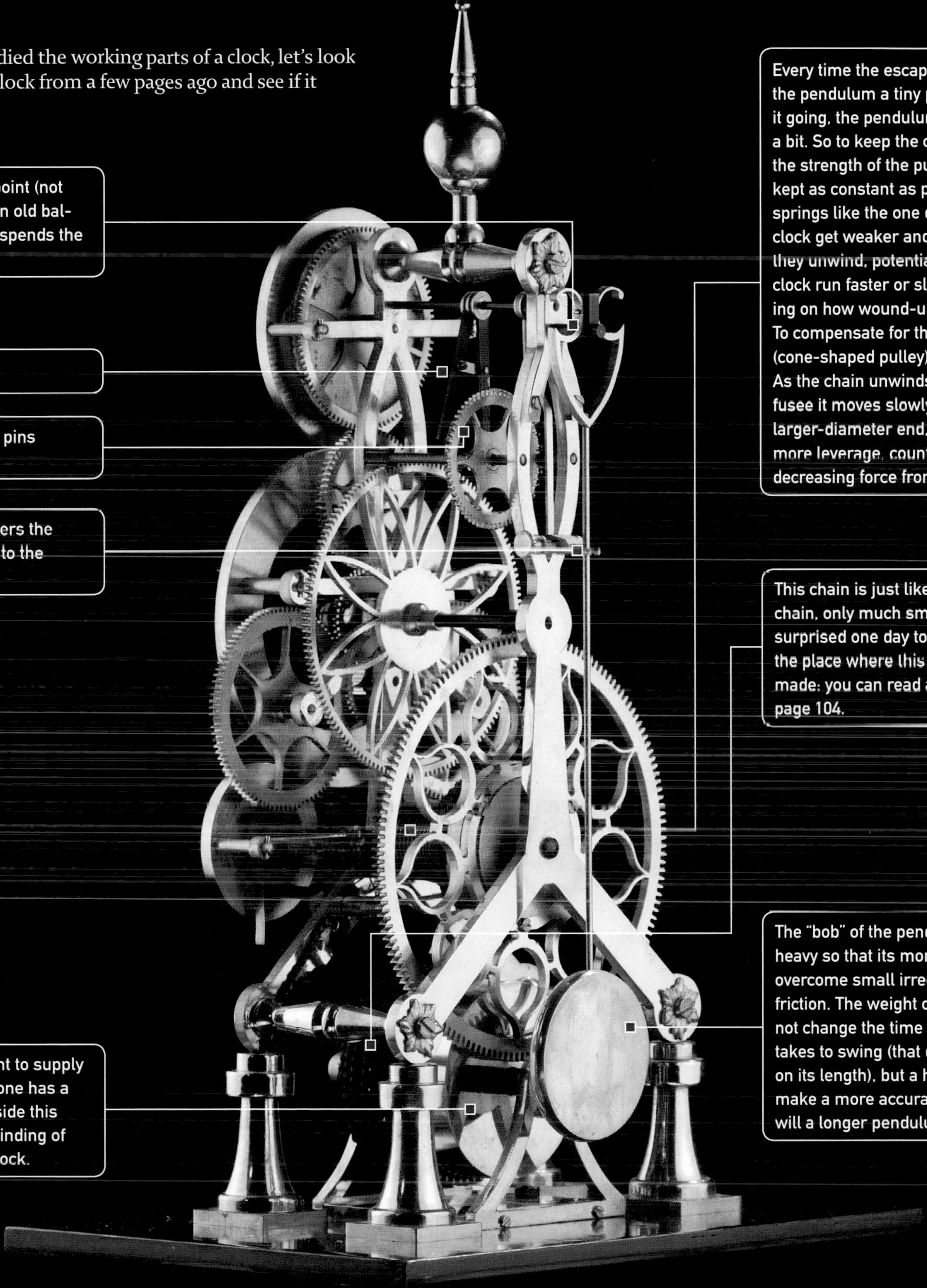

A knife-and-anvil pivot point (not unlike what you find in an old balance scale, page 168) suspends the pendulum.

Escapement anchor

Escapement wheel, with pins instead of teeth

This pin delicately transfers the motion of the pendulum to the escapement anchor.

Instead of a falling weight to supply energy to the clock this one has a coiled spring (hidden inside this protective can). The unwinding of the spring powers the clock.

Every time the escapement gives the pendulum a tiny push to keep it going, the pendulum speeds up a bit. So to keep the clock accurate, the strength of the push must be kept as constant as possible. But springs like the one driving this clock get weaker and weaker as they unwind, potentially making the clock run faster or slower depending on how wound-up it is.
To compensate for this, a "fusee" (cone-shaped pulley) can be used. As the chain unwinds from the fusee it moves slowly toward the larger-diameter end, where it has more leverage, counteracting the decreasing force from the spring.

This chain is just like a bicycle chain, only much smaller. I was surprised one day to find myself in the place where this very chain was made: you can read about that on page 104.

The "bob" of the pendulum is heavy so that its momentum can overcome small irregularities in friction. The weight of the bob does not change the time the pendulum takes to swing (that depends only on its length), but a heavier bob will make a more accurate clock, and so will a longer pendulum.

The Grandfather Clock

A PROPER GRANDFATHER clock, even a modern one like this (made in 1996), is a thing of beauty. Lovely woodwork, ornate dial, polished brass pendulum, and chimes every 15 minutes—all are part of the deal for such a clock—as is a pendulum about 3 feet, or a meter, long.

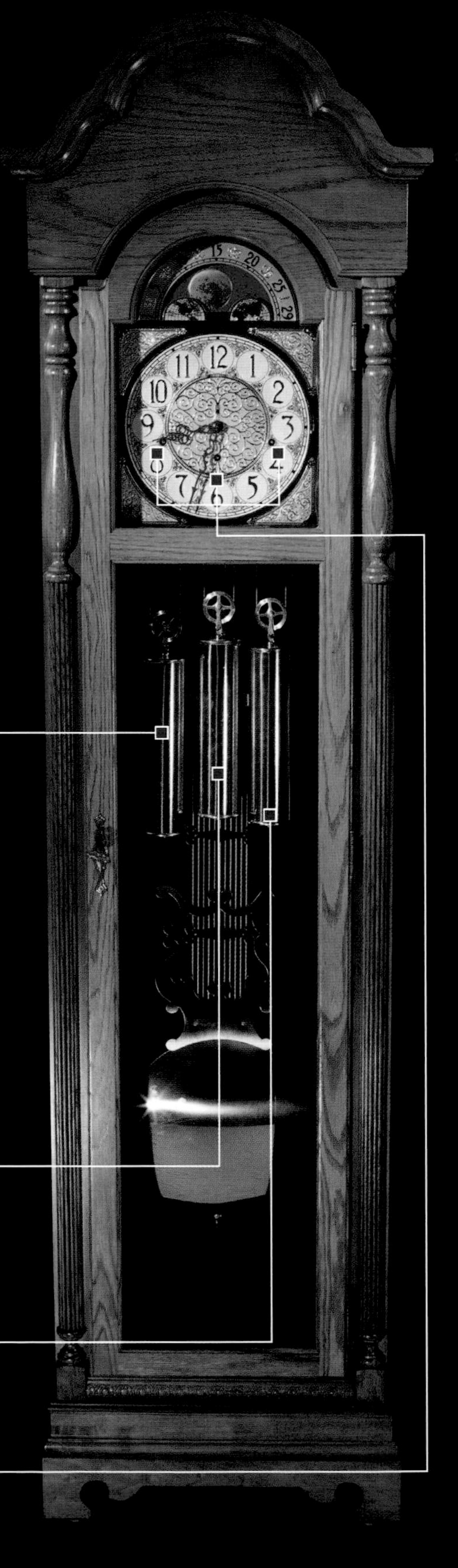

◀ The chimes (like bells) in this clock are long spring steel rods that are struck by little hammers driven by the complex clockwork inside. I've studied the gears for quite a while and still really don't get how they are able to execute the complex 4 to 16 note sequences every 15 minutes, and the chimes that count out the hour on the hour (three bongs at three o'clock, etc.).

▼ As we will see throughout this chapter, clocks and fakery go hand in hand. Unfortunately not even this beautiful grandfather clock is an exception. No, its pendulum is pure nonsense.

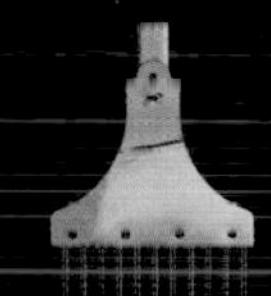

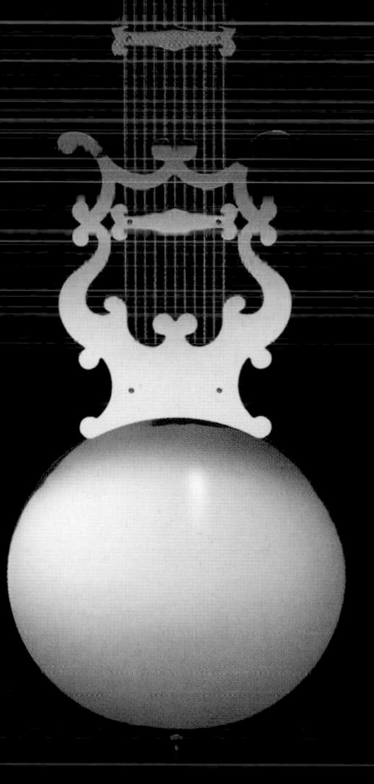

◀ Somewhere in this mess of gears is the explanation for how the chimes work.

Fancy Pendulums

WE LEARNED THAT the swinging time of a pendulum depends on its length. When metals (and most other materials) heat up, they expand slightly. So if you have a pendulum with a rod made of metal, your clock will slow down when it gets warmer. Especially in times before central heating and air conditioning, clocks experienced pretty wide swings of temperature, making it impossible for a pendulum clock with a simple metal pendulum rod to keep consistently good time all year.

One solution is to use alternating brass and steel rods connected in a clever way that cancels out both their thermal expansions. The "gridiron"-style pendulum from my grandfather clock appears to be made with alternating brass and steel rods—and it is. But the rods are not connected to each other in the necessary clever way. They are just stacked up decoratively next to each other, serving no function. People who have such clocks, and even the people who make them, often have no idea that this pattern of brass and steel, very commonly seen on such clocks, isn't random, but is instead a degraded echo of something that used to be functional.

Genuine temperature-compensating gridiron pendulums are almost impossible to find. In fact, I was unable to find one, so I made this one myself from brass and stainless steel rods. As the temperature goes up, *all* the rods get longer, but the brass ones get longer by almost twice as much as the steel ones (the coefficient of thermal expansion for brass is about twice that of stainless steel). The rods are connected in such a way that growing steel rods lower the pendulum bob, while growing brass rods lift it back up again. The lengths of the rods are not arbitrary: they are carefully calculated so they exactly cancel out each other's thermal expansions.

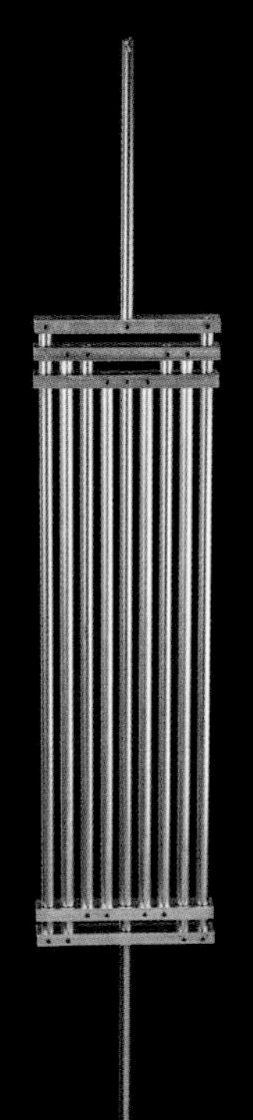

For clarity, I made the main image on this page with brackets that are only as big as they need to be to hold the rods they are connected to. In a real gridiron pendulum, all the brackets extend the full width, with extra pass-through holes for the rods they are not connected to. That prevents the rods from bending forward or backward.

0 mm pivot point

The pendulum hangs from this pivot point.

Where you see a black set screw, the rod is locked to the bracket.

0.15 mm

0.2 mm

0.5 mm

The scale here refers to the amount by which each point moves up or down with a 100°C change in temperature.

The center of the pendulum bob (a heavy weight) would be attached at this point at the bottom of the rod. The length of the rod between this point and the hanging point at the top stays almost exactly the same regardless of changes in temperature.

0.17 mm

0.5 mm

Where you see an oversized hole with no set screw, the rod is free to slide through the bracket.

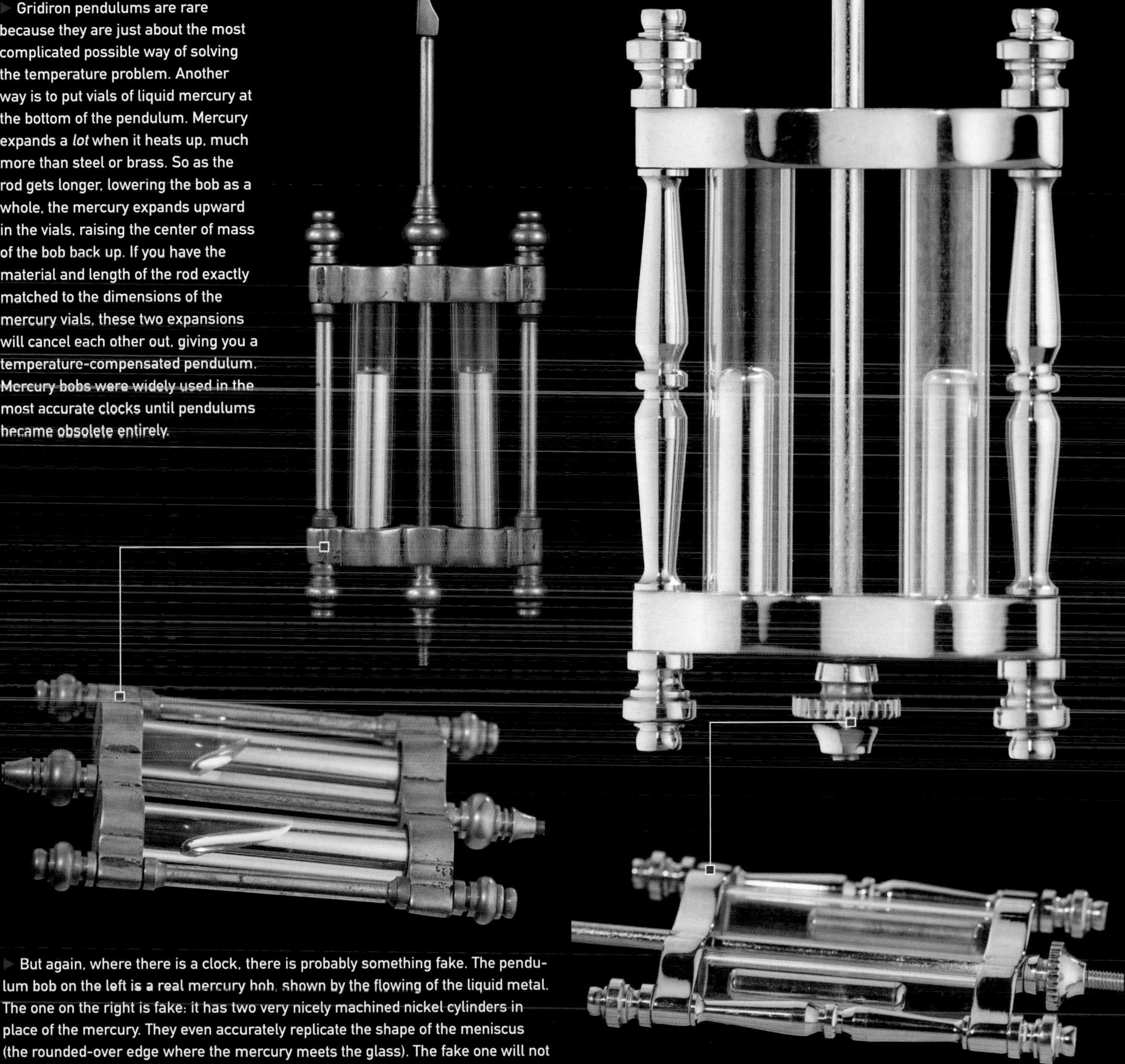

Gridiron pendulums are rare because they are just about the most complicated possible way of solving the temperature problem. Another way is to put vials of liquid mercury at the bottom of the pendulum. Mercury expands a *lot* when it heats up, much more than steel or brass. So as the rod gets longer, lowering the bob as a whole, the mercury expands upward in the vials, raising the center of mass of the bob back up. If you have the material and length of the rod exactly matched to the dimensions of the mercury vials, these two expansions will cancel each other out, giving you a temperature-compensated pendulum. Mercury bobs were widely used in the most accurate clocks until pendulums became obsolete entirely.

But again, where there is a clock, there is probably something fake. The pendulum bob on the left is a real mercury bob, shown by the flowing of the liquid metal. The one on the right is fake: it has two very nicely machined nickel cylinders in place of the mercury. They even accurately replicate the shape of the meniscus (the rounded-over edge where the mercury meets the glass). The fake one will not do anything to improve the accuracy of a clock it's on, but on the plus side, if you break it, you don't have to call a hazmat team to clean up a toxic mercury spill.

A VISIT TO THE 河北昱昌古典钟表有限公司 CLOCK FACTORY

CHINA IS FULL of amazing time machines that take you backward and forward by two hundred years at a time. They're called "trains."

The ancient capital of Beijing and the commercial centers of Shanghai and Hong Kong are among the most advanced cities on earth—and they are tame compared to the new-technology megacities of Shenzhen and Guangzhou in the southern province of Guangdong. The inhabitants of these places truly are living in the future compared to the rest of China and to the rest of the world.

An hour or two on one of the smooth-as-silk high-speed trains will take you to places where older ways of life survive, not as museums, but as real cities, villages, farms, and people. In these places life as it was known in past centuries is not a memory, but a lived experience, a real place. *河北昱昌古典钟表有限公司* is one such place. (That's pronounced Hebei Yuchang Gudian Zhongbiao Youxiangongsi and translates as The Hebei Yuchang Antique Clock Company.)

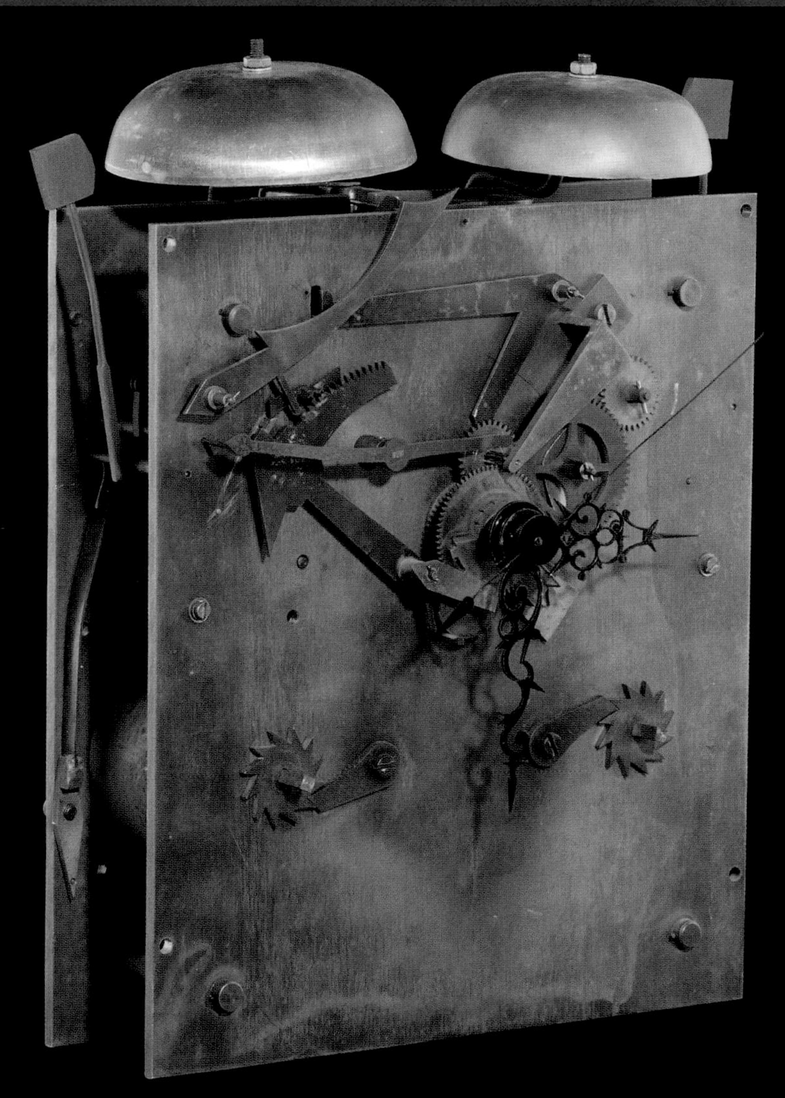

There are a lot of clocks in the Yuchang clock factory. By their reckoning, and by their own proud admission, they make about 90 percent of the "fake" antique clocks in the Chinese market. But what does it mean to be fake? These clocks are detailed copies of antique clocks that should be hundreds of years old, yet they are brand new (made from scratch in this factory). That makes them fake, right?

Throughout this chapter you will find many clocks that I dismiss as ugly fakes. This is not one of them. This movement is beautifully hand-made of twelve pounds (5.4 kg) of brass. The side plates are over 1/8-inch-thick (3 mm) solid machined brass. The gears are lathe-cut brass, not stamped steel. The escapement is an authentic verge escapement, a design that has remained unchanged since the fourteenth century, and that has not been used in clockmaking since the mid-1800s. Every detail of this movement is as it should be. It is honest to a fault (i.e., it's not a very accurate clock by modern standards). Does the fact that it was made last week mean it's fake? I don't think so.

▲ The verge escapement was the first type of mechanical escapement devised, well over six hundred years ago.

▲ Clock shop in Beijing's Silk Street tourist market

▼ Walking into this place is like entering a time warp. It's as if they have re-created the working methods of a nineteenth-century French or English workshop for the purpose of re-creating nineteenth-century French or English clocks. Aside from a few modern intrusions, the tools and techniques are largely the same.

While I don't think these clocks are fake, they can easily *become* fake when misrepresented as actually old. Yuchang's clocks are for sale in shops all over Beijing, including in the famous Silk Street tourist market. I've gotten different stories in different shops. In the one shown here they were completely honest: "This is a Chinese-made copy of an antique clock, and look at how nicely made it is, solid brass, heavy marble!" (which is absolutely true on all counts). Other places the story is half true: "This is a reproduction made in Italy" (about a clock that I know for an absolute fact was made by Yuchang). And sometimes they leave the clear impression that the clock is old, which is often false.

▲ A closer look reveals why you should never judge a mess by its messiness. Just look at these beauties, with their elaborate enameled metal ornamentation and fluted brass columns! Shine it up, put it in a box, and you have a first-rate commercial product that sells for a good price.

▶ Every clock is a collection of parts, and every part has to be made in its own way. To make this one, colored glaze is placed in a thick layer between raised lines cast into the metal of the part. After firing melts the glaze into a glass-like enamel, the part is machined smooth and polished.

▼ If you look closely, the lines in this fancier version of an enamel inlaid part are much crisper. This one is made with a more time-consuming technique in which thin strips of brass are carefully bent to shape, then applied to the surface of a metal cup or dome. This is genuine "cloisonné" enamel inlaying, while the cast version to the left is a cheaper imitation. Both are made by Yuchang.

▲ Of course the most important parts of any clock are its gears. These are mass-produced by a combination of stamping and lathe cutting.

Earlier I describe the fine chain used to keep the torque in a clock mechanism constant as its spring unwinds. I never thought I'd find myself looking at the very place where the chain in that clock came from: this old heart-shaped metal tin filled with *tiny* chain links.

The brass template is used to line up three layers of bowtie-shaped links, with the middle layer shifted by one notch.

Next tiny *tiny* pins are dropped into the holes in the bowtie links.

The whole time I was watching the template being loaded up, I was wondering how on earth they were planning to hammer the pins in without all of them bouncing out. Sure enough, there's a clever solution. A thin metal strip with a notch at the end holds all the loose links in as the worker delicately hammers the pins one by one (which spreads them out enough to lock them into the holes in the links).

A little more hammering to straighten it out and this segment is finished. Later these short pieces will be linked together to make the 3-foot (nearly 1 m) chains used in the clocks.

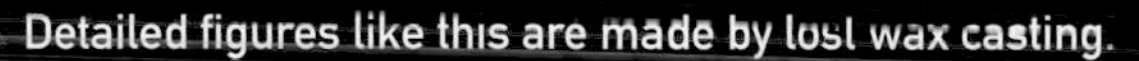

Detailed figures like this are made by lost wax casting.

The process starts here with a skilled craftsperson painstakingly carving a block of wax into a three-dimensional shape that will end up being made of metal.

Once the hand-carved original is finished, it is coated in many layers of latex rubber, with cloth or other reinforcing materials added after the details have been captured in pure latex. Latex is amazing stuff: it's incredibly tough, and it can capture nearly microscopic details. When it's hardened, the latex mold is carefully cut into two or more parts that allow it to be disassembled from around the wax original.

The latex mold is used to make potentially many hundreds of copies of the figurine in wax. These workers are ladling molten wax into the hollow molds, swishing it around, and then pouring out the excess. After several cycles of this, a hollow wax copy has been made and can be removed from the mold.

Just as an aside, let me say that this room smells *amazing*. It reminds me of candle making in grade school in Switzerland. It's a good smell, not sweet, but rich and warm. It's not what I was expecting a foundry to smell like. Frankly, foundries have a reputation for making people sick from fumes, not nostalgic for their childhood making candles at Christmastime.

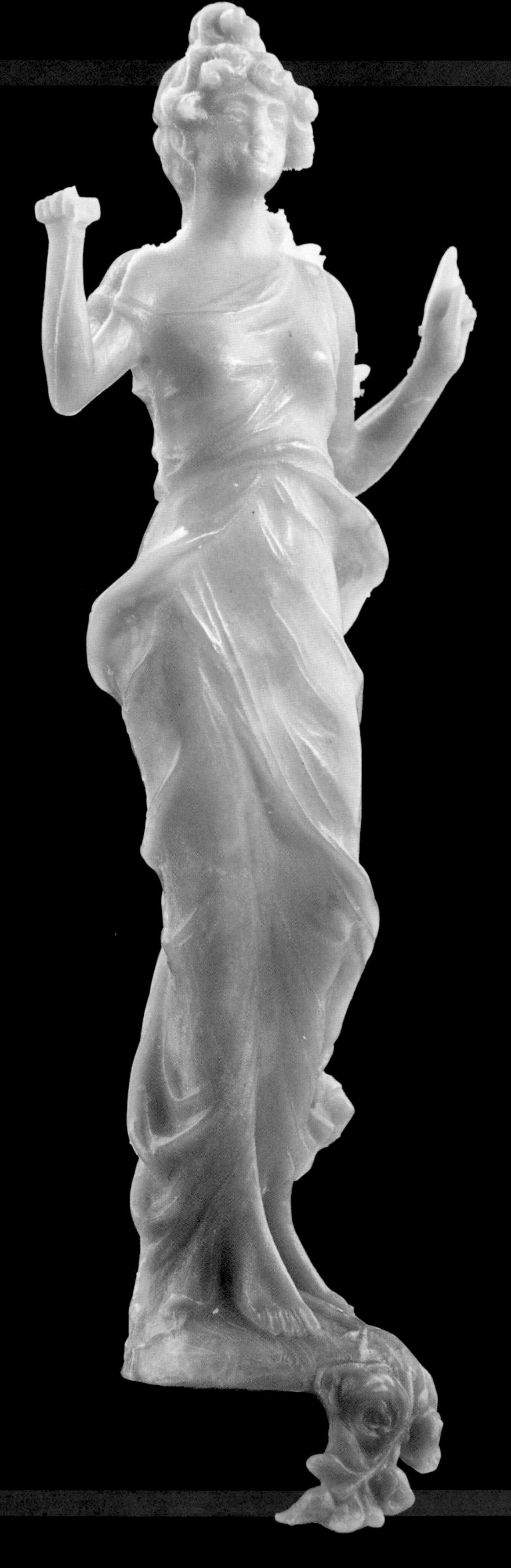

This is what one of the wax copies looks like right out of the mold. It's not perfect: there will always be little marks where the mold is split, tiny bubbles, or other defects. But it's got the basic shape and all the important details. The defects can be fixed far more easily now in wax form than when they have been copied into metal, so time is put into fixing the wax at this stage.

Wax is very forgiving. This worker is using a hot iron to gently melt small sections of a wax copy to remove flaws. They will also add "sprue" channels made of wax that will direct the metal to all sections of the mold (see page 108 for what the sprue looks like after it's turned into metal).

▼ Once the wax copy is perfect, workers coat it with "slip," a form of plaster that hardens around the wax. The sprue end is left open so molten metal can be poured in later. After several layers of plaster have been built up, coarser aggregate is added to further strengthen the mold.

▲ This looks like a giant soup pot, and that's not too far off. It's a steaming chamber used to melt the wax copy out from inside the plaster mold. The molds are stacked up, open end down, the lid is closed, and steam is directed in until all the wax has melted and flowed out the bottom. (Beeswax is expensive: as much as possible is recovered and reused.)

▲ The brightly glowing liquid metal is poured into the fresh-out-of-the-oven molds. It hardens in a few seconds.

▲ Here we are at the heart of the foundry: the casting floor. On the left you see the mold firing oven: after the majority of the wax has been melted out, the molds are put in this oven and fired at high temperature until every last trace of moisture and wax residue has been burned out. The glowing hole below is a sunken crucible containing hundreds of pounds of liquid brass—an alloy (mixture) of copper and zinc. Between them is a section of gravel floor on which the actual casting will be done. It's always gravel, because the metal always spills. If it spills on concrete, the concrete may shatter or the metal may stick, or flow too far. On gravel, no problem, the metal spilled so we pick it up, dust it off, and put it back in the pot.

◄ In a larger foundry, one capable of producing castings of a hundred pounds or more, there would be an overhead crane that carries a tiltable pot with a pouring spout over to the molds. But this foundry specializes in relatively small castings, so workers simply dip a big soup spoon down into the liquid metal and carry over enough for one or two castings.

▲ Oh, I feel your pain. This happened to me *so many times* when I was casting lead, zinc, and aluminum as a kid. I'm really happy to see that it happens to the professionals too. They, however, had a better solution: using a long spatula to slather on some damp mud to cool the metal enough to stop the leak. I don't think they lost this casting (and if they did, the metal can just be recycled, though it wastes the time used to prepare the wax copy and mold).

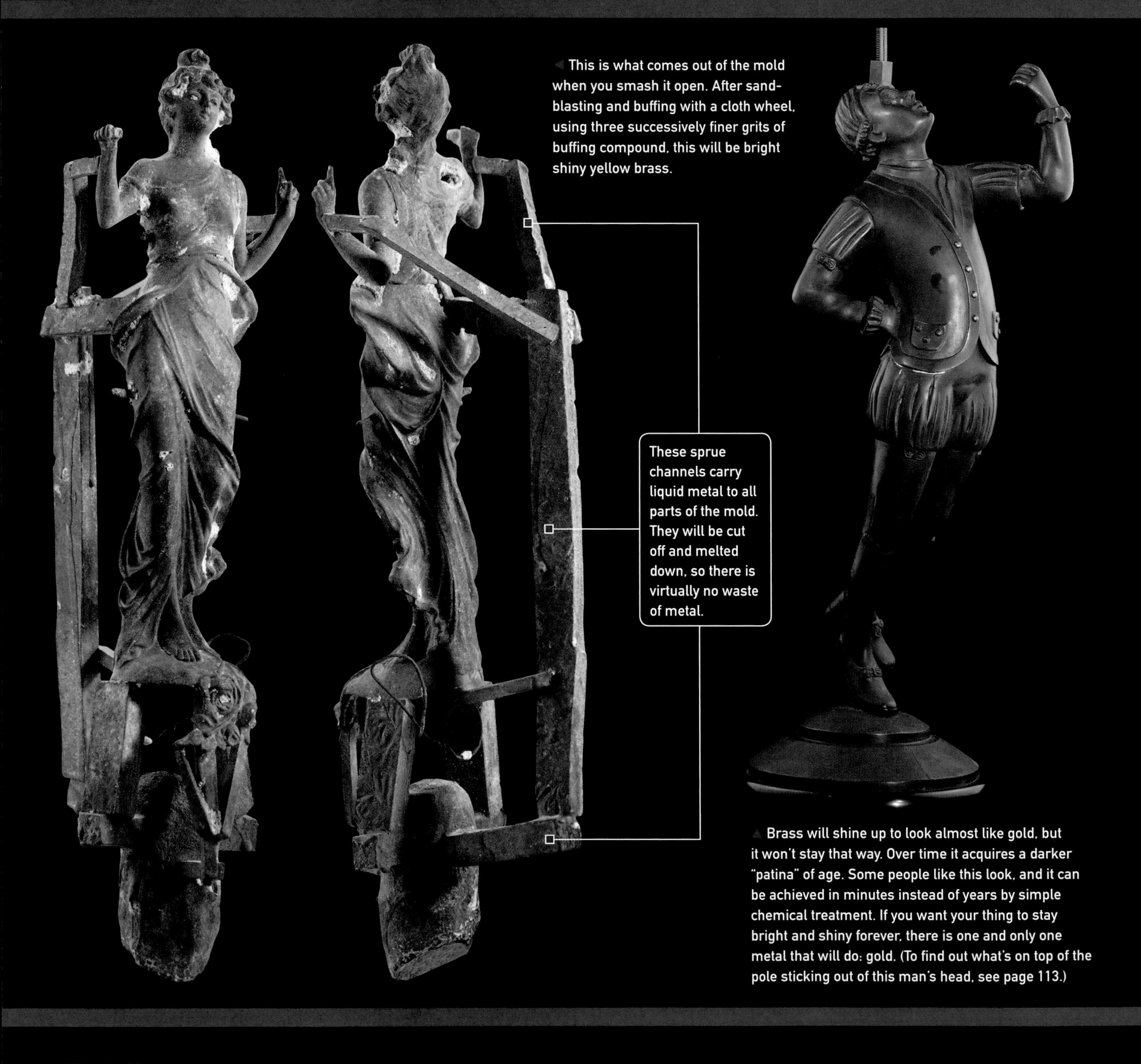

This is what comes out of the mold when you smash it open. After sand-blasting and buffing with a cloth wheel, using three successively finer grits of buffing compound, this will be bright shiny yellow brass.

These sprue channels carry liquid metal to all parts of the mold. They will be cut off and melted down, so there is virtually no waste of metal.

Brass will shine up to look almost like gold, but it won't stay that way. Over time it acquires a darker "patina" of age. Some people like this look, and it can be achieved in minutes instead of years by simple chemical treatment. If you want your thing to stay bright and shiny forever, there is one and only one metal that will do: gold. (To find out what's on top of the pole sticking out of this man's head, see page 113.)

The first plating bath deposits a relatively thick (but still microscopically thin) layer of nickel, which protects the part from corrosion if the gold layer is scratched, and provides a hard, mirror-smooth, "sticky" surface for the gold layer to adhere to.

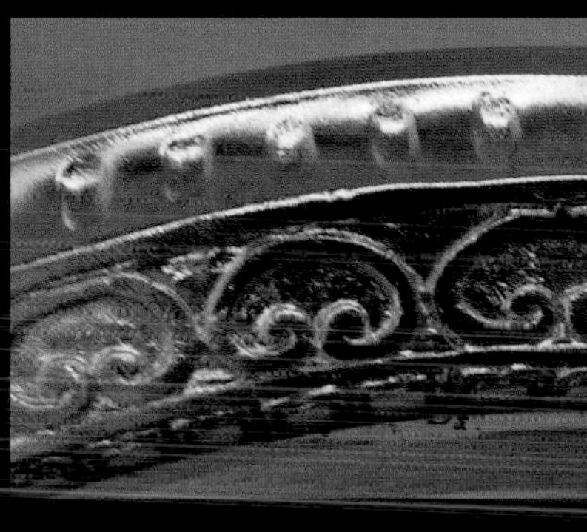

Electroplating works by running an electric current through a water-based solution in which the metal (nickel or gold in this factory) has been dissolved. The current converts dissolved metal ions into solid metal on the surface of the object being plated (which is connected by a wire to the power supply, completing an electrical circuit through the bath).

The second plating bath adds a *very* thin layer of gold. After being rinsed to remove the plating solution, the part is ready to go.

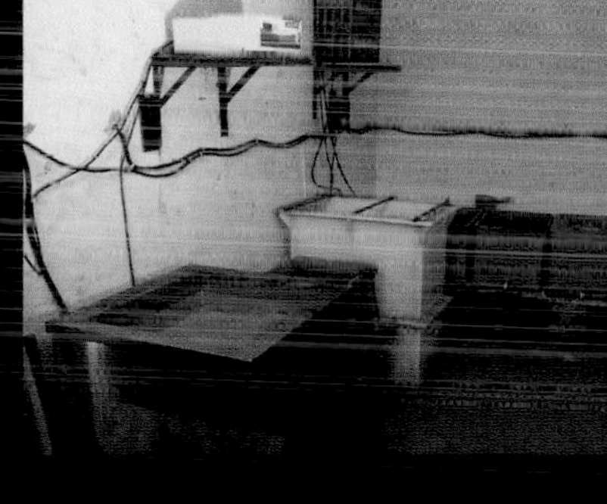

Electroplating is a poisonous business. These bins contain a sequence of detergent baths, ultrasonic cleaning baths, acid etching baths, and two electroplating baths where metal is deposited. Most of them are either toxic or highly corrosive.

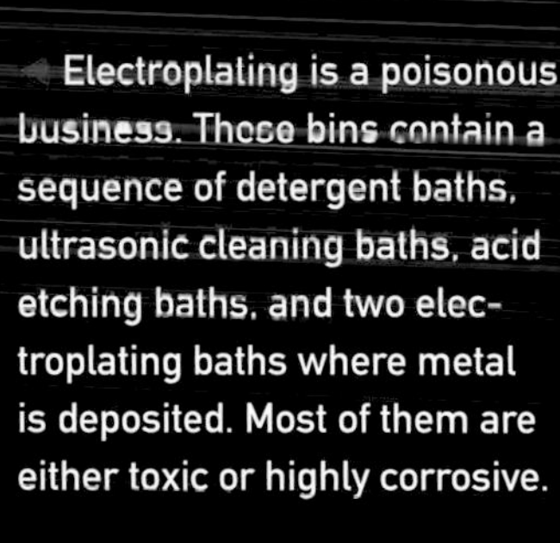

There must be at least some satisfaction in seeing things of such beauty being pulled out of a poison vat every day. Unlike most manufacturing processes, plated objects are at their most beautiful immediately as you pull them out of the bath—no buffing or polishing required.

GOLD IS GOLD, no two ways about it. No other metal looks like it, and no other metal will stay that color literally forever (like age-of-the-universe forever, as long as it's not exposed to aqua regia or a few other nasties not likely to be found in the average home).

Even a genuine antique clock is not likely to be made of solid gold. Except for a few examples of the most extravagant excesses of kings and emperors (each worth millions), no matter how old a thing like this is, the only gold in it will be a very, very thin layer on top of a base metal casting. Fortunately, because of its absolute resistance to corrosion under all normal circumstances, a very thin layer is all it takes.

In old times this layer would have been applied by gilding, a process in which incredibly thin sheets of beaten out gold leaf, only about 500 atoms thick, are glued to the surface (Yuchang actually did it this way up until just a few years ago). Today it's done by electroplating, in which a similarly thin layer of gold is deposited by an electric current running through a liquid bath in which the figurine is submerged.

Gold is not exactly easy to dissolve: the powerful acid mixture known as aqua regia—"royal water"—is so named because it's the only acid that will touch gold. So instead of trying to dissolve their own gold, the factory buys its gold in the form of a salt, which has already been converted to a soluble form. This is how you arrive at the perhaps slightly scary place of paying on the order of $3000 for a small 100 gram (3.5 ounce) bottle of white powder that looks *nothing* like gold. (It would, however, be fairly easy to tell if the powder was fake: while it does not share gold's color, it does share gold's astonishing density. A true gold salt will weigh far more than you'd expect of your average expensive white powder.)

Three Fundamental Functions of Clocks

I HAVE A total of eleven clocks from the Yuchang Company (which, given that I have no history of collecting clocks, seems like a lot to me). We've already seen the beautiful "French" skeleton clock and the impressively heavy brass movement. I picked the following clocks because each of them demonstrates a different purely mechanical way of solving one of the three fundamental design requirements of a clock: something has to beat with a steady rhythm, something has to supply the energy needed to keep that thing moving, and something has to display the current time for people to read.

KEEPING THE BEAT

LET'S START WITH a look at some variations on the theme of pendulums, the mechanism that beats the steady rhythm within the clock.

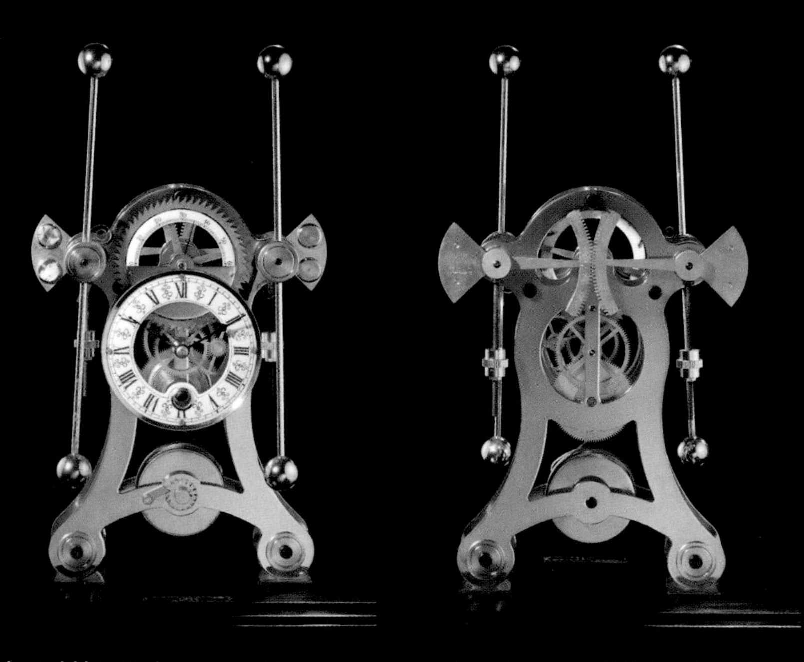

▲ I would love to have a genuine Harrison sea clock (see page 93), but the only existing ones are on display at the Greenwich Observatory. This is a compromise: a modern sort-of reproduction. It has a genuine grasshopper escapement, but the twin linked ball-and-rod pendulums are false imitations of those in Harrison's clock. They are out of balance, with slightly more weight on the bottom than the top, and they don't have the balance springs of the original. In other words, they are gravity-based pendulums in disguise, and would not work at all on a boat. But on the plus side, the clock was cheap, and I don't have a boat anyway.

▲ There are many different pendulum designs that have been tried over the years in an attempt to make clocks more accurate. This is a huge step in the other direction. The "pendulum" is a small metal ball hanging from a fine thread. The ball swings to one side, wraps around a peg until it's all wound up, then unwraps, then wraps around a second peg on the same side, then unwraps and finally swings over to the other side where it repeats the double wrap-and-unwrap motion. The whole cycle takes 10 seconds. Mechanical clocks normally go to great lengths to eliminate sources of variation in the timekeeping mechanism (see page 98). The thread in this clock is just a mess of potential errors: humidity, stretching, wear, and so on. In short, this clock is just for fun: at no point in history was this design a good idea. But what's wrong with a little fun in life?

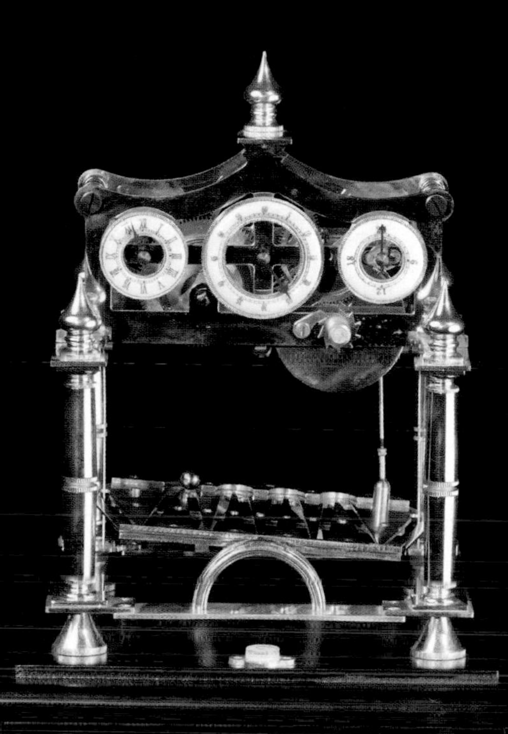

Continuing the theme of pendulum alternatives that cost more and make the clock less accurate, this one has a grooved plate that can tilt to the left or the right. It takes 12 seconds for the ball to roll from one side to the other, at which point the plate is triggered to flip the other way, sending the ball rolling back where it came from. Fun to watch, bad for timekeeping.

This is a clever, pointless weight-saving idea. There are two heavy things in a spring-wound pendulum clock: the movement (all the gears and the plates that hold them together) and the pendulum bob (the heavy weight at the bottom of the pendulum). In this design the movement doubles as the pendulum bob, cutting the total weight of the clock in half! For no reason, since pendulum clocks can't be made portable.

This clock also has what appears to be, but isn't, a gridiron pendulum rod (see page 98). I was delighted to be able to confirm my suspicion about these things: the owner and engineer at the Yuchang clock company, who designed and made this clock, had no idea why they were using this particular pattern of rods. They were just copying an old one, which was itself probably not built right. I showed them the picture on page 98 to enlighten them as to why they are making their clocks this way.

This "anniversary clock" isn't from the same company, but it illustrates another interesting variation of pendulum. Its mechanism is actually closer to that of Harrison's later sea clocks, or a modern wristwatch, than it is to a hanging pendulum clock. A set of heavy balls rotates slowly back and forth around a vertical axis. The restoring force that causes them to change direction is provided by a very fine coiled spring called a balance spring. Since the balls don't rely on gravity, and are perfectly balanced around their point of rotation, this clock should run at a pretty constant rate even on a moving ship.

Energy to the Beat

NOW LET'S TAKE a look at some different ways energy can be supplied to a clock without electricity. There are fundamentally two ways it's done: springs and gravity. We'll start with a look at the back side of the anniversary clock, where we find a monster of a spring.

This clock is currently wound all the way down, so the spring is entirely on the bottom half. When you wind it up with the supplied crank, the spring—a long, flat steel strip—gets wound up on this spool in the opposite direction from the one it wants to be coiled in. Fully wound-up a spring like this holds a *lot* of energy: never take apart a clock like this with a wound-up spring!

◀ This is called an anniversary clock because you only have to wind it once a year, for example on the date of your anniversary, if you got it as a wedding gift. (Or on the day after your anniversary if you forgot the date and had to make up for it with a really nice present to smooth things over, in which case I would recommend a more expensive model.) The spring is nearly identical to the spring in the wind-up radio we saw previously, but it lasts far longer because running a clock takes only a tiny fraction of the energy required to run a radio.

◀ The vast majority of spring-driven clocks use coiled spiral springs because they can store a lot of energy in a small space, and their two end points don't change position as the spring is unwinding. (One end stays put and the other rotates, which is perfect for driving a gear.) But there's no law that says you can't build a clock that runs on a linear extension spring instead, which I suppose must be a good enough reason to do it, because there is nothing else to recommend this design.

▲ As discussed on page 95, springs have an inherent flaw: as they unwind, the force they supply to the clock decreases, potentially introducing timekeeping error. The spring tension also changes with temperature. Clocks powered by (slowly) falling weights don't have these problems: the weight supplies a constant amount of force. As the weight goes down, gravitational potential energy is converted into kinetic energy in the movement of the gears.

Why are the weights hanging from pulleys off a doubled-up cable, instead of just hanging from a single strand of cable? This allows the weight to be twice as heavy as the clock mechanism would otherwise be strong enough to support, letting the clock run twice as long (seven days in this example). You can see the same pulley system in the grandfather clock we looked at earlier.

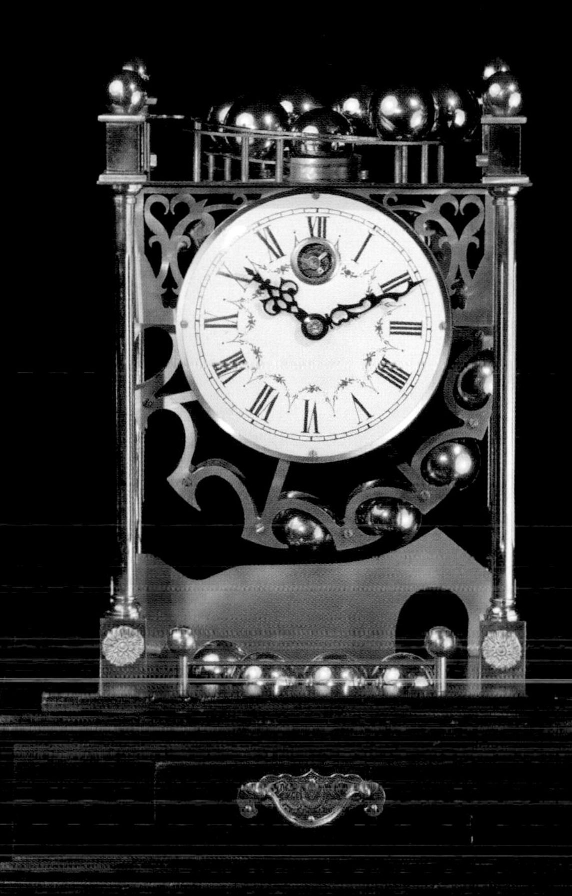

▲ The inventiveness of people who come up with unnecessary clock designs knows no bounds. This one also uses the clock movement as the drive weight, but instead of sliding down a pole, it rolls ever so slowly down an inclined plane.

▲ This clock is also powered by slowly falling weights, but they are not hanging from anything. Instead it has a hopper full of heavy steel balls on top, which feed down into a sort of waterwheel that powers the clock as it slowly rotates. Every four hours a ball drops out the bottom and a new one enters from the top. Instead of rewinding the clock, you lift the balls back up to the top every couple of days. (The energy you put into lifting the balls is what powers the clock.) This design has the fun property that you could make the clock run for pretty much as long as you like just by adding a larger hopper on top and filling it with more balls!

▲ From the front it looks like the clock should just roll down immediately, but inside we see that the movement is isolated from the outer wheel, and weighs much more on the bottom than the top. That keeps it from rolling freely, and keeps the clock face upright as the outer wheel rolls down the slope. Power to run the clock is taken from the rotation of these two parts relative to each other. (See the transparent clock on page 117 for an example of a similar principle at work.)

► On page 111 we saw a clock that used itself as its own pendulum bob. Here is a clock that uses itself as its own driving weight. Over the course of three days the whole clock moves down the stick, powering itself with its own gravitational potential energy. No reason you couldn't put it on a 10-foot pole and have it run for weeks. Where is the pendulum in this clock? It doesn't have one, instead it uses a balance wheel. Could you make a clock that is both its own drive weight and its own pendulum bob all at the same time? I hope not, because that might cause some kind of implosion of the space-time continuum. (You can see what's holding up the stick on page 108.)

▼ To make this clock run longer, all you need is a longer inclined plane, which they actually have available! Or set it on the side of Mt. Fuji and come back next year.

Displaying Time

FINALLY, LET'S LOOK at different ways of displaying the time. There's really only two systems that have commonly been used: digital and analog (dial), with a few variations possible within each category.

Dials we've seen dozens of already, and this is the only other common mechanical method: digital flipping displays. Plastic or metal leaves are mounted on a slowly rotating drum. Once per minute (on the right) or once per hour (on the left) the front-most leaf on the top moves down far enough to clear the clip at the top, allowing it to flip down, and revealing the next number underneath. (These kinds of flipping displays used to be universal in train stations and airports for displaying departure times and gate numbers. It was quite a thrill back in the day watching a whole board full of hundreds of these things update simultaneously as the positions of all the trains shifted up by one row. Flip flip flip flip flip flip, oh man, I better hurry to track 7!)

This isn't an old-school mechanical watch, nor it is a modern electronic watch. It's a middle-aged-school LED watch, which displays the time in segmented digital form, but using the old technology of LEDs (light-emitting diodes). LEDs today are incredibly bright and used for area lighting, replacing incandescent and fluorescent lights nearly everywhere. But back in the 1970s, they were available only in red, and they were only just bright enough to act as indicator lights. This watch is from a flea market in Moscow, where it is just as obsolete as everywhere else.

CASIO WATCH LOVE

I REMEMBER THIS watch. Oh boy, it's not often the feels hit you like this. It would have been around 1980, I was walking to school, just past the dome of the Kreuzkirche, down a fairly steep section of Dolderstrasse, a few blocks before I would reach the Kinderspital where my sister was to die later that year. Under the shade of the Linden trees a man stopped and asked if I had a new watch—he'd seen me checking it over and over as I was walking along. I was of course shy, but also tremendously proud because, yes, I did have a new watch! This watch! (Actually this model of watch. The one I had is long lost, and this is a replacement from the amazing memory restorative that is eBay.)

It's got everything: a full-month calendar display, a bunch of different melodies it can play, dual time zones. It blew my little mind that such things were possible. That moment, standing at the crest where the street began to slope down, the sun bright, the day young, and me with the coolest thing I could ever imagine strapped to my wrist. Just give me a minute.

OK, now, the reason the watch is here is because it demonstrates how inescapable analog watch hands are for displaying time. Even in a watch that was going out of its way to be hypermodern, the best they could come up with was to replicate analog watch hands in digital LCD form. (The hands are not real: they are just segments in an LCD display, whose operation is explained in detail on page 127.)

▲ So strong is the pull to display time on a conventional clock face that this method is extended even to the extreme case of a clock for the blind. This braille clock is no different than any other clock, except there are raised dots around the dial, and the hands are stronger, to allow the time to be read out by feel. (These days a more common solution is a talking clock, but that's an electronic solution and in this section we are only talking about mechanical clocks.)

I could go on with more crazy clocks, but I ran out of room in my seven suitcases.

▶ After I bought all those clocks at Yuchang they said I could have the dog for free.

Beyond the Pendulum

PENDULUM CLOCKS WERE the first accurate clocks, and are still the most accurate of all purely mechanical clocks, but they are seriously inconvenient. Portable clocks need a more compact and motion-insensitive timekeeping solution.

This balance wheel replaces the pendulum. Instead of swinging like a pendulum it oscillates back and forth around its axle, like a miniature version of the balls in the anniversary clock we saw earlier.

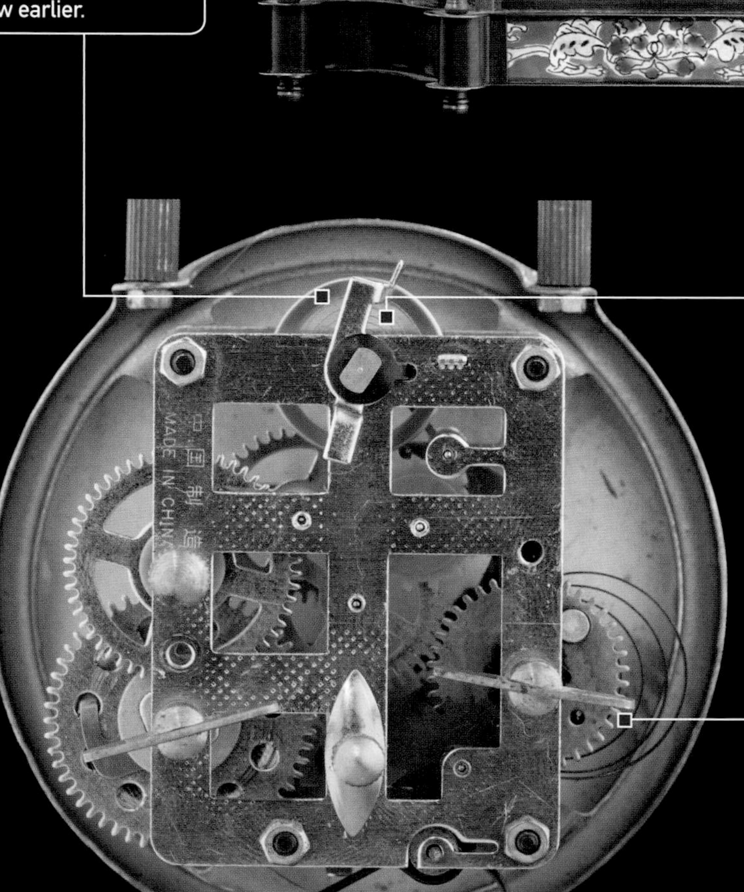

Instead of gravity pulling on a pendulum to get it to reverse direction at the end of its cycle, this tiny spring pulls the balance wheel back and forth.

The clock is rewound by turning the inner core of the mainspring with a key.

▲ Compact clocks use a coiled spring for power, and a "balance wheel" instead of a pendulum. These parts can be made very small, and they can be made nearly immune to movement of the clock. That's important if you want to put a clock on a ship, on your wrist, or just on your desk.

The beauty of the spring-and-balance-wheel design is that it can be made almost arbitrarily small. This wristwatch has all the same components as the desk clock on the previous page, just in miniature form. It's called a "skeleton" watch because the case is see-through, and the frame is cut down to the minimum necessary to hold the gears in place.

This tiny "regulator" lever lets you adjust the speed of the watch by slightly changing the tension in the balance spring.

Balance spring

The balance wheel in this watch spins back and forth three times per second. A smaller, faster balance wheel makes for a more accurate watch, and a smoother-running second hand. (The fastest commercial watch movements beat 10 times per second, with some rare mechanical stopwatches going as fast as 100 per second.)

Run down

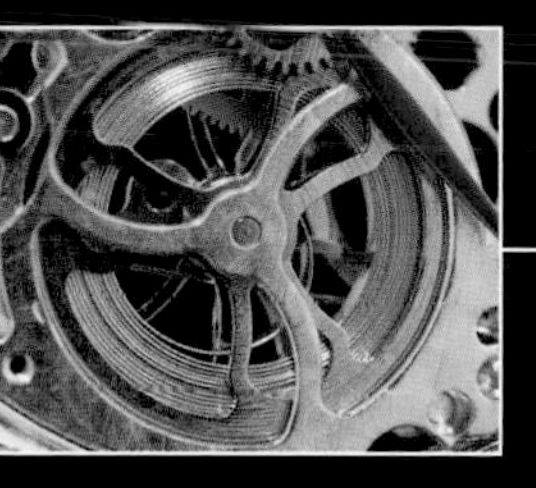

Half wound

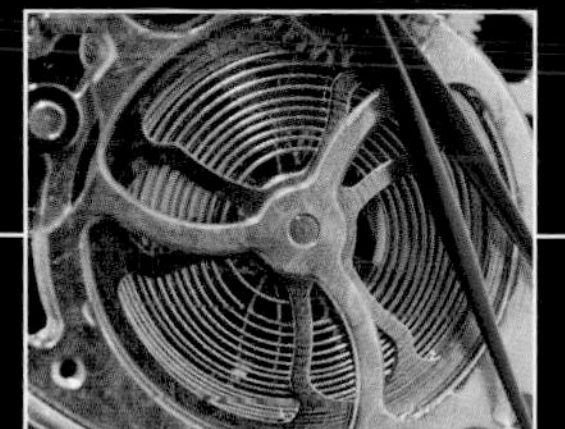

Fully wound up

This tiny escapement has sapphire pallets, which is typical of all but the cheapest watches. Sapphire is much harder than steel and moves with less friction, allowing this escapement to run for many years without oiling. Sapphire or ruby are also used in the axle bearings of some of the gears in better-quality watches.

RUSSIA IS HOME to the world's largest clock movement, installed in the atrium of the children's mall on Lubyanka Square in central Moscow. Notice I said largest clock *movement*, not largest clock: there are other clocks with bigger faces and longer hands, but this one has the largest mechanism. This, of course, makes it far more interesting than some church tower with a really big clock face.

Measuring the length of the pendulum does not require getting to the clock: it can be done simply by timing the duration of its swing. This pendulum takes 6 seconds to make a full back-and-forth cycle, which means, according to the formula for the period of a pendulum, it must have an effective center of mass 30 feet (9 m) below its pivot point (overall, it's 42 feet, 13 meters, from top to bottom).

The gears on this thing are truly immense, some at least 10 feet (3 m) in diameter. Normally gears this big would be made of tons of steel, and would be transmitting huge amounts of force through their massive teeth. But these gears are in a clock, not a ten-story crane, and they need to be light enough to hang over the heads of carefree shoppers. The engineers chose to do something I have never seen or heard of in any other machine: they put individual ball bearings on each and every tooth of each of the gears. The design is beautiful, and the execution is outstanding. A true celebration of the clock as mechanical art in motion.

▲ The world's largest clock movement is located in a children's mall in Moscow.

▲ Ordinary wristwatches are already very small, but this one is just crazy small—yet inside is a complete mechanical watch movement just like the one we saw on the previous page, with fifteen jewel bearings. It is claimed to be the smallest ever made in Russia (back when it was the USSR). For reference, the dime on which it is sitting is the smallest US coin in circulation at just under 0.7 inches (18 mm) in diameter.

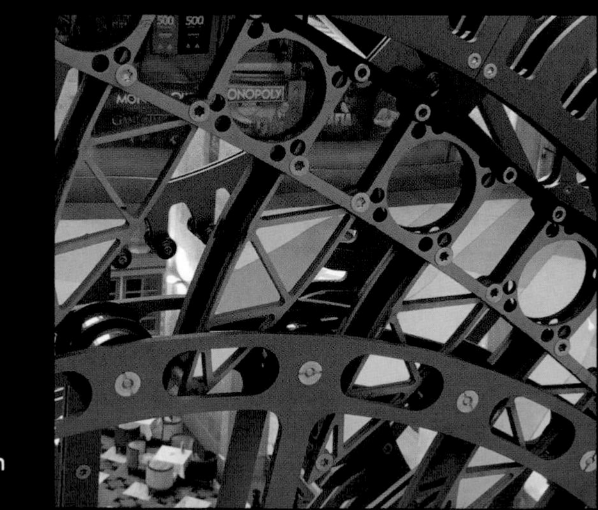

▶ Individual ball bearings roll along the teeth of the huge gears.

▲ This lower quality skeleton watch has all the same moving parts as the one on page 117, but is marred by also having fake gears engraved on its case. It's trying to look fancier than it is, which is a shame, because machines are beautiful for what they are, not for what they pretend to be.

▲ Mechanical wristwatches can get *ridiculously* expensive. Some brands routinely sell for upward of $300,000 and watches in the $10,000 range are considered "reasonably priced" in some circles. (No, that's not actually reasonably priced.) I didn't get any answer to my emails politely asking expensive watch companies for permission to use photos of their watches in this book, so instead we took a

Cheaper Is Better

THE IRONIC THING about ridiculously expensive mechanical watches is that, as beautiful and expensive as they are, they kind of suck at being watches, if you measure their performance against the standard set by dollar store electronic watches. This watch literally cost a dollar, and keeps time as well as almost any mechanical watch. (Actually I'm exaggerating on the price. It was actually 99 cents, with free shipping to America. Within China they go for 30 cents each, retail.)

The accuracy of cheap quartz watches comes down to one thing: what's oscillating to keep time. Instead of a pendulum or balance wheel they have a little metal can with a tiny quartz crystal inside.

LCD display

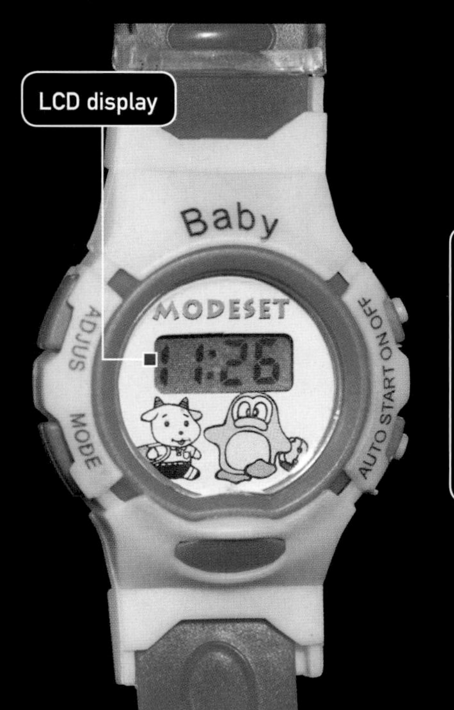

Underneath this dot of glue (called "potting compound") is the silicon chip that controls all the functions of the watch. The actual chip is much smaller than the dot of glue: you could barely see it, yet it contains the equivalent of all the gears in a mechanical clock.

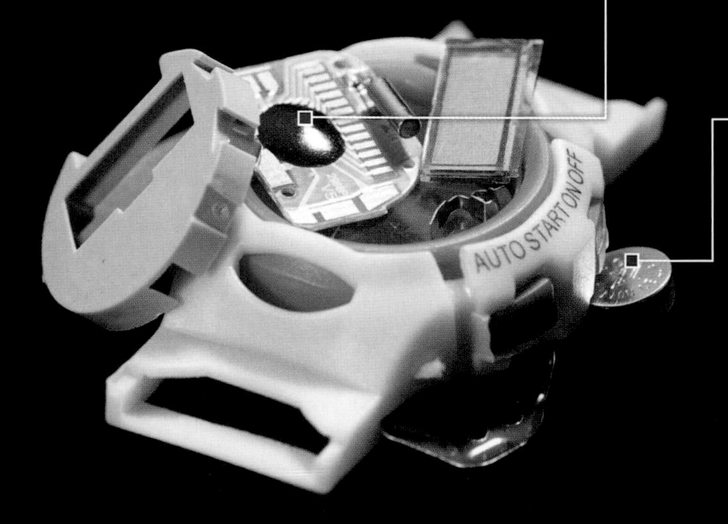

A tiny "watch battery" supplies enough energy to run the watch for at least a year. Some watches will run for ten years on a single battery.

LCD display

When the watch is assembled, this "elastomeric connector" is squeezed between the bare copper contacts on the circuit board and the invisible, transparent electrical contacts on the back side of the LCD display. It is a very fine stack of alternating layers of conductive and insulating materials, which allows it to conduct electricity only in the direction between the circuit board and the display. This is a common technique for transferring a lot of electrical signals from one board to another without having to connect individual wires to each contact. There are so many conductor/insulator layers that it doesn't matter how exactly they line up with the contacts.

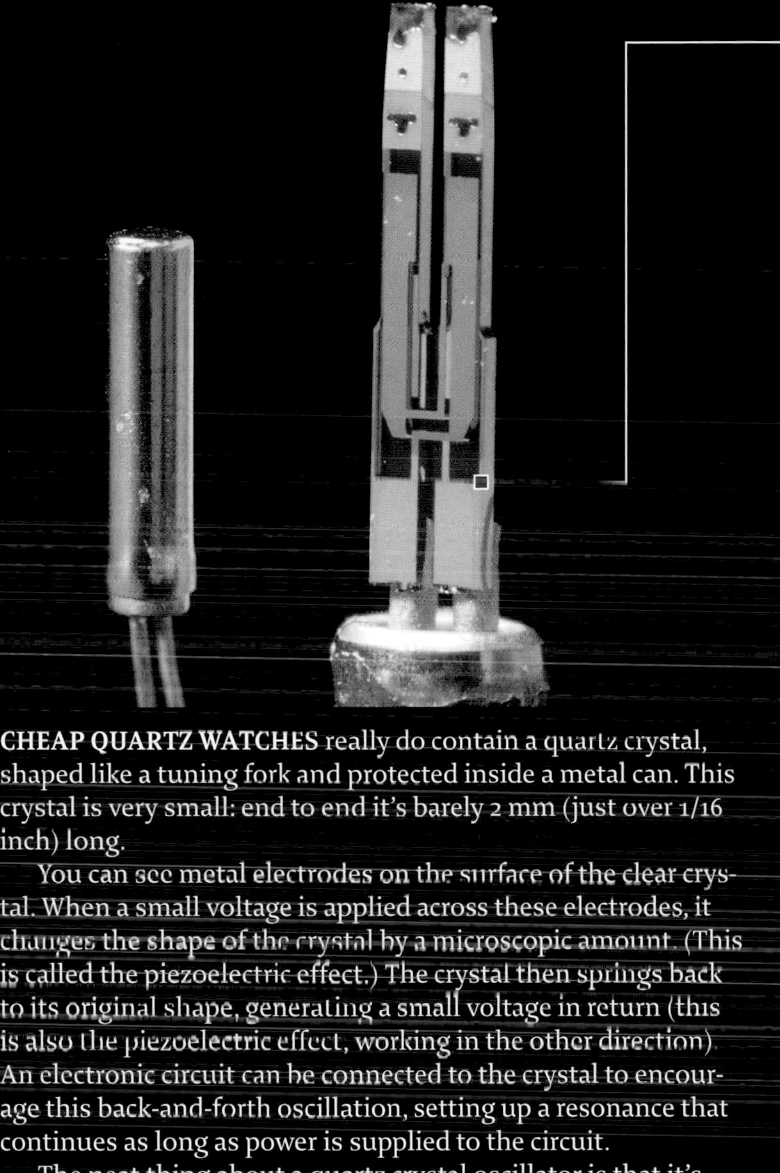

CHEAP QUARTZ WATCHES really do contain a quartz crystal, shaped like a tuning fork and protected inside a metal can. This crystal is very small: end to end it's barely 2 mm (just over 1/16 inch) long.

You can see metal electrodes on the surface of the clear crystal. When a small voltage is applied across these electrodes, it changes the shape of the crystal by a microscopic amount. (This is called the piezoelectric effect.) The crystal then springs back to its original shape, generating a small voltage in return (this is also the piezoelectric effect, working in the other direction). An electronic circuit can be connected to the crystal to encourage this back-and-forth oscillation, setting up a resonance that continues as long as power is supplied to the circuit.

The neat thing about a quartz crystal oscillator is that it's very, very stable. Even the cheapest quartz watches, unless they are actually defective, will keep time within a dozen seconds per month. Yes, there are purely mechanical watches more accurate than that, but only the most delicate and expensive models match dollar-store quartz watch performance. And there are a number of techniques used to make quartz watches significantly more accurate, with reasonably inexpensive models being more accurate that even the most expensive mechanical watches.

Most quartz watches use a crystal tuned to oscillate at 32,768 cycles per second (32.758 KHz, a tone somewhat above the range of human hearing). Why this strange number? Actually it isn't strange at all, it's a number you see all the time in digital circuits: 2 multiplied by itself 15 times. 2 × 2 × 2 × 2 × 2 × 2 × 2 × 2 × 2

IT IS VERY easy to build an electronic circuit called a "binary divider," which takes an alternating up/down signal and produces an alternating up/down signal going at exactly half the rate. That is, for every two up/down cycles of the input, the output does a single up/down cycle. This is the digital electronic equivalent of a pair of gears with a gear ratio of 2 to 1. With gears you can make any integer ratio you want (within reason) but with digital dividers it's much easier to divide by 2 than by any other ratio.

Put fifteen of these binary dividers in a row and you can divide the 32,768 cycle per second signal from the crystal down to a one-cycle-per-second signal you need to drive the seconds display in a digital watch. (Similar, but somewhat more complicated, dividers then cut the signal down further for the minutes and hours displays.)

These days the complexity of the digital circuitry needed to divide by any other number would be completely trivial. Even a 99-cent watch could have a thousand times more circuitry than what's needed to divide by any arbitrary number without raising its cost. But in the early days of quartz watches the size, power requirement, and cost of digital circuits were all much higher, so the frequency of the crystal was chosen to make the circuits as simple as possible.

◀ Binary divider chain

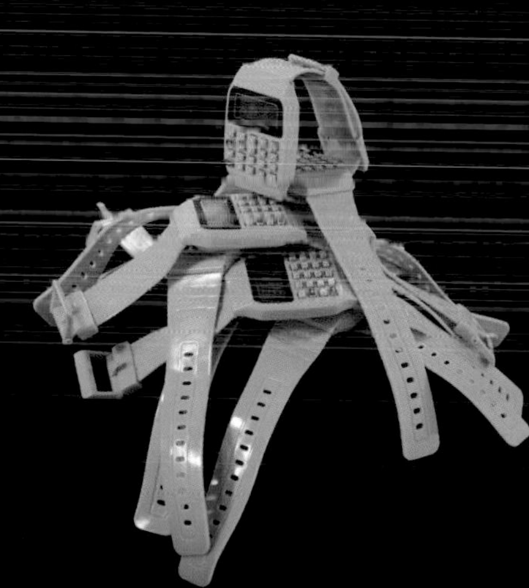

▲ A pile of cheap watches

Just to drive home the point that digital circuitry is as good as free these days, here is a pile of 99-cent *calculator* watches. Not only do they have all the circuitry of a quartz watch, they can also add, subtract, multiply, and divide 8-digit numbers, no extra charge.

Of course this is as commonplace to me as to any of you, but I do like to remind myself from time to time that there is living memory—my memory—of a time when such things did not exist. I told my parents in third grade that it was pointless for me to learn my multiplication tables, because if I ever needed to do that sort of thing, there would be cheap portable machines to do it for me. At the time I was talking about machines that didn't

THE PUREST DIGITAL watches use LCD displays, but there are also lots of quartz watches that use old-style hands to display the time. These watches use the same quartz crystal oscillator and digital dividing chain to create a signal that switches once per second. But instead of sending that signal to an LCD display, they send it to a tiny "solenoid," a kind of electromagnet with a coil of wire that generates a magnetic field when electricity is sent through it. The magnetic field pulls in a tiny iron rod, which in turn moves a gear by one notch. This is roughly the equivalent of the escapement in a mechanical watch, but the timing is controlled by the crystal instead of a pendulum or balance wheel.

The crystals used in wristwatches are tiny! Here are ten 32,768 Hz crystals fitting comfortably on a penny.

This has got to be some kind of high point in phony watchmaking. It's a standard cheap quartz watch with electro-mechanical movement just like the one on the previous page. But this movement, with its real gears, solenoid, battery, and crystal, is completely hidden behind a stamped metal, completely fake "skeleton watch" design. None of the "gears" you see actually move. The clear quartz watch is much more beautiful, because it lets you see real things doing real work. This one hides the naturally beautiful quartz mechanism behind an embarrassing wall of fake. Oh well, at least it only cost $5.

Quartz crystal oscillators are found in nearly all digital electronic devices, not just in watches. They come in many sizes and designs, with frequencies from a few KHz (thousands of cycles per second) up to many MHz (millions of cycles per second). You will find at least one in every cell phone, radio, computer, digital camera, and talking doll. In each of these devices the crystal provides the "clock signal" that synchronizes the operation of all the parts of the device. In anything with a microprocessor (a computer processor chip) the crystal times the execution of the instructions in the program that the chip is running. In a radio the crystal provides the reference frequency for transmitting and receiving signals.

A raw quartz crystal oscillator is already a very accurate timekeeper, but it does have one weakness: the frequency depends a bit on the temperature of the crystal. There are two ways to improve accuracy. The cheap way, commonly used in clocks and watches, is to include a temperature sensor mounted near the crystal. This doesn't help keep the frequency more constant, but if you know the effect of temperature on the frequency, you can use the measurement of the temperature to correct the signal coming from the crystal. This integrated, temperature-corrected crystal oscillator module is accurate to a few parts per million, at least 10 times more accurate than a raw crystal.

Now this is a brutally beautiful watch. It's like the nerd equivalent of those incredibly oversized bling watches you see on insecure jock types. The case is made of multiple layers of laser-cut acrylic, which you assemble yourself with a few tiny screws, after soldering together the circuit board. The LED (light-emitting diode) display uses so much power it can only be switched on for a few seconds at a time.

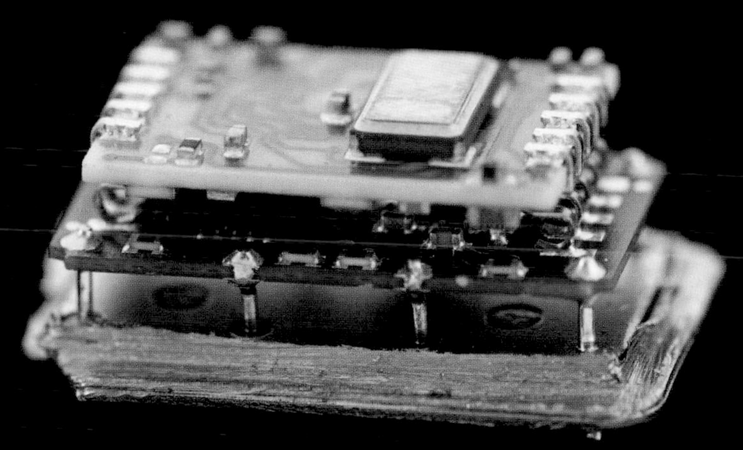

An even more accurate, but more expensive and *much* more power-hungry solution, is to put the crystal into a thermostat-controlled "oven" that keeps it at a constant temperature somewhat above room temperature. This expensive ($50) high-precision oven-controlled quartz oscillator has a long-term accuracy of twenty parts per billion (about one second in two years). But you could never use it in a wristwatch: they are practical only in devices that have access to a lot more power than you can get from a tiny battery. And they are also far more accurate than you'd normally need in a watch, especially since there is an alternative that will give you basically perfect accuracy with very little extra power required.

THE PENULTIMATE WATCH

QUARTZ OSCILLATORS ARE great, but they are not even close to being the most accurate clocks in the world. Later we'll learn about *really* accurate time, but you don't have to wait that long to encounter an inexpensive watch that is nearly as accurate as the most accurate clocks anywhere. This $40 watch is able to keep time to within one second in something like *a hundred million years*.

It does, however, cheat to achieve this level of perfection: the timekeeping mechanism is not actually in the watch. It has a common quartz crystal mechanism that keeps the time accurate over the course of a day or two, but it also has a tiny radio receiver permanently tuned to radio station WWVB broadcasting out of Fort Collins, Colorado. This station does one thing and one thing only: it broadcasts a time signal derived from the collection of super-accurate atomic clocks that together define the standard called Coordinated Universal Time.

This is basically the same as the system used in the 1800s by the Greenwich Observatory to communicate its time standard to the rest of London. Instead of a giant red ball there's a radio station, but the principle is the same: use a more accurate national time standard to regularly reset a less accurate local clock.

The downsides are that the watch works only within a few thousand miles of Colorado, and that it will stop working (or at least stop being so accurate) if civilization falls apart and WWVB stops broadcasting. On a positive note, you never have to set it to the correct time—not even once. Instead you tell it what *time zone* you're in, and the actual time is received from the radio signal and set automatically.

Believe it or not, even at one second in a hundred million years, this watch is not the most accurate one you can carry with you. That honor goes to your cell phone.

MOST CHEAP WATCHES are not temperature compensated or radio controlled, so they may drift more than usual when kept at extreme temperatures. One way to keep such a watch accurate is simply to wear it most of the time: quartz watches are calibrated to run at the correct speed when they are at the temperature they typically reach when in contact with a human arm. In other words, they are using your constant-temperature arm the same way an oven-controlled crystal oscillator uses its constant-temperature oven. Sorry undead zombies, your watches will run a few seconds slow per month. (Unless, hypothetically speaking, an entire zombie watch industry were to develop making watches calibrated to work best at lower temperatures. Why is this never discussed in zombie movies?)

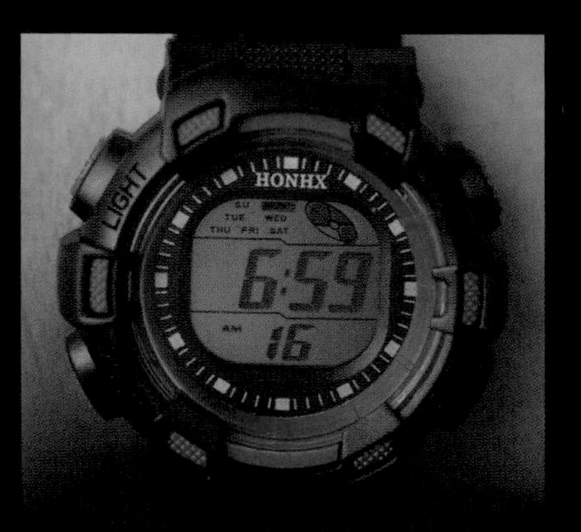

▲ Some quartz watches run most accurately when worn.

▶ This is the end of a large (8 inch, 20 cm long) perfect quartz crystal from which crystal oscillators are cut. They are grown in laboratories from molten silica (silicon dioxide), which is basically purified sand.

▶ Even though there is no pendulum, balance wheel, or escapement in a quartz clock, most of them still tick because the solenoid inside pulls very suddenly on the second hand's ratchet gear. Wall clocks can be quite annoyingly loud in a quiet room. My dad always *hated* ticking wall clocks when he was trying to eat breakfast. I remember him trying out all kinds of different battery operated quartz clocks until he finally found one that was absolutely silent. That clock is long since lost, but it looked a lot like this one, which too is gloriously silent.

Silent quartz wall clocks use a "sweep" mechanism that contains a tiny synchronous electric motor instead of a solenoid. The trade-off is that this mechanism uses more power than a solenoid-based mechanism. You buy silence at breakfast with a new battery every few months instead of every few years.

The core of the silent clock mechanism is a completely self-contained combination timekeeping circuit and synchronous motor.

A gear chain cuts down the speed of the rotor by 480 times to reach the speed of the second hand. After that are the same 60-to-1 and 12-to-1 gear chains found in any other clock with hands.

Instead of dividing the high frequency of the crystal down to 1 pulse per second, this clock divides it down to 32 pulses per second, which are fed to this coil of fine wire, which creates an alternating magnetic field from the electric current. All those pulses are why this mechanism uses more power than a solenoid that pulses only once per second.

Quartz crystal in can

These metal plates transfer the magnetic field generated by the windings over to the rotor.

A tiny permanent magnet in this rotor responds to the alternating magnetic field by spinning at a rate exactly in sync with the oscillations of the field (hence the name "synchronous motor").

Clear Face Clocks

THERE IS A category of clock that pulls a clever visual trick: hands that appear to float in air with no visible mechanism. I have seen five quite different ways in which this effect is achieved, ranging from pure beauty to sublime cleverness to sad faking.

▸ This clock is in the sublime cleverness category. At first I was baffled how it could work at all. It uses a clear disk (glass instead of plastic, because it's not a tacky bar clock). But there is only *one* disk driving both hands. How can you have both the minute and hour hands turning at different rates mounted to the same piece of glass? The secret is in plain sight from the back: a pair of gears and a small weight. Wait, a weight?

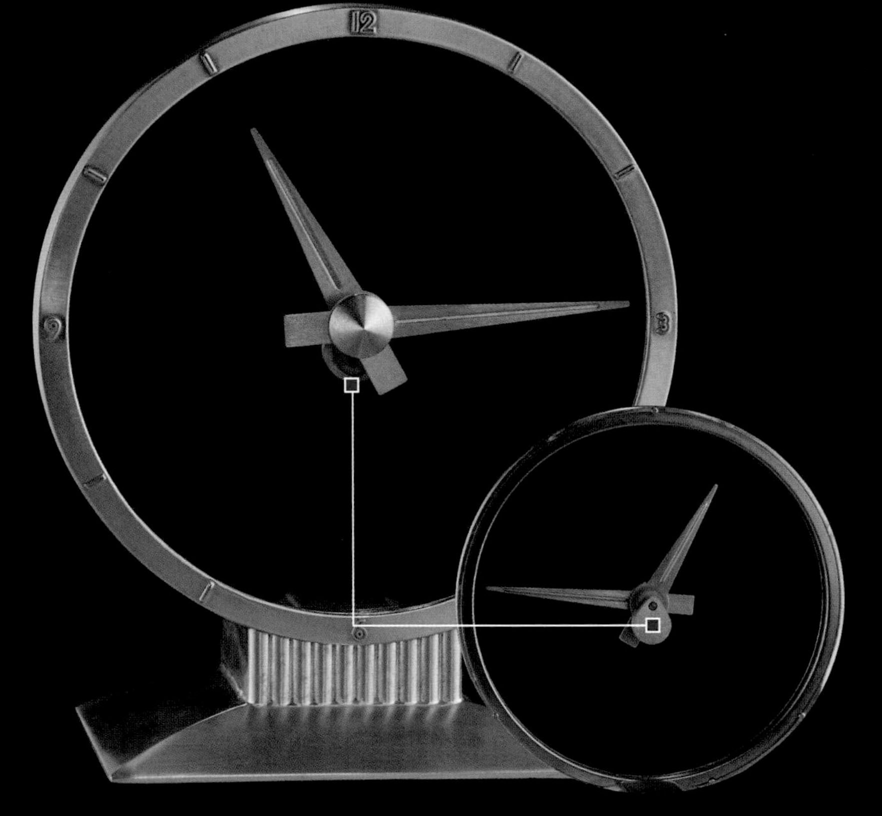

▴ The minute hand is mounted directly to the glass disk, which makes one complete turn every hour. As it is turning, a small gear locked to the glass plate is turning a larger gear whose axle is mounted to the dangling weight. The large gear itself is not locked to the weight, only the axle on which the gear is mounted. The hour hand is locked to this larger gear, turning once every twelve hours.

Because the weight is hanging down from the center of rotation of the disk, it does not turn as the disk turns. Instead it stays stationary relative to the rest of the clock. In this way it provides the necessary second reference point that allows one gear to turn against another. Without the weight (and the gravity needed to keep it pulled down) the whole mechanism in the center of the disk would just turn as a unit, and the hour hand would also go around once per hour, instead of twice a day as it should.

▴ In the pure beauty category is this bar clock, which is also an advertisement for a brand of beer (that's not the beautiful part).

◂ This is the sad fake example of a floating-hands clock. It's just a normal battery-powered clock with hands that are longer than the mechanism. You can tell this is a low-effort floating-hands clock by the large size of the hub in the middle. Compare this to the beer clock that has literally nothing in the middle.

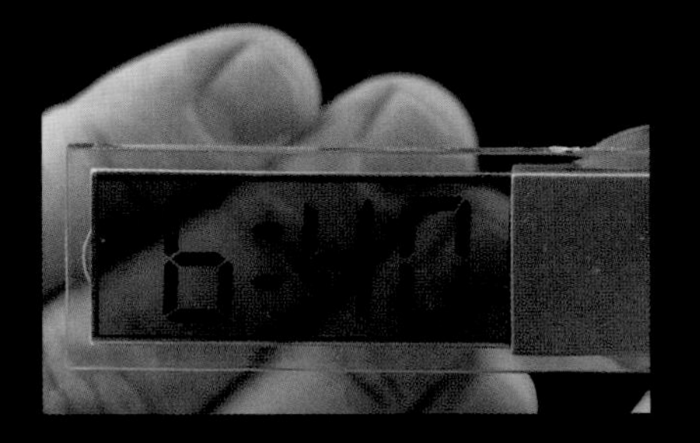

▴ The beauty of this clock is in its pure implementation of a simple idea: mount the hands of a clock on clear plastic disks, then drive those disks invisibly from their outer rims. This leaves the center of the clock completely empty: other than the hands themselves, nothing obscures the view through the face of the clock right back to the cheap whiskey behind the bar.

▸ This clock has numbers instead of hands, but the idea is the same: time is floating on a completely clear face. Clear is possible here because of the way LCD technology works. An LCD display has a layer (the liquid crystal) that is able to rotate polarized light, sandwiched between two polarizing filters. Let's unpack these words.

Anatomy of an LCD Display

LIGHT IS A wave, but it's not a wave *in* something, like an ocean wave is a wave in water. Instead it is a self-sustaining oscillation of electric and magnetic fields that races through empty space at the speed of light. (Because it is, literally, light.) Every light wave has an orientation: its electric field is oscillating back and forth in a particular direction relative to the direction it is moving. This light wave, for example, has its electric field oscillating up and down. (I'm not showing the magnetic wave in this diagram because that would get way too confusing in the diagrams to come over the next few pages.)

Normal light is made up of waves with their electric fields randomly oriented in all directions, all superimposed on top of each other. For simplicity, in the diagrams that follow I'm going to show a beam of light with all the waves going in the same direction in a regular grid. (In reality of course things would be a lot more random.)

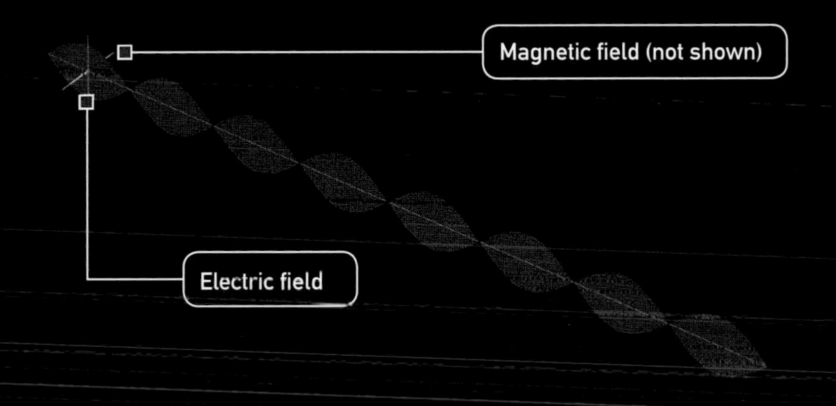

▼ A polarizing filter is a piece of special plastic that forces (nearly) all the light passing through it to oscillate in one particular direction. This one is oriented vertically, so all the light coming through has its electric field oscillating in the up/down direction: that's called polarized light. (A good quality polarizer will give you about half the total amount of light you started with, about 98 percent aligned in the desired direction.)

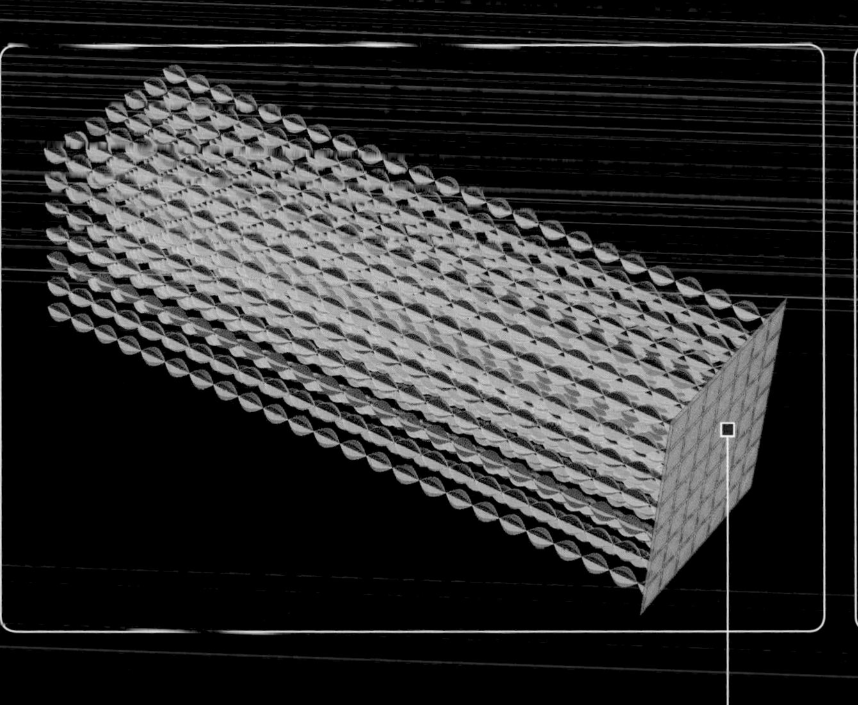

This represents a "screen" that the beam of light is shining on, so we can more easily see when parts of the beam are blocked. Here we see the screen all lit up.

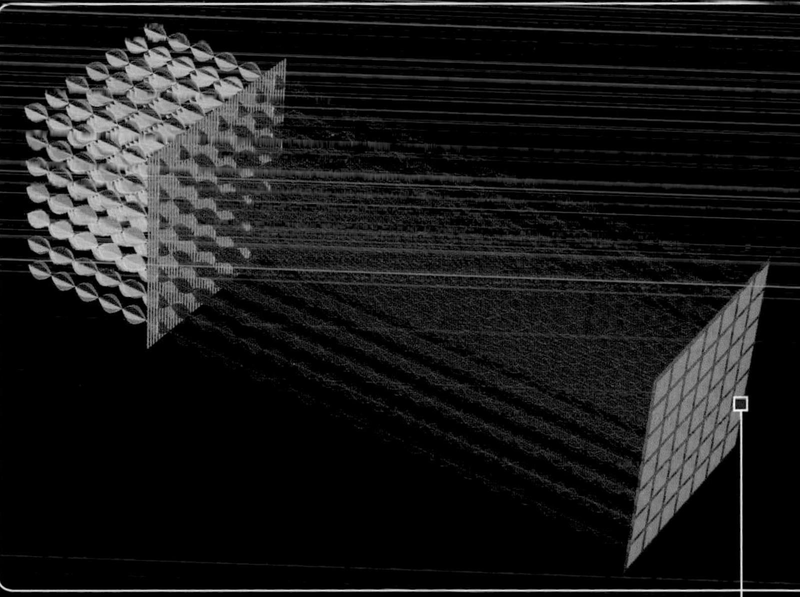

The screen is still all lit up, because it's still getting hit by a good amount of light everywhere.

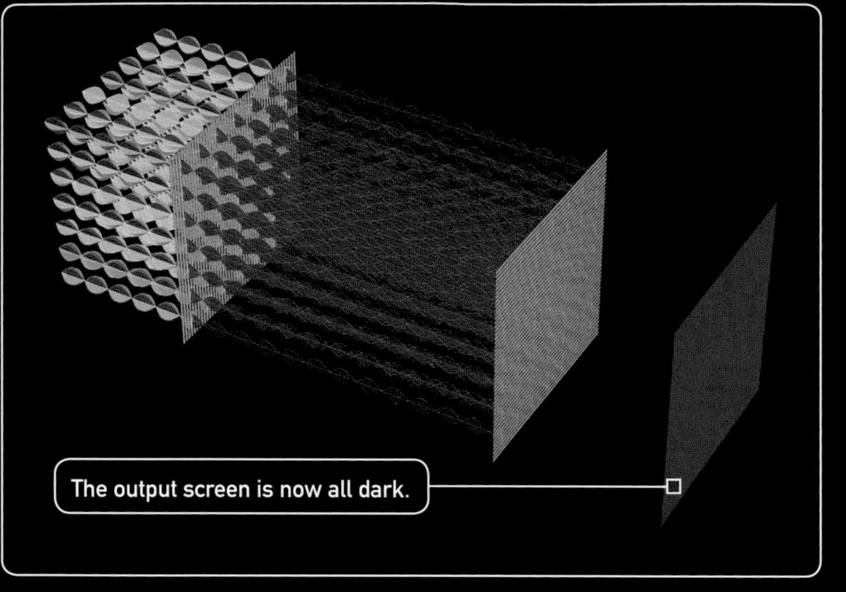

▲ **1.** If we insert a second polarizer rotated 90 degrees (1/4 turn) from the first one, all the light reaching the second one is oriented in exactly the wrong way, and none of it goes through. (With good quality polarizers, as little as one-thousandth of the light will leak through.)

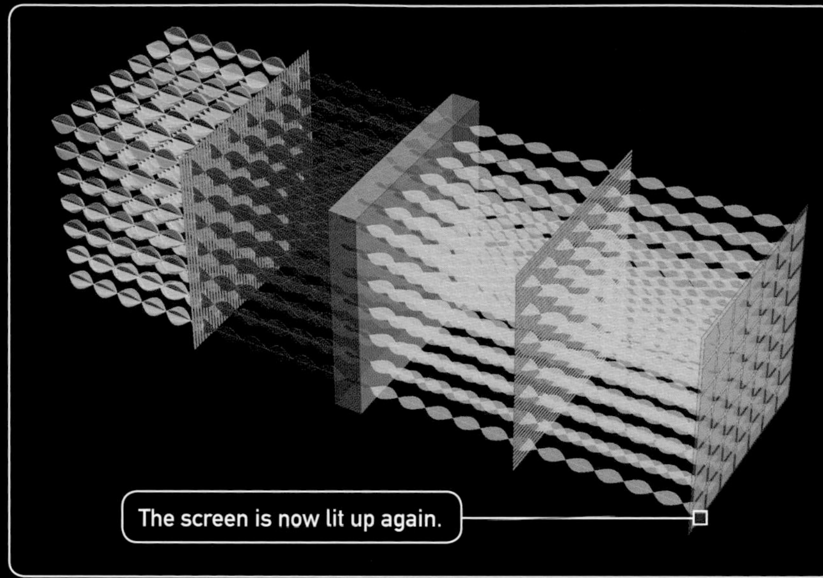

▲ **2.** Here we've got the same two polarizers at 90 degrees to each other, but we've inserted between them a sheet of material that rotates the polarization of the light by 90 degrees. This time when the light reaches the second polarizer, it's been rotated to the correct orientation to pass through.

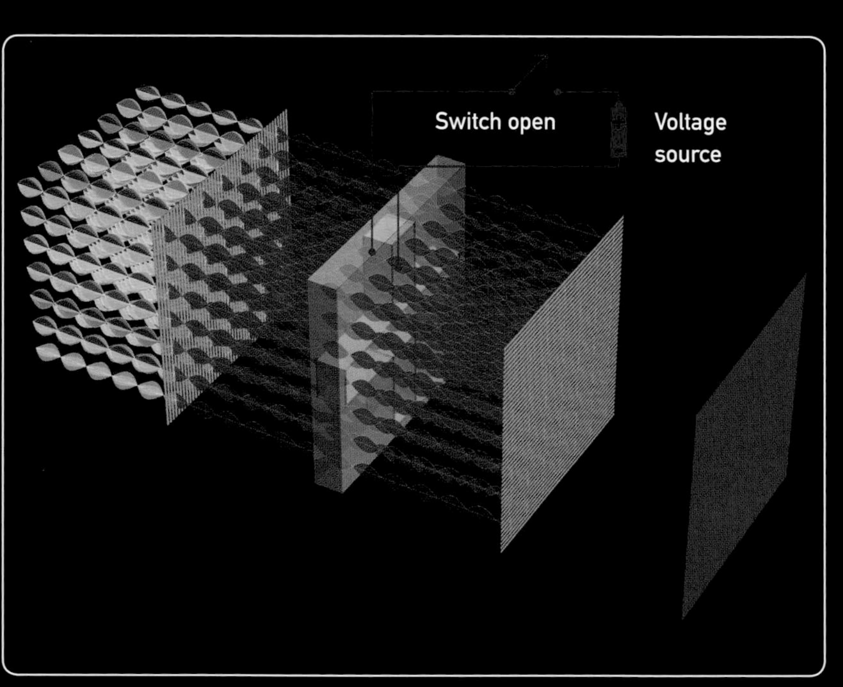

▲ **3.** Now imagine what you could do if you had a material that either does or does not rotate polarized light, depending on whether there is a voltage applied to it. That's exactly what liquid crystals can do. This liquid crystal sheet is split into separate zones, with a transparent electrical connection made to the plus-shaped patch in the middle. With no voltage applied, none of the light is being rotated, so none of it makes it past the second polarizer and the output screen is dark.

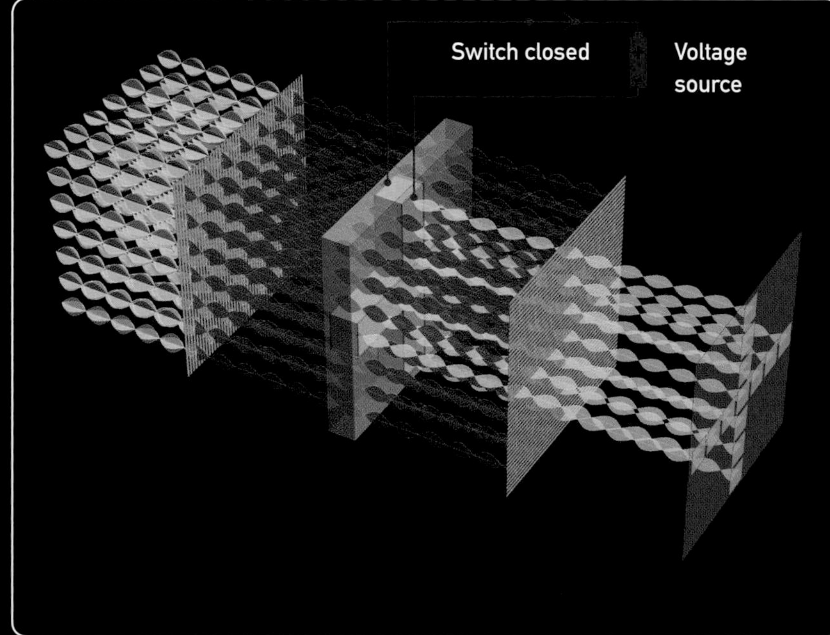

▲ **4.** If we apply a voltage to the plus-shaped segment, it starts rotating the light that goes through that part of the sheet, lighting up that part of the screen. This is an LCD display that shows a plus sign when activated.

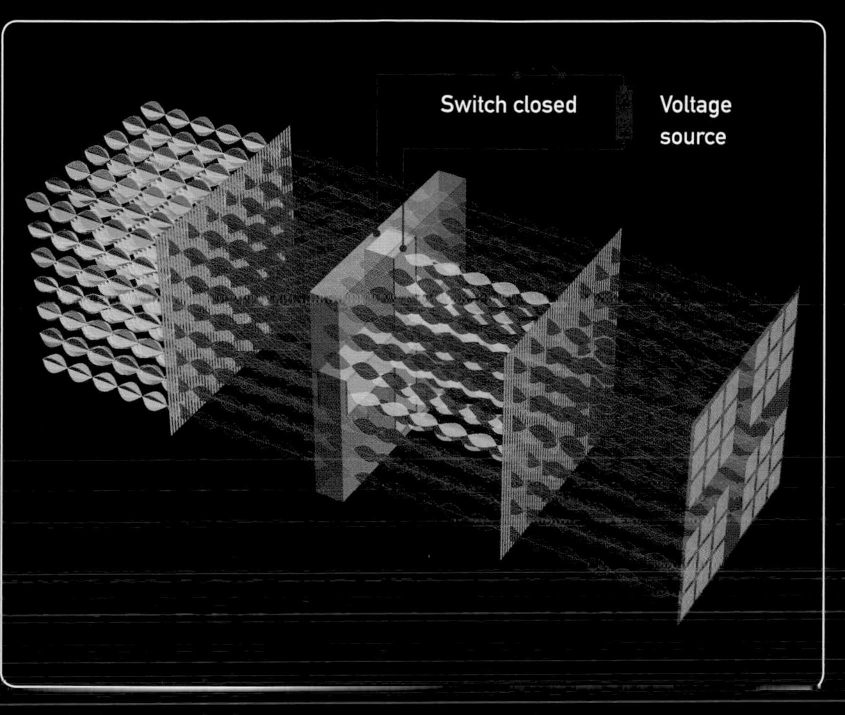

There's one final trick to get us to the transparent clock that got us started on this journey into polarized light. You can easily change the light vs. dark behavior of any LCD display by rotating the second polarizer 90 degrees. In the previous diagram we had a bright plus on a dark background, but here we have a dark plus on a bright background. The only thing that's changed is that the second polarizer is now aligned vertically, the same as the first polarizer.

If you leave off the backlight supplying bright light behind the whole display, then "bright" turns into "transparent" and "dark" turns into "opaque" (blocking). With both polarizers aligned in the same direction and no backlight, you have a display that is transparent everywhere except where voltage is applied to the liquid crystal segments.

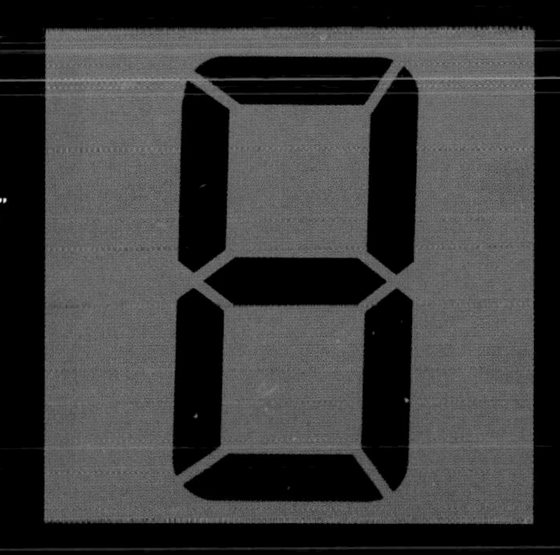

To make a more complicated display you just need to arrange little patches of electrodes in the shapes you want on a sheet of liquid crystals. This set of shapes is standard for "seven segment" displays, which can show any number. A wonderful property of these displays is that no significant amount of current needs to flow through the segments to keep them activated. They just need to have a voltage difference from one side to the other. That means a display like this uses very close to zero power (as long as you don't have a backlight). Watches with LCD displays can run for years on a tiny battery. (For those of you familiar with electrical components, an LCD segment is a bit like a capacitor. Once you charge it up, no further current needs to flow to keep it activated.)

LCD displays in computers and cell phones don't have patches of specific shapes, instead they have a huge number of tiny squares (millions of them) so they can display any image they want by turning on and off individual squares (called pixels). LCD displays usually have a backlight behind the first polarizer to make the whole display light up, but in principle any LCD display can be made transparent. Here I've removed the backlight from a cheap cell phone screen so you see my hand behind the image on the screen. The polarizers in this case are crossed at a 90 degree angle, so the display becomes transparent where the image it's trying to show is bright.

Computer and phone displays have two further refinements beyond what's shown in the diagrams above. First, each tiny square is actually split into three separate zones with color filters—red, green, and blue—in front of each third of the square. Second, the voltage applied to the liquid crystal can be varied over a range, which varies how much the polarized light passing through is rotated. When the light is rotating less than the full 90 degrees, only a proportional fraction makes it through the second polarizer. That's how the display can show a continuous-tone image where each square has a different brightness level in each of its three colors. A nice cell phone with a full-HD display has over six million individual patches (two million of each color), each of which can be set to one of 256 different brightness levels.

▲ Here is the fifth and final way I'm aware of to make a floating transparent clock face: spinning LEDs on a fan blade. The face of the clock looks mostly transparent because the blades are spinning so fast that you don't see them. Fancy versions can display not only the time, but also text and images you download to them from your phone by Bluetooth.

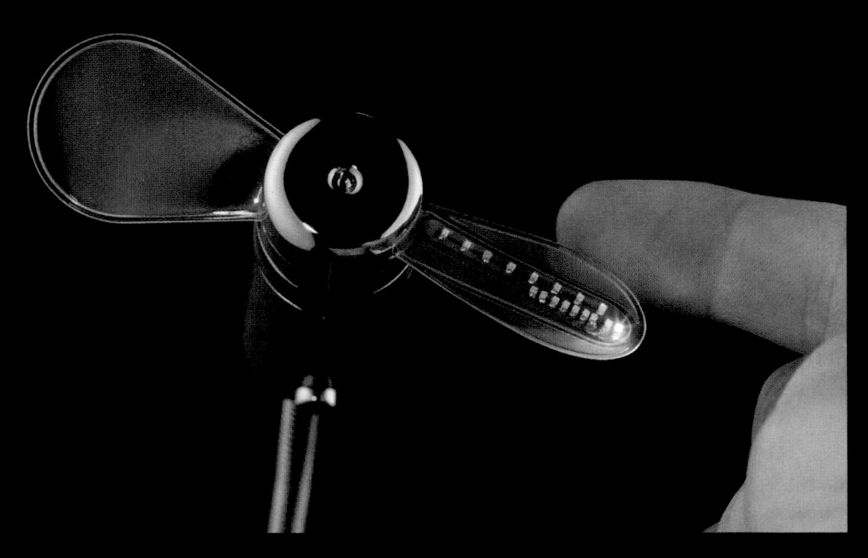

▲ If you stop the spinning fan blades, you see that there is actually only one radial line of LEDs. At any given moment it is showing a meaningless pattern of on and off dots.

▲ If you take a medium-long-exposure picture of the fan/clock, you see that the single line of LEDs is painting out a pattern as it is swept around the clock face on its fan blade. LEDs can do this because they can be switched on and off very, very fast (billions of times per second in fact, though for this application just ten thousand times per second is fast enough).

Really Accurate Time

THIS CUTE LITTLE clock may look like a toy, but—in common with every plug-in electric clock—its timekeeping ability is extraordinarily good. If this clock were battery powered, the speed of its motor would be controlled by a quartz crystal oscillator, and it would have pretty good accuracy. But because it plugs into the wall, it can instead be controlled by the alternating current (AC) that comes from the power company. The frequency of that power is maintained at an extremely accurate 60 cycles per second, synchronized to the global UTC (Coordinated Universal Time; see page 133), the most accurate time standard that has ever existed. If civilization were to last another hundred million years, and this clock stayed plugged in, and you never forgot to pay your power bill, it would still be accurate to within a second. (Power is kept accurately synchronized so that multiple power-generating plants can feed power into the grid without fighting each other with out-of-sync voltages.)

This metal plate transfers the oscillating magnetic field generated by the current flowing through the wire coil over to the rotating part of the motor, from where the motion is carried forward by a standard clock gear chain.

You can just see the rotor inside the motor housing.

You can see that this coil of wire is connected directly to the incoming 120V wires from the wall plug. There is literally no electronic circuitry at all in this clock, because it doesn't need any; the timekeeping function takes place back at the power plant.

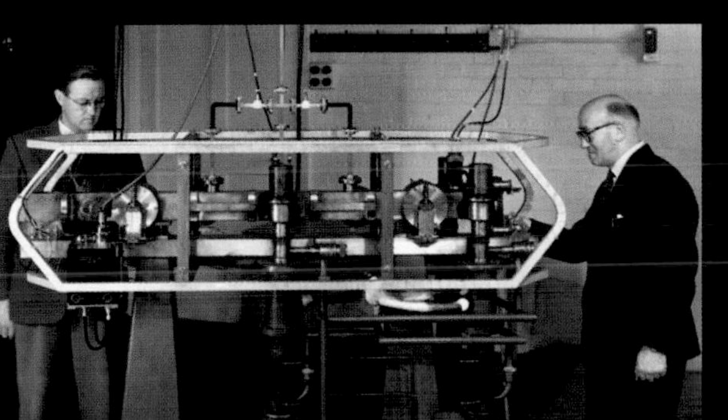

The first cesium atom clock. Descendants of this clock are what control the frequency of the AC power grid.

WE'VE NOW ARRIVED at the subject of really *accurate* time. Earlier I said that up until 1955 a glorified sundial was the most accurate clock in existence. Another way of saying this is that we were using the rotation of the earth as the reference standard for time, because we didn't have anything better.

Every day, the time told by the world's best clocks was compared to the exact moment that a set of reference stars passed directly overhead at the Greenwich Observatory in London. If there was a deviation between when the clocks said the stars should pass overhead and when they actually did, we trusted the stars over the clocks, and reset the clocks. Based on everything we knew about the earth and the clocks, there was no reason to think that the clocks were a more accurate measure of time than the rotation of the earth.

It's a bit of a logical puzzle, actually: how do you know if a clock is accurate? There's an old saying: "A person with one clock always knows what time it is. A person with two clocks is never quite sure." The point is that if you have two clocks and they go out of sync with each other, how do you know which one is right? All you can say for sure is that at least one of them is wrong. There is no absolute way to know which one it is.

You might say that the more accurate one is the one that predicts the moment of noon more consistently over a period of months. But that's just using the rotation of the earth as a reference clock. If your clock gets out of sync with the earth, your clock might be wrong, but then again, the earth might be wrong instead.

Two clocks, even if one is the earth, are not enough. But if you have *three* clocks, and you find that two of them stay exactly in sync while the third one drifts away, now you've got some evidence that the third one is the problem. If you have a hundred clocks spread all over the world, built by different people using different methods, and all but one of them agree with each other, then you can be pretty sure that the outlier is the problem, not the other ninety-nine clocks. In 1955 the earth became that outlier—not because anything changed about the earth, but because we suddenly got much better at building clocks.

What changed in 1955 was the invention of the cesium atomic clock. Now suddenly we had multiple clocks that all stayed very, very, *very* closely in sync with each other, and they all agreed that the speed of the rotation of the earth was changing. Using these clocks, and the better ones that followed, we now know that with each passing day the earth takes on average 0.00000005 seconds longer to make one revolution than it did the day before. Every century the length of a day gets about 2 milliseconds (0.002 seconds) longer. Every 500 years the day gets a whole second longer. The ancient Egyptians had an especially hard time when building the great pyramids because their days were almost nine seconds shorter than ours!

Over even longer spans of time the slowing of the earth has been dramatic. There is good evidence that about six hundred million years ago the day was only 21 hours long (by today's definition of the hour, of course). A few billion years ago the day was only 18 hours long.

The length of a day is getting steadily longer *on average*, but day to day there is fluctuation both up and down in the length of a day. Our clocks, and our measurements of the rotation of the earth, are now so good that we can measure these individual fluctuations and deduce from them fascinating things about what's going on in and around the earth. This graph shows what the length of a day has looked like over the past sixty years (since such measurements became possible).

You can see regular seasonal cycles. That's because the earth slows down a bit when, on average, there is more water evaporated up high in the atmosphere vs. down on the ground in lakes and oceans. (For the same reason—conservation of angular momentum—that a spinning ice skater goes faster when they draw their arms in closer.)

In 2004 the massive Indian Ocean earthquake pulled in the "arms" of the earth by enough to immediately decrease the length of a day by 2.68 microseconds (0.00000268 seconds).

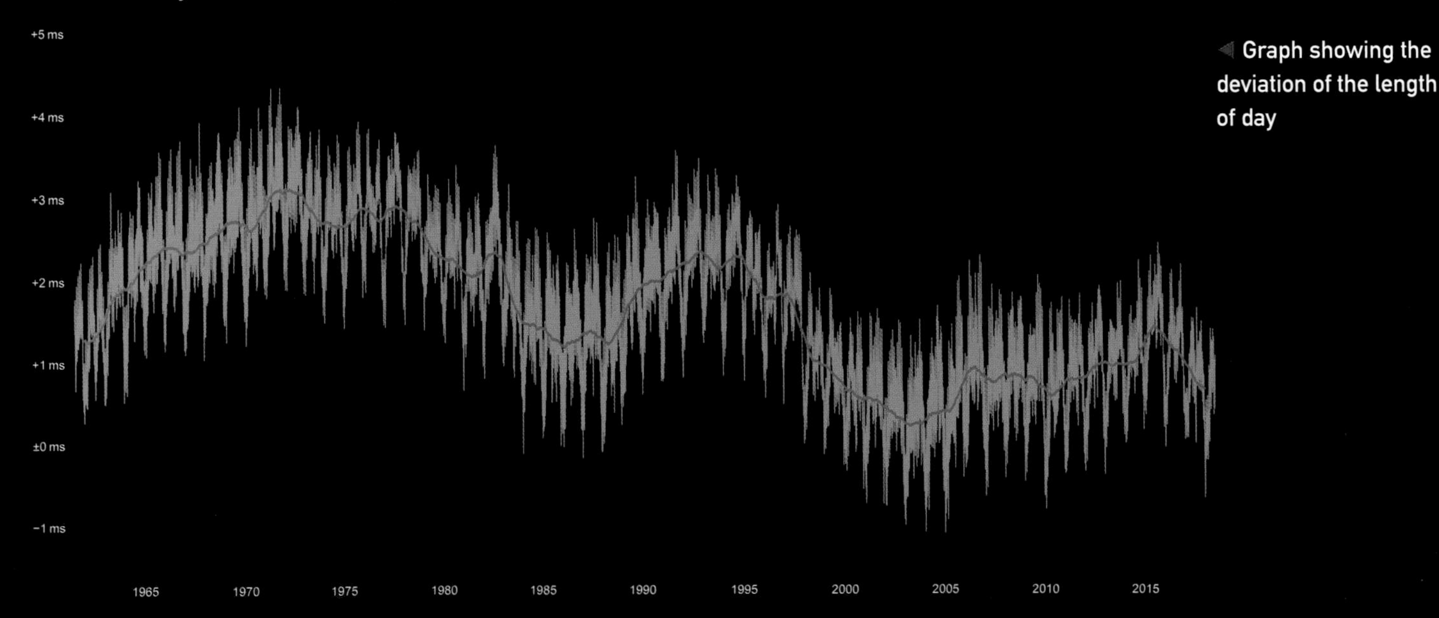

Graph showing the deviation of the length of day

TODAY'S BEST CLOCKS are so far removed from the pendulums, balance wheels, and even the quartz crystal oscillators of everyday clocks that they are scarcely recognizable as clocks at all. This, for example, is a so-called cesium fountain clock. It uses the resonant frequency of a certain quantum mechanical state change within cesium atoms as a measure of time. This frequency is in theory perfectly constant, but if the cesium atoms you're measuring bump into each other, or into the walls of a container, or experience any other force, the frequency will be slightly disturbed.

What you want for timekeeping, ideally, is individual cesium atoms free-floating in an otherwise perfect vacuum. But if the atoms were floating free, they would fall down and bounce off the bottom of the vacuum chamber, messing up the measurement. What to do? This is called a cesium fountain clock because it solves the problem by gently throwing little puffs of cesium atoms vertically upward into an evacuated column. Like balls thrown up into the air, the cesium atoms fly upward, slowly come to a stop, and then fall back down again.The resonant frequency measurement is taken while they are in freefall at the top of their arc, untouched by external forces.

Other clocks have been proposed and built that are more accurate than cesium atomic clocks, but they are still research projects. The internationally accepted definition of a second, as of 2019, is still in terms of the cesium atom. Officially, "The second is the duration of 9,192,631,770 periods of the radiation corresponding to the transition between the two hyperfine levels of the ground state of the cesium 133 atom."

The world's absolute time standard, called International Atomic Time, is set by a network of about four hundred of the best clocks, most of them cesium atomic clocks, maintained in dozens of countries all over the world and synchronized with each other by radio signals. International Atomic Time is an absolute standard: it counts up one second every second, period. That's useful for long-term comparisons of spans of time, but unfortunately it means that this time standard is getting progressively farther and farther out of sync with the earth itself—because the earth is wrong (currently by a cumulative 37 seconds since International Atomic Time was first defined).

To deal with the steadily growing error in the earth's speed of rotation, a second time standard is more commonly used: Coordinated Universal Time. (Which is abbreviated UTC, not CUT, for political reasons: UTC does not make sense as the abbreviation in any language, so it was acceptable to the French.) UTC ticks off seconds at the same rate as International Atomic Time, but it's kept in sync with the rotation of the earth by the introduction of "leap seconds" every couple of years, as needed. (The earth is sufficiently unpredictable in its rotation that the time of these leap seconds has to be determined on a case-by-case basis, and is typically announced a few months before they happen.)

▲ Modern cesium clock

CELL PHONES AND smartwatches that get their time by association with a companion smartphone have access to absolutely the most accurate available time, even more accurate than what you can get from the WWVB radio signal mentioned on page 124. The problem with WWVB isn't in the signal it broadcasts, it's in the time it takes that signal to reach you. Radio waves travel at the speed of light, which is about one nanosecond (billionth of a second) per foot (30 cm). If you're, say, about 500 miles (800 km) from the transmitter, the time signal will take about 0.0027 seconds (2.7 milliseconds) to reach you. If you're using a watch like the one on page 124 it has no idea where it is—it could be right next to the transmitter or it could be 2000 miles away in a boat somewhere in the Atlantic. That means its time could be off by as much as, gasp, 0.01 seconds (10 milliseconds)!

Cell phones take care of this problem by using their GPS (global positioning system) receivers to get the current time. GPS time is based on Coordinated Universal Time, as is WWVB, but in a much more sophisticated way.

A set of GPS satellites in orbit around the earth is constantly broadcasting signals that encode within them the exact time and the exact *location* from which they were broadcast. When your phone gets one of these signals, it compares the time encoded in the signal with the current time, and from that it can calculate how long the signal took to

▶ Smartwatch

LIGHT TRAVELS AT 186,282 miles per second (299,792 km per second), or about one nanosecond (billionth of a second) per foot (30 cm). That sounds pretty fast, until you consider that a modern computer chip has an instruction cycle time of maybe 1/3 of a nanosecond (in a 3 GHz processor). In other words, during the time it takes a typical chip to go through one step of executing a program, light has time to travel only about 4 inches (10 cm)—not even all the way from one side of the computer to the other.

0.ns 0.1ns 0.2ns 0.3ns

reach the phone. Knowing the time delay, the phone can calculate how far away it is from the satellite.

But wait, how does the phone know what time it is accurately enough to measure a few microseconds of delay in the satellite signal? The trick is that the phone isn't just listening to one satellite, it's listening to half a dozen or more of them at once, each broadcasting from a different location in the sky. By cross-referencing the time and location signals, the phone can deduce what absolute time must be. And by triangulation of the distances, it can calculate exactly where on earth it must be. And from that it can deduce exactly the delay from each satellite, and thus the absolute time. Circular, but it works.

It's very clever, and potentially allows your phone to display the time to within a few nanoseconds of true Coordinated Universal Time. (I say "potentially" because cell phone makers don't really care about nanosecond accuracy, and the layers of software between the GPS receiver and the time you see on the screen might allow the time to be off by whole thousandths of a second. Specialized GPS time-standard receivers achieve a consistent accuracy of about ± 2 nanoseconds.)

▼ The fountain clock on the previous page is an extreme example of a cesium atomic clock, but smaller ones are readily available. This is my favorite example of a *much* smaller cesium clock: the whole unit including glass cesium chamber, heater, and antennas is barely a quarter of an inch (5 mm) high. This picture shows only those components: in operation they would be connected by fine wires to some microcircuitry and a power supply.

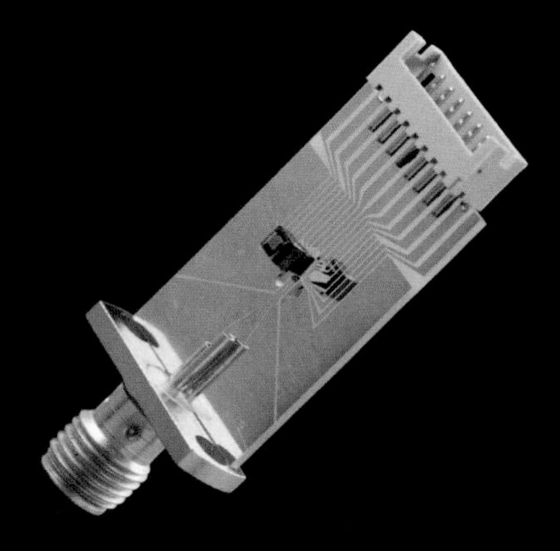

A Clock Joke

SUPER-ACCURATE TIME IS a serious topic, but fortunately we don't have to end the chapter this way, because I know a clock joke!

First, you have to understand that in Switzerland there are twenty-six cantons, and the population of every one of them (except Canton Zürich) has a unique flaw or quirk of personality that requires it to be made fun of by everyone else. These flaws have been memorized by every citizen by the age of six. My family, fortunately, is from the canton of Zürich where the people are perfect (though I now live in Illinois, which is so hopeless the Swiss don't even have jokes about it).

Zürich is known for the huge clock faces on its many church towers, particularly the great clock of the Fraumünster cathedral. From time to time this clock face requires painting, and one year the fine and upstanding people of Zürich, who definitely are not moneygrubbing, penny-pinching, stuck-up bankers, hired a low-bid painter from the canton of Bern. Big mistake there—frankly they really should have known better. After a whole week of work he'd made almost no progress! Asked why, he explained that every time he put his hand up to start painting, the darn hour hand would come spinning around and whack the brush right out of his hand! If you're Swiss you immediately know why: those people from Bern are just *so slow*. (Trust me, it's funny when told in Swiss dialect, without the explanation, to a six-year-old. I can hear my mother's voice right now.)

Cesium clocks are the gold standard for time (appropriate since cesium is the only other gold-colored element), but very similar clocks based on rubidium atoms are, for various reasons, cheaper, so they are often used in applications that require local clocks that are better than quartz crystal clocks, but not as good as cesium atomic clocks. The principle of operation is exactly the same, just a different element in the glass chamber.

SCALES

PEOPLE HAVE BEEN weighing things for thousands of years, mainly in the service of money and commerce. If you're paying for potatoes or gold, you want to know that you got what you paid for. More often than not, this is defined by the weight of the thing you're buying—ten pounds of potatoes, a thousand tons of iron ore, a troy ounce of gold, and so on.

When you hear farmers complaining about the price of corn (which they always do, no matter what it is), they are complaining about the price per bushel. If you look up the word "bushel" you'll find it defined as 8 gallons (about 35 liters). Gallons and liters are measures of volume, not weight, so that's saying a bushel is a measure of volume. You might think this is a counterexample to what I just said about people using weight when selling this, but it turns out that for large commercial transactions, they actually don't use this definition of bushel anymore. A commercial "bushel of corn" is 56 pounds of corn with a moisture content of 15.5 percent. In other words, it's a measure of weight.

Weight is the preferred way to quantify for commerce because it's hard to fake. It's also the most consistent, invariant way of measuring how much of something you have. Cereal might settle in the box, making the box appear only partly full, but its weight doesn't change. How much corn fits into 8 gallons of volume depends on the shape and orientation of the individual kernels. If you mill 8 gallons of corn into cornmeal, you might get a bit more or a bit less volume each time. But if you mill 56 pounds of corn, you will always get 56 pounds of meal. A pound of feathers is always a pound of feathers, no matter how fluffed up or packed down they are.

Because weight is so important in commerce, and commerce is central to nearly all aspects of public life, scales have played a central role in human civilization for a very, very long time.

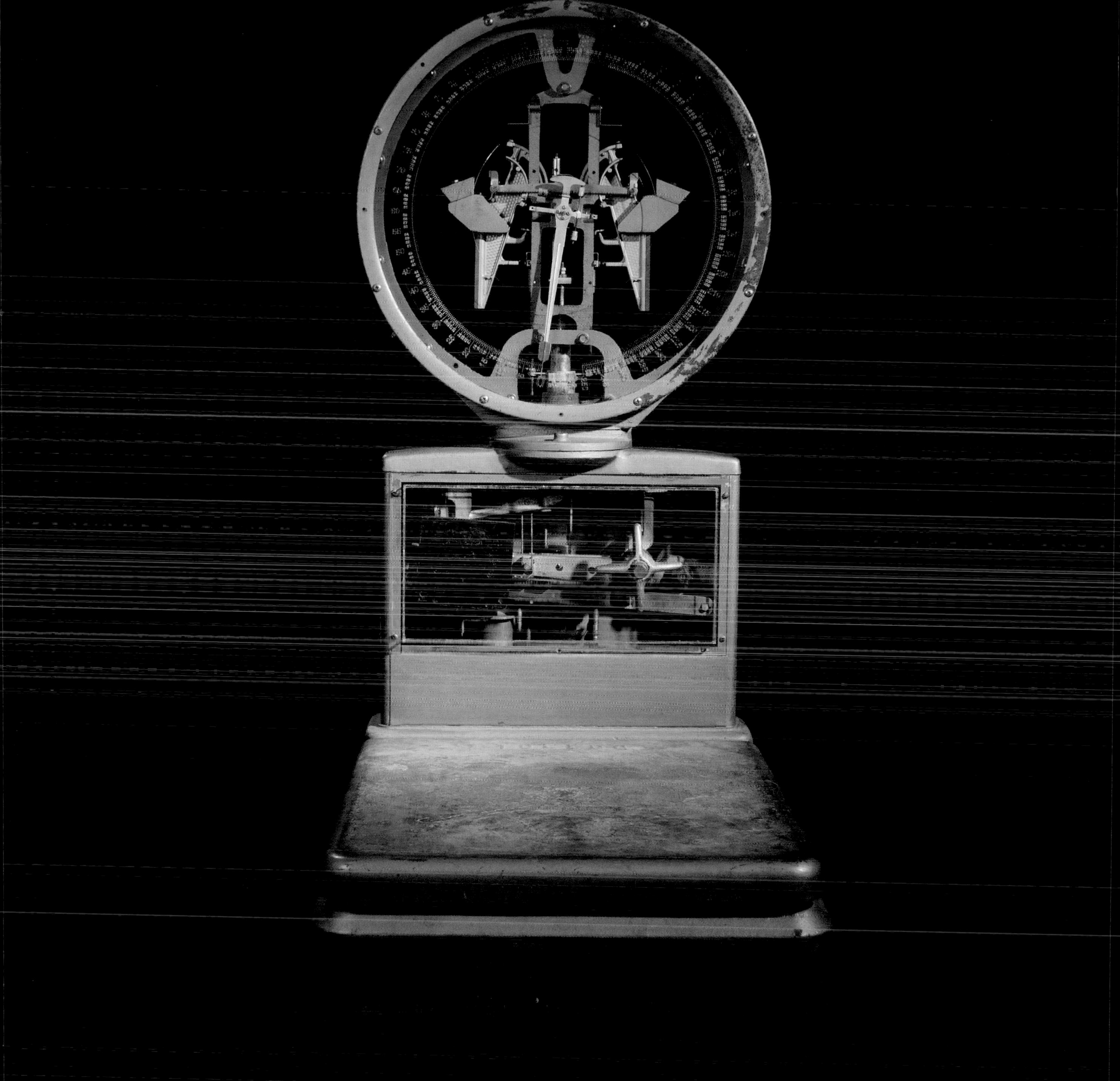

The First Balances

THE FIRST BALANCES, which date back at least four or five thousand years, were probably no more complicated than this one I made from a popsicle stick and a match. All you need to weigh something is a beam (like a stick or rod) balanced on a pivot point (the fulcrum). Beyond that it's just refinements to make the scale more convenient or more accurate.

This scale is just accurate enough to do what I wanted it to do: tell the difference between pennies made of solid copper (pre-1982) and pennies made of copper-plated zinc (anything after 1982). Copper pennies weigh more, so by putting a zinc penny (plus a bit of tape) on one side, I could easily tell whether a test penny is heavier, or about the same, as the reference penny. This is much faster than reading the date on each one.

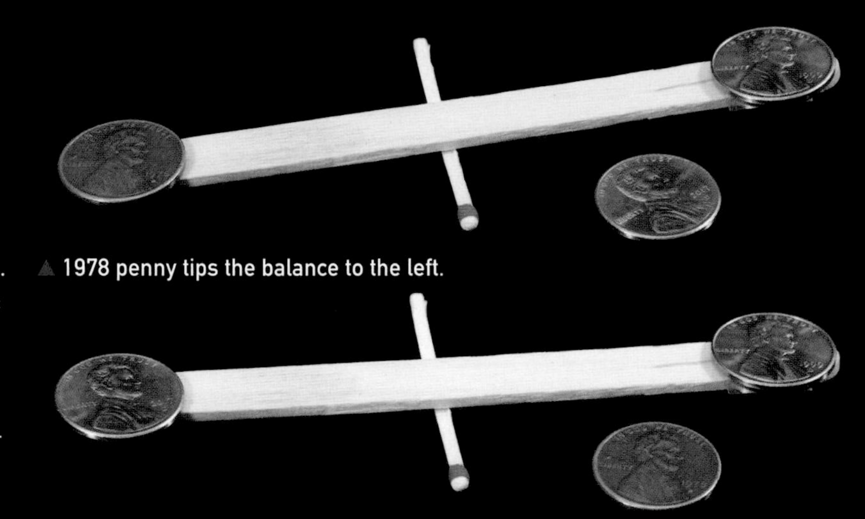

▲ 1978 penny tips the balance to the left.

▲ 2017 penny leaves the balance tipped to the right.

◀ Even things you might think are sold by volume (by the gallon, liter, bushel, cubic foot, etc.) are often actually sold by weight. This label, found on many boxes of breakfast cereal, is meant to assure you that even though the box you're buying doesn't appear to be full, you're still getting the full promised weight of the cereal.

▶ The ring of ridges around the edge of some coins, called reeding, is designed to make shaving easy to spot. This isn't really a big issue today since no coins in general circulation are made of precious metals, but the reeding stays by tradition.

Actual coin balances are a bit fancier than my version and go back a long way. This brass model is from around the time of the "Great Recoinage of 1816," when British coins were redefined and put back on a gold standard after the disastrous wars with Napoleon. One new "gold sovereign" coin was supposed to contain exactly 160/623 of a troy ounce (about a quarter ounce) of 22K gold. As with all coins whose value is based on their precious metal content, these new coins were vulnerable to wise guys shaving off small amounts of gold, or drilling and hiding tiny holes in them, or even shaking bags full of them in order to collect the gold dust that wore off. (Enough of that and you could maybe collect enough gold to make a whole new coin.)

This special-purpose balance was designed to verify three things about a sovereign or half-sovereign coin: that it had the correct diameter (by fitting in the circle); that it had the correct thickness (by fitting through the slot); and, of course, that it had the correct weight (by exactly balancing the fixed counterweight).

Why would you care if the coin is larger than it should be, as long as it has the correct weight? Because gold is a much denser metal—that is, it weighs more per unit of volume—than any of the cheaper metals (available at that time) that might replace it. If you have a coin that weighs the right amount but is too big, that is proof that it has been made with too much of some metal other than gold.

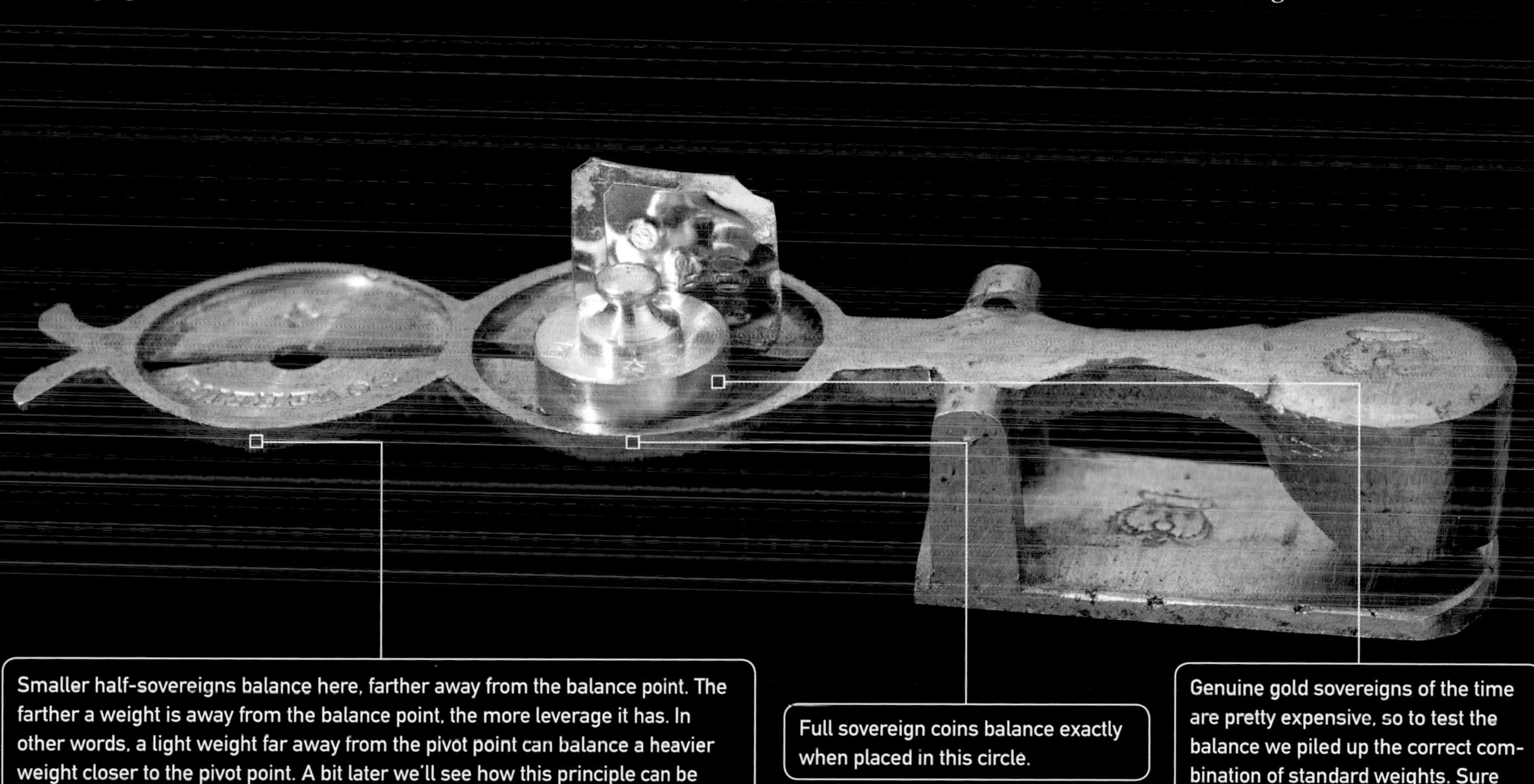

Smaller half-sovereigns balance here, farther away from the balance point. The farther a weight is away from the balance point, the more leverage it has. In other words, a light weight far away from the pivot point can balance a heavier weight closer to the pivot point. A bit later we'll see how this principle can be used to make a convenient general-purpose balance.

Full sovereign coins balance exactly when placed in this circle.

Genuine gold sovereigns of the time are pretty expensive, so to test the balance we piled up the correct combination of standard weights. Sure enough, they balance perfectly.

AS WE'VE JUST seen, it makes a difference where along the balance beam you place the object to be weighed. The sovereign balance works because it's designed to weigh only specific things, which fit in certain exact spots on the beam. But to make an accurate, general-purpose scale, you need to find a way to make sure the weights on both sides are always centered exactly the same distance from the balancing point. Hanging pans from both sides of the balance beam solves this problem. The pans automatically swing to exactly center the weight under the point from which they are hanging, regardless of where on the pan you place the object.

A balance doesn't have to be fancy to be quite accurate. This cheap plastic one can detect a difference of just 20 to 30 milligrams (thousandths of a gram), the weight of a few dozen grains of salt.

For many thousands of years, up until the middle of the 1700s, this kind of two-pan, equal-arm balance with hanging pans was pretty much the only method used for accurate weighing.

A problem with two-pan balances is that they can only compare two weights and tell you if they are equal or not. If you want to determine the weight of an unknown object, you need a set of accurate reference weights that add up to the weight of the unknown. There are countless examples of these sorts of weight sets. Some are intended to be used for weighing goods, others are calibration sets used to check the accuracy of scales, or the accuracy of other weight sets.

These sets usually come in units that are divided just like units of money (in nearly all countries except the United States). That is to say, in units of 1, 2, 5, then repeated ten times larger—10, 20, 50, 100, 200, 500, and so on. For example, England has 1 p, 2 p, 5 p, 10 p, 20 p, 50 p, and 100 p (£1) coins, and this set has 1 g, 2 g, 5 g, 10 g, 20 g, 50 g, and 100g weights (as well as smaller divisions of 0.5 g, 0.2 g, 0.1 g, 0.05 g, 0.02 g, and 0.01 g).

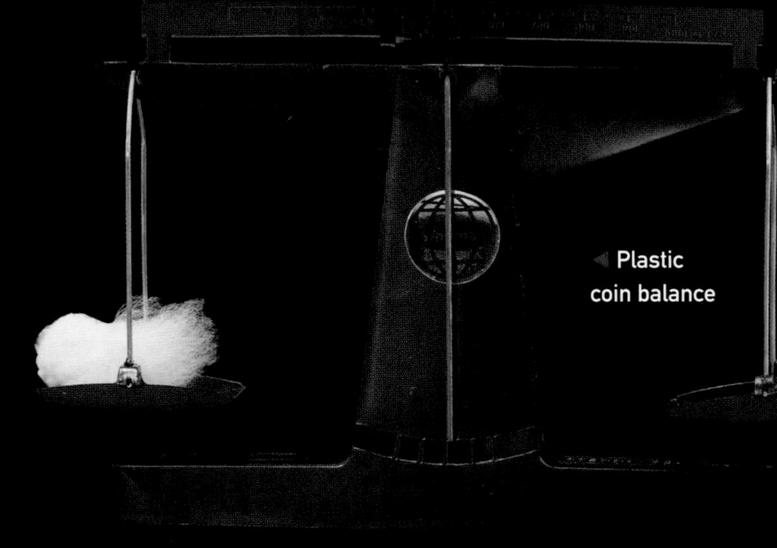

◄ Plastic coin balance

▲ Modern reference weights for hanging scales

▲ The Egyptian god Anubis had the important job of weighing the hearts of those seeking to enter the afterlife. To do this he used, of course, a two-pan balance. The earliest mentions of Anubis and his scale go back five thousand years, but this particular illustration, in the *Book of the Dead* compiled for a scribe named Ani, is only about 3,250 years old. (This section of the *Papyrus of Ani* is about two feet (67 cm) wide. The whole thing is 78 feet (24 m) long, and is currently held in the British Museum. According to the hieroglyphic text, Mr. Ani passed the test.)

▲ These reference weights all nest together like Russian dolls.

▲ Modern reference weights up to 100 g

▶ Decorative reference weights came in many fun shapes.

A modern weight set has two of each of the "2" units, so two 0.02 g (20 mg), two 0.2 g (200 mg), two 2 g, and two 20 g weights. Why? Because you need two of those in order to be able to get all possible weights in between. For example, if you want a set of weights that adds up to 96.59 grams, you need 50 g + 20 g + 20 g +5 g + 1 g + 0.5 g + 0.05 g + 0.02 g + 0.02 g, which includes two each of the 20 g and 0.02 g (20 mg) weights.

For those of you who like mathematical puzzles, can you figure out a more efficient set of weights that would let you reach every possible value with the smallest possible number of separate weights? Since I hate mathematical puzzles, I'll tell you the answer: if you're allowed to put the weights only on one side (the side opposite the thing you're trying to weigh), then you want weights in successive multiples of two, as in 1, 2, 4, 8, 16, 32, 64, 128, and so on. These weights might seem incredibly inconvenient to use, what with all the math you'd need to do, but on page 170 we will encounter a type of scale that uses exactly this system, and does the math for you—mechanically.

Steelyard Scales

THIS TYPE OF "steelyard" scale has a big advantage over two-pan balances: A single counterweight can be moved along the beam, varying its distance from the balance point. That lets one weight balance against any unknown weight over a wide range. You read off the measured weight from the marks scribed into the beam. For example, if a 10-pound weight balances an unknown when it's hanging twice as far from the balance point as the unknown, then the unknown must weight twice as much, or 20 pounds.

Auncel (steelyard) scales are compact and convenient, but unfortunately they are vulnerable to human nature. A statute issued by King Edward III of England in 1350 emphatically bans, on pain of "grievous punishment," anyone from using an uneven-beam auncel balance for weighing goods being sold or traded (see facing page). There's even language that implies this is a criminal, not just a civil, offense (a civil offense could be pursued only by the wronged party, but the statute says the party *or* the king can pursue the offender).

The problem is that it's very easy to cheat an auncel scale by making small adjustments to the short arm of the balance, or by tampering in other ways with the ratio of lengths.

With a two-pan balance where both arms are supposed to be exactly the same length, it's much harder to cheat. As long as the reference weights are accurate, it's generally safe to trust the balance even if it's owned by a merchant who would happily cheat you if he could.

(Keeping the reference weights accurate was the subject of another statute issued three years later, which ordered that "Weights shall be sent to all the Shires" and "The Clerk of the Market shall carry with him all his Weights and Measures signed.")

▶ Examples of brass, iron, and stone steelyard scales

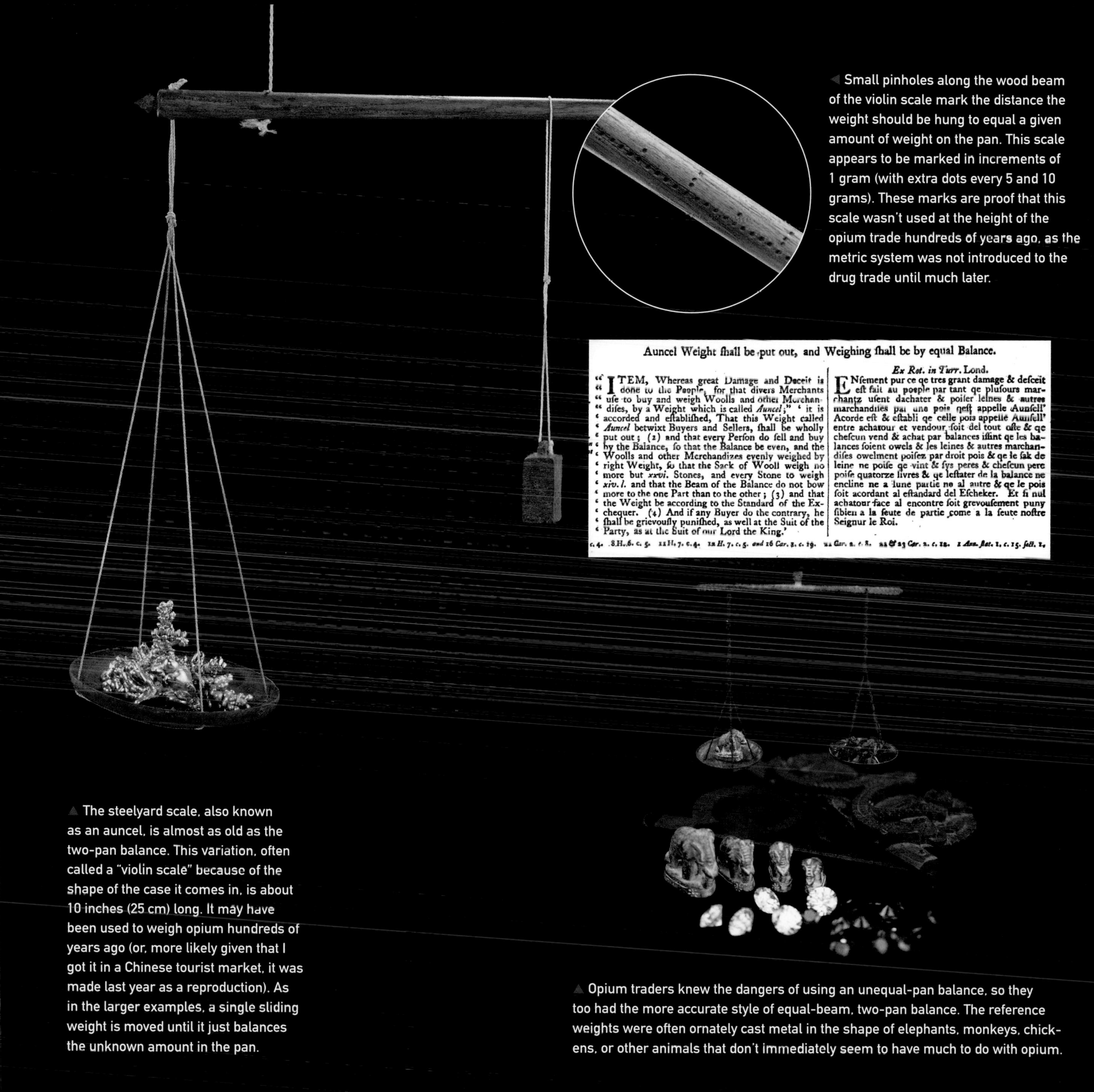

Auncel Weight ſhall be put out, and Weighing ſhall be by equal Balance.

"ITEM, Whereas great Damage and Deceit is done to the People, for that divers Merchants uſe to buy and weigh Woolls and other Merchandiſes, by a Weight which is called *Auncel*;" 'it is accorded and eſtabliſhed, That this Weight called *Auncel* betwixt Buyers and Sellers, ſhall be wholly put out; (2) and that every Perſon do ſell and buy by the Balance, ſo that the Balance be even, and the Woolls and other Merchandizes evenly weighed by right Weight, ſo that the Sack of Wooll weigh no more but *xxvi.* Stones, and every Stone to weigh *xiv. l.* and that the Beam of the Balance do not bow more to the one Part than to the other; (3) and that the Weight be according to the Standard of the Exchequer. (4) And if any Buyer do the contrary, he ſhall be grievouſly puniſhed, as well at the Suit of the Party, as at the Suit of our Lord the King.'

Ex Rot. in Turr. Lond.

ENſement pur ce qe tres grant damage & deſceit eſt fait au poeple par tant qe pluſours marchantz uſent dachater & poiſer leines & autres marchandiſes par une pois qeſt appelle Aunſell' Acorde eſt & eſtabli qe celle pois appelle Aunſell' entre achatour et vendour ſoit del tout oſte & qe cheſcun vend & achat par balances iſſint qe les balances ſoient owels & les leines & autres marchandiſes owelment poiſez par droit pois & qe le ſak de leine ne poiſe qe vint & ſys peres & cheſcun pere poiſe quatorze livres & qe leſtater de la balance ne encline ne a lune partie ne al autre & qe le pois ſoit acordant al eſtandard del Eſcheker. Et ſi nul achatour face al encontre ſoit grevouſement puny ſibien a la ſeute de partie come a la ſeute noſtre Seignur le Roi.

c. 4. 8 H. 6. c. 5. 11 H. 7. c. 4. 12 *H.* 7. *c.* 5. *and* 16 *Car.* 2. *c.* 19. 22 *Car.* 2. *c.* 8. 22 & 23 *Car.* 2. *c.* 12. 1 *Ann. ſtat.* 1. *c.* 15. *ſect.* 1.

◀ Small pinholes along the wood beam of the violin scale mark the distance the weight should be hung to equal a given amount of weight on the pan. This scale appears to be marked in increments of 1 gram (with extra dots every 5 and 10 grams). These marks are proof that this scale wasn't used at the height of the opium trade hundreds of years ago, as the metric system was not introduced to the drug trade until much later.

▲ The steelyard scale, also known as an auncel, is almost as old as the two-pan balance. This variation, often called a "violin scale" because of the shape of the case it comes in, is about 10 inches (25 cm) long. It may have been used to weigh opium hundreds of years ago (or, more likely given that I got it in a Chinese tourist market, it was made last year as a reproduction). As in the larger examples, a single sliding weight is moved until it just balances the unknown amount in the pan.

▲ Opium traders knew the dangers of using an unequal-pan balance, so they too had the more accurate style of equal-beam, two-pan balance. The reference weights were often ornately cast metal in the shape of elephants, monkeys, chickens, or other animals that don't immediately seem to have much to do with opium.

YOU MIGHT THINK that weights and measures regulations are always designed to ensure a *minimum* quantity. "One pound" of butter should always be *at least* a certain amount or else you're getting cheated. But many of the early laws are actually the other way around: they specify a *maximum* amount that a certain measure can be. For example, back when a bushel was measured by filling a standard-sized tub or barrel, the same statute quoted earlier also said "the Quarter shall contain Eight Bushels by the Standard, and no more. And every Measure of Corn shall be stricken without Heap." The statute is saying that after the measuring container is filled with corn, it should be "stricken off," or wiped across the top, rather than left heaping. Why would they want to ensure that a Quarter of corn isn't more than the maximum allowed? The answer is in the next part of the sentence: "saving the Rents and Ferms of Lords." The owner of the land, the lord, was paid a tax based on the number of bushels harvested. If the farmers, with a nudge and a wink to the merchants, measured off their crops in heaping units, the total number of bushels recorded would come out less on paper, "cheating" the lord out of his "fair" share of tax money.

▼ Striking off a bushel of corn

▲ Throughout history, one of the most useful and legitimate functions of kings and governments has been standardizing, inspecting, and enforcing units of measurement. We see this today in, for example, the seals stuck on gas station pumps and

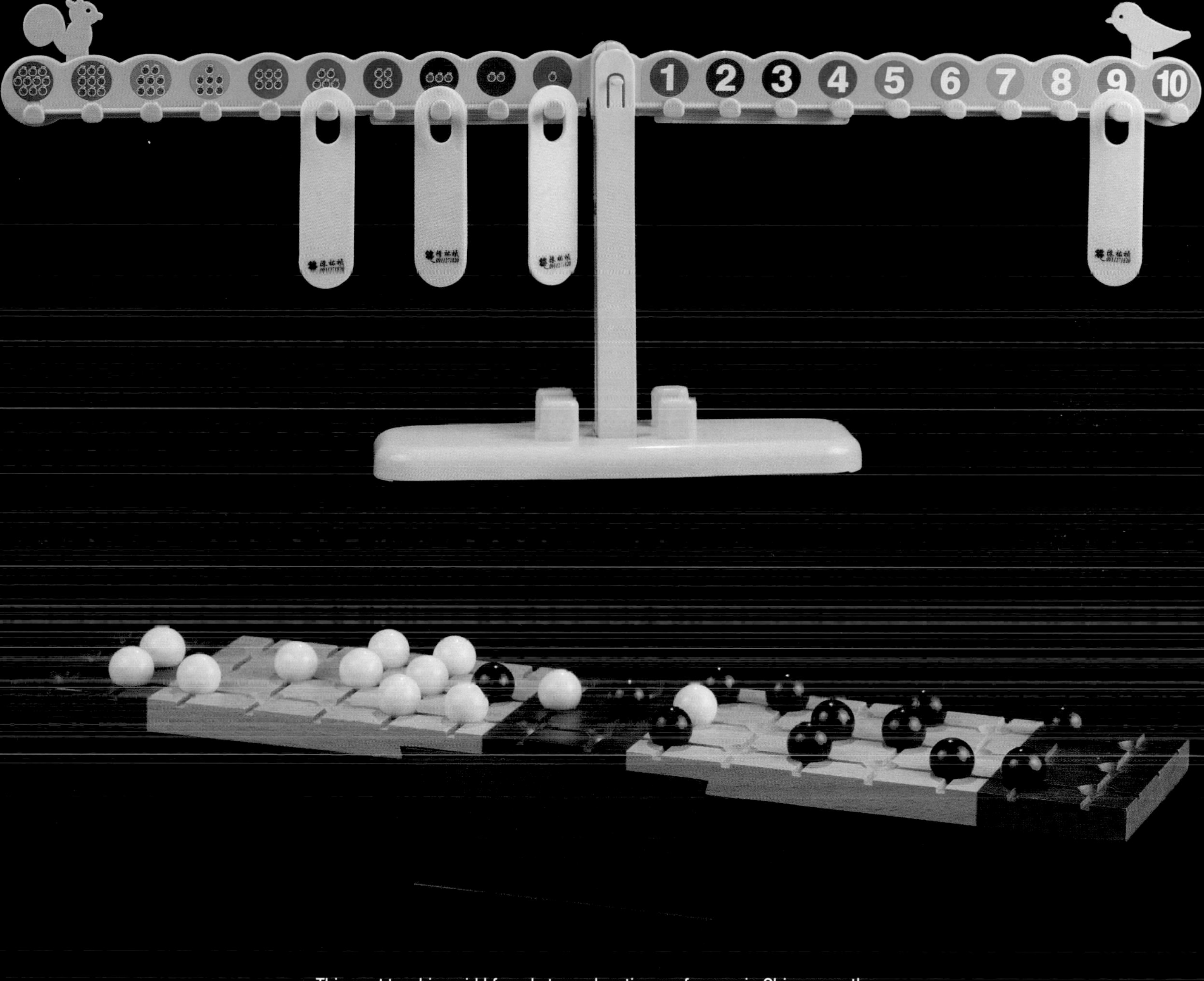

▲ This neat teaching aid I found at an education conference in China uses the same principle as the steelyard scale—the fact that a weight has more leverage the farther out it is on the balance beam—as a tool for teaching math. If you hang plastic weights from differently numbered pegs, the scale will only balance if the sum of the numbers on each side is the same. The marble-balancing game works on the same principle: each marble becomes more and more important the farther out it is from the pivot point.

Toward a More Convenient Balance

SINCE THE INVENTION of balances, people have been trying to make them more accurate and more convenient. You generally can't have it both ways—super-convenient and super-accurate at the same time—so we're left with many different styles of balances that try to optimize one or the other. In the following pages we're going to look at simplified, schematic scale designs that take us step by step through the key ideas that lead to modern scales appropriate for a range of situations.

The ancient two-pan hanging balance is simple, accurate, and a pain to use, clean, move, and store. One constant issue is that the strings or rods holding the pans get in the way of large objects. Wouldn't it be great if the pans were on top?

The version below seems like a great balance. It's compact, has only one moving part, and you can put objects of any size on the pans. Too bad it's a terrible design. The problem is that it matters where on the pan you put the weights. If you move the weight toward the edge of the pan, it will tilt the balance in that direction, even though the weight in the pan hasn't changed. You can use this style of balance only in special situations where it's possible to ensure the exact location and type of object being weighed. But in general, this is, in fact, a no-good, very bad balance.

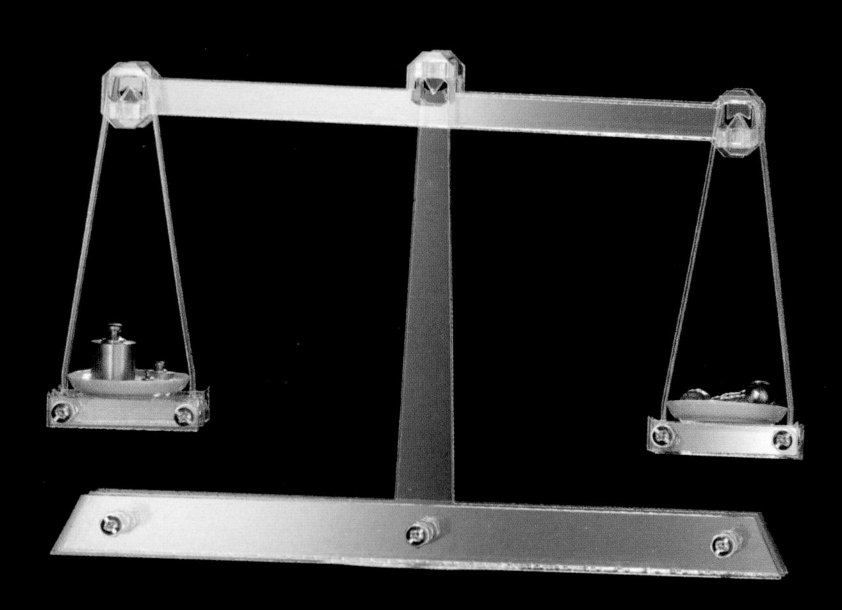

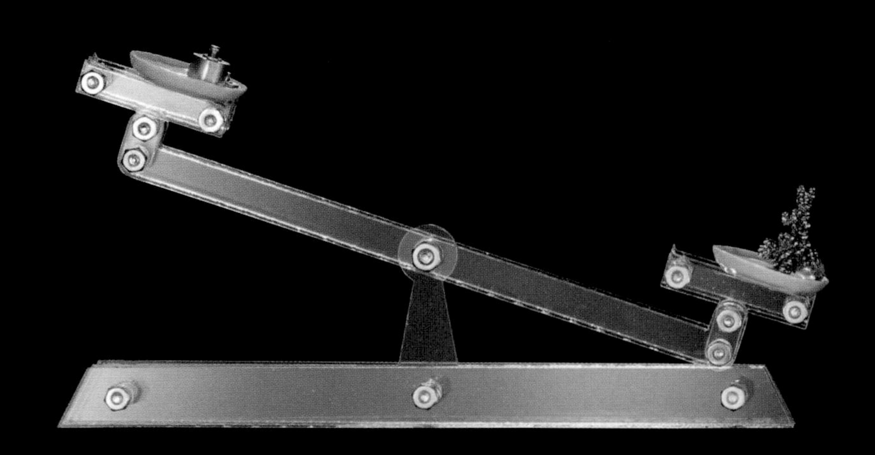

▲ Acrylic model of a hanging balance

▶ Acrylic model of a top-pan balance. It doesn't work very well.

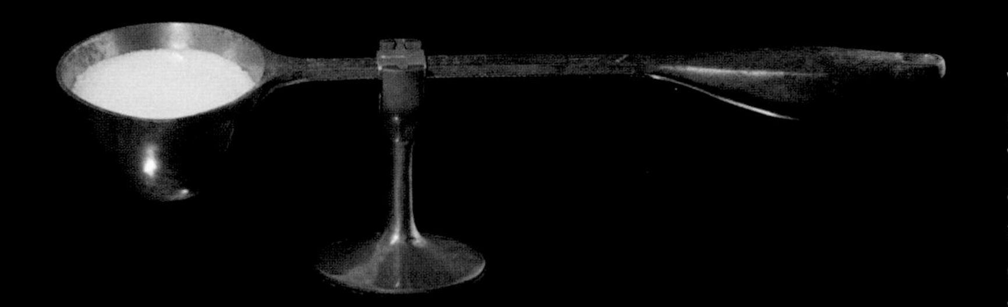

◀ This is a clever example of a super-simple, single-pivot balance: an automatic weighing spoon. You slide the pivot point along the handle of the spoon up to the graduated mark showing the number of grams you want to weigh. Then you pour water, oil, flour, or whatever you like into the spoon until it balances. This works as long as the thing you are weighing flows uniformly into the spoon, thus always finding a consistent position relative to the balance point. If you tried to weigh solid chunks of something, it would matter where in the bowl you put them, and the weight would not be accurate.

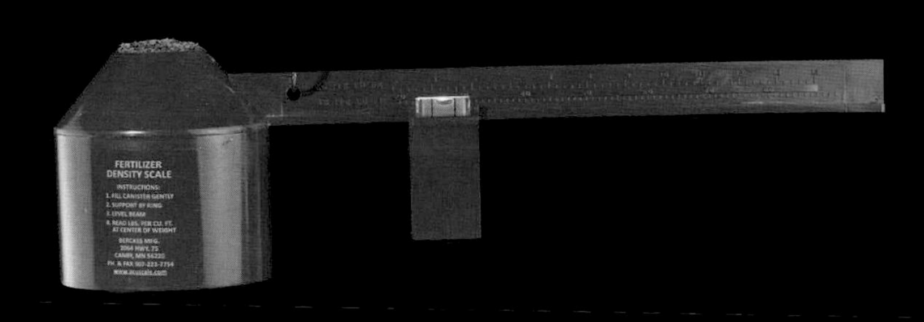

This is another example of a scale that, given specific circumstances, can get away with being with what would, in other situations, be an inaccurate design. It's meant to be used only to weigh powders and fine granular material (specifically fertilizers) that completely fill the integrated measuring cup. It's basically a steelyard design, but with two clever twists. First, there's a bubble level mounted on the balance arm, which lets you accurately determine when it's level (an obvious enhancement that I'm frankly surprised isn't used more often). Second, because the measuring cup has a calibrated volume, this scale can be used to determine not only the weight, but also the density (weight per unit volume) of the thing you're testing. This is apparently something you want to know about fertilizers.

This is an example of a scale not really intended to weigh anything. Yes, it's a traditional two-pan balance, and even comes with a set of weights, but the pivot point is a metal hook that doesn't allow the balance to tilt freely. The four weights have pretty random masses, not the nice integer multiples you'd expect for proper weights. This is just a knickknack to put on your shelf. (It's from the Panjiayuan market, so this is no surprise.)

THE BALANCE THAT GOT AWAY

THIS IS A very rare style of scale, so rare that it does not actually exist, even though I've seen one. Let me explain. I found a scale of this design in a small antique shop in the giant Panjiayuan flea/antiques/fakes market in Beijing. The shopkeeper insisted it was a valuable Italian antique, hundreds of years old. She would not budge from the thousands she wanted for it, but I would not budge from my opinion that it was an obvious fake.

A balance of its supposed quality would have had sharp metal or stone pivoting knives as its balance points (see pages 168 and 169 for a description of balance knives). This one instead had crude screws poked (off-center) into holes drilled into a brass ball. I think it was made by someone who had seen pictures of, but didn't really understand, or didn't care, how balances work—so they improvised badly what is actually the crucial thing to get right in a real balance. I almost wish I'd paid the ridiculous amount they wanted for it, fake or not, because since then I have been unable to locate even a photograph of any balance like it. (And when I went back six months later, it was gone. It's now forever the scale that got away.) Scale experts I've consulted say this is what's called a "fantasy" scale, and may have been made for the interior decorator market. No real examples, made to actually be used as a balance, are known to exist, at least not by me or the experts I've consulted.

But...if it were built properly, it could in theory work. This model is my best recollection of the one that got away. It solves exactly one problem with the hanging pan balance: the problem of the strings getting in the way. The heavy pendulum weights under each pan keep the center of mass of whatever you put on the pan exactly centered over the pivot point, which means the balance is accurate no matter where the objects are placed. Other than that, it's just as much of a hassle to use, transport, or buy from a Chinese antique shop as any hanging pan balance. Except with the added bonus that if you put something a bit too heavy on top, the pan flips over and dumps out your stuff. Which may explain why no one ever made scales like this for serious use.

Mathematics Makes a Contribution

THE FIRST REAL advance in above-the-beam pan balances came in 1669, thanks to a mathematician who took the time to analyze the physics of the situation. As we saw in the last section, the fundamental problem with this type of scale is how to keep the center of mass of the objects being weighed a fixed distance away from the balance point.

There are two ways of looking at why the distance of weights from the balance point makes a difference: leverage and potential energy. One of these two ways is the key to a more convenient balance.

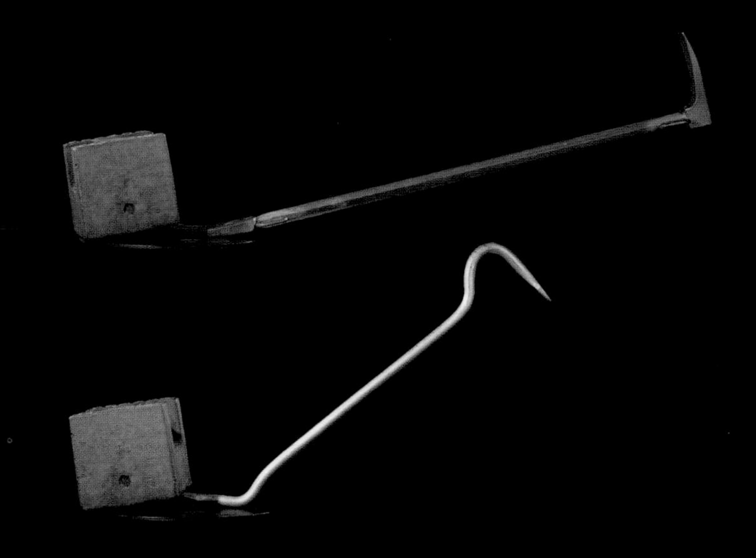

▲ Leverage is what allows you to turn a small force into a large force. Common experience tells you that a longer lever makes it easier to lift a weight. If you double the length of the lever, thereby creating more leverage, you cut in half the amount of force needed to lift the thing on the other side.

Potential energy is the energy of things that are high up. The higher a thing is, the more potential energy it has. When the thing falls, its potential energy is converted into kinetic energy—the energy of things in motion. Imagine a weight that gets dropped on your foot. The higher up that weight is dropped from, the more it's going to hurt, because the weight will have more kinetic energy when it lands on your toes.

The number one rule of energy is that it's *conserved*. In other words, if you want to increase the potential energy of an object, you can do that by moving the object higher up. But in order to move the object higher you have to do work. The heavier an object is, the more work (energy) it takes to lift it. And the higher you want to lift something, the more energy you have to add to it.

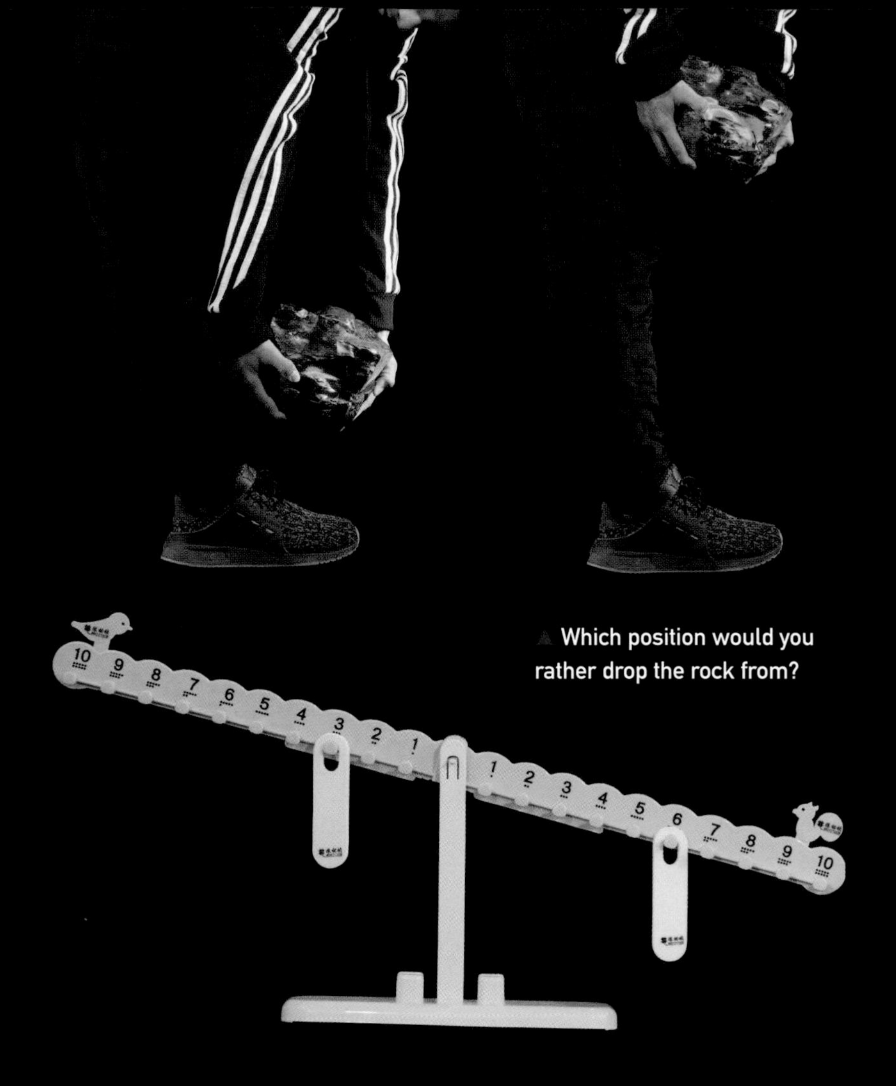

▲ Which position would you rather drop the rock from?

A balance is just a type of lever: It's got a fulcrum (balance point) in the middle, and a force pushing down on each side. So think about this balance in terms of leverage. Two equal weights are placed at a different distance on either side of the balance point. The one farther out (on the longer beam) has more leverage, so it wins, tilting the balance down while the weight on the other side is lifted up.

Now consider the same situation in terms of potential energy. If the beam were to tilt so that the longer side went up, the weight on the longer side would rise up by more than the weight on the shorter side went down. One would gain more potential energy than the other one lost, so overall the pair of them would have more total potential energy. The only way that could happen is if you added energy—for example, by doing the work of pushing down on the short side with your finger. Instead, what will of course happen is that the balance will tip so the long side goes down. Overall, the total amount of potential energy between the two weights will decrease, being converted into the kinetic energy of the movement of the beam.

Looked at in terms of potential energy, the problem isn't so much that one weight has more leverage than the other, it's that one weight moves a larger distance than the other. In 1669 the mathematician Gilles de Roberval realized that this was a solvable problem.

Roberval Balance

THIS CLEVER ARRANGEMENT of beams, to this day known as a Roberval balance, solves the problem of the weights moving a different distance by keeping the two pans *level* regardless of what position the balance is in. The weight on the right is farther away from the pivot point, but because the pans stay level, it travels the same distance up or down (in the opposite direction) as the weight on the left. Therefore, by conservation of energy, it must balance regardless of where the weight is placed.

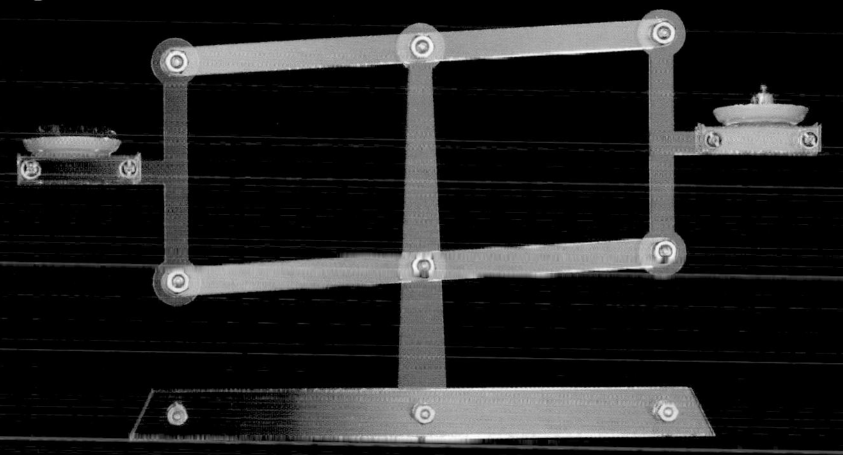

Model of a Roberval balance

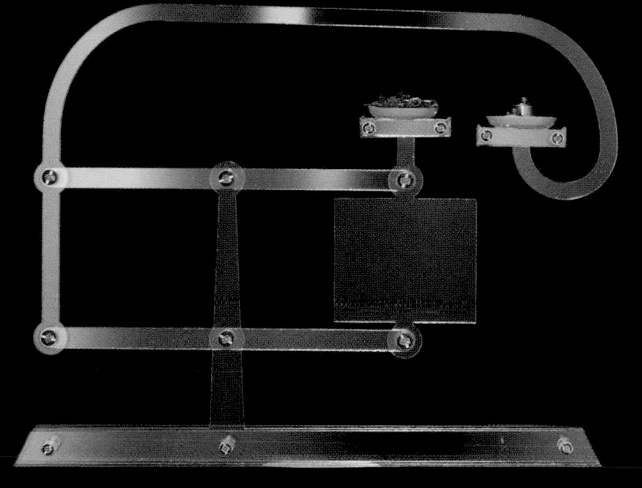

Model of a Roberval balance with an overarm

Something really surprising about a Roberval balance is that it works, and balances correctly, *even if one pan is shifted all the way over onto the other side*. How can this be? Well, conservation of energy demands that it must be so, and conservation of energy is a principle that never lies. Understanding why it can be so in terms of torque and leverage is much harder. It has to do with sideways forces, I suppose. Honestly, I find it as baffling as you probably do, but it does work: no cheating in the photo. As in many other situations, conservation of energy lets you cut through a lot of complicated calculations and confusing geometry to see what must be so.

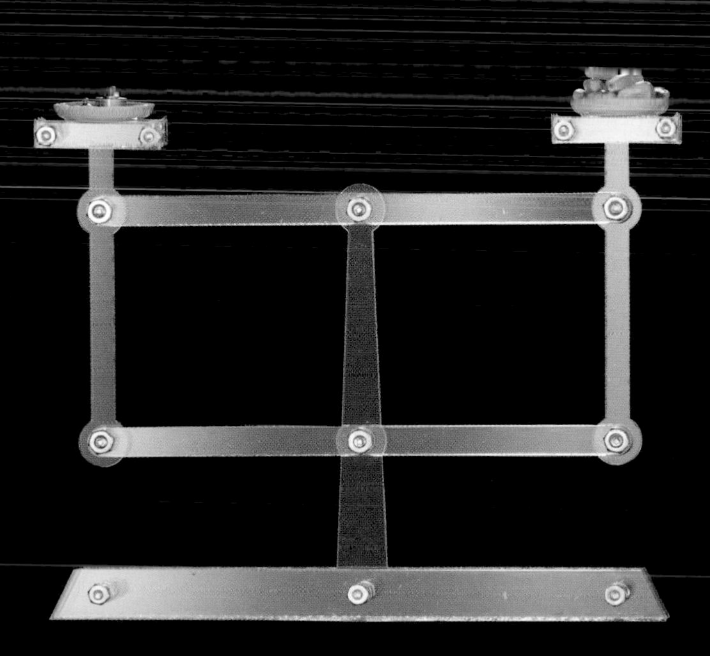

Model of Roberval balance with top pans

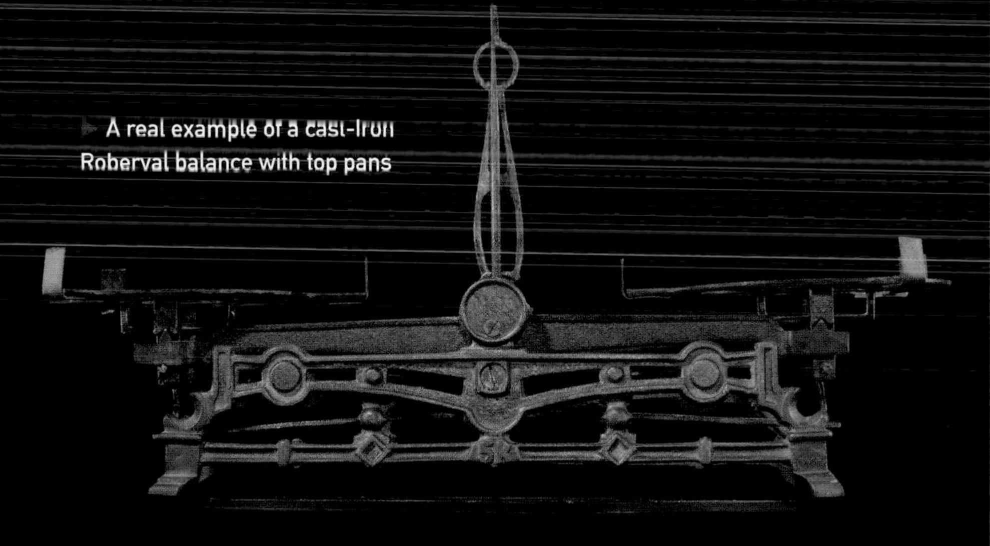

A real example of a cast-iron Roberval balance with top pans

Normally the pans are placed above the top beam, for compactness and convenience. But the principle is the same: use a second beam to keep the pans level. The top beam is the main one: its pivots bear the full weight of the things being weighed, and need to be precise and friction-free. The bottom beam can have simple, loose-fitting pivot points, because all they have to do is keep the pans from tipping over. They bear only a small amount of sideways force.

▼ Nearly all old-school, two-pan kitchen and commercial balances are Roberval balances of one style or another. Often the second beam is hidden inside the base, but you know it must be there because otherwise the pans would tilt.

▶ Sometimes both beams are hidden, so all you see is the pans sticking up out of the case. But inside, it's still a Roberval mechanism as you can see on the following page.

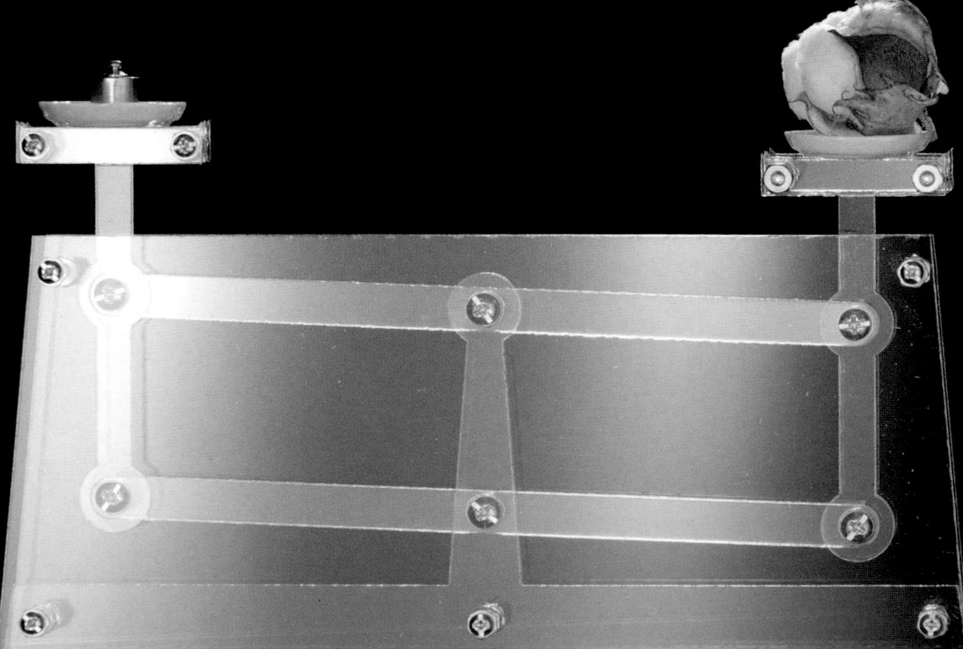

RUSSIAN ROBERVAL BALANCE VARIATION

OPEN-FRAME, cast-iron Roberval balances are pretty common flea market and eBay finds. This is because (1) a lot of them were made over a hundred-year span, (2) they are nearly indestructible, and (3) they're apparently not the kind of thing that anyone would throw away, even long after they stopped being a sensible tool to use for weighing. They survive as decor in rustic-themed restaurants, as knickknacks in trendy apartments, and as doorstops or dust collectors in countless antique shops and market stalls.

Case in point, this odd variation of the Roberval design that I found deep in the heart of the biggest flea market I've ever seen, the Izmailovsky Market in Moscow (not the tourist section at the front of the market, the crazy-stuff section that stretches for miles out back). Instead of a single lower beam that moves together with the upper beam, this scale has two independent lower beams anchored at the outside edges of the scale. I am puzzled as to why this more complicated design is better, but presumably it must have some advantage or they wouldn't have made it this way. Just another little mechanical mystery to keep things interesting—until someone tells me the answer.

▲ This variation of a Roberval balance was found in a Russian flea market. I have replaced the front and back cast-iron frame with clear acrylic so you can see the mechanism more clearly.

▼ Hidden beam art deco scale

One Weight, Many Angles

ROBERVAL AND STEELYARD scales are more convenient than hanging two-pan balances, but they still require the user to manually add, remove, or slide weights until balance is achieved. It would be a lot simpler if the scale would just tell you the weight. Such scales do exist, and have existed since long before the digital era. By the simple trick of *bending* the balance beam, it's possible to make a scale that automatically shows the measured weight with a pointer on a dial.

▶ Small hanging postal scale

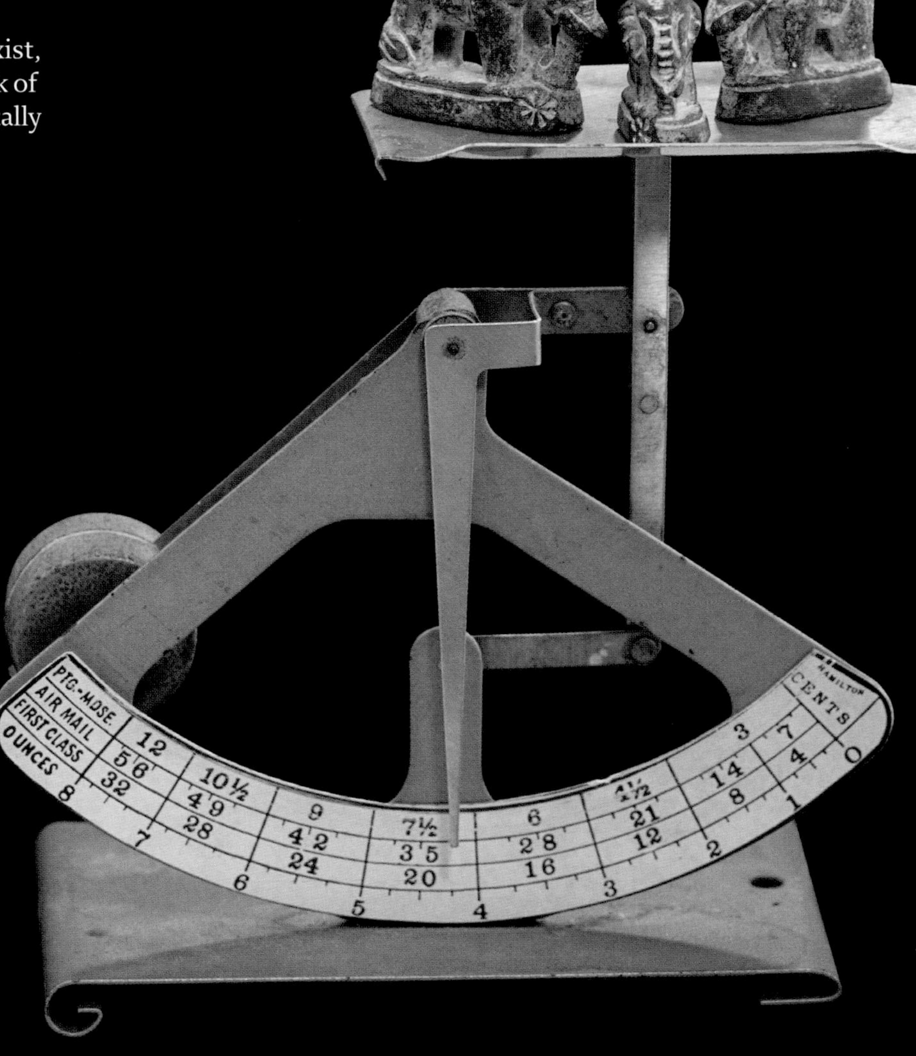

▲ This small postal scale is used for weighing letters up to a few ounces. The weight of the object causes the counterweight to swing away from the fulcrum, increasing its distance away from the balance point, and thus increasing its leverage. The trick here is that the balance beam is bent at the pivot point. If the beam were straight, as in an equal-beam balance, then the rotation angle wouldn't matter, and the balance would stay in equilibrium at any angle. But if the beam is bent, then as the beam rotates, one side gets closer to the balance point while the other one moves farther away. This shifts the balance of leverage between the two sides. Rotation continues until the weight multiplied by the leverage on each side is equal. The pointer then shows the angle of rotation, calibrated in ounces and/or grams.

▲ This pan balance uses the same bent-beam design as the hanging postal scale. As the pan moves down, the counterweight swings out to balance an ever-larger amount of weight on the pan. (This particular scale belonged to my mother. I remember it well from many years ago and am pleased to be able to give it further life in this book.)

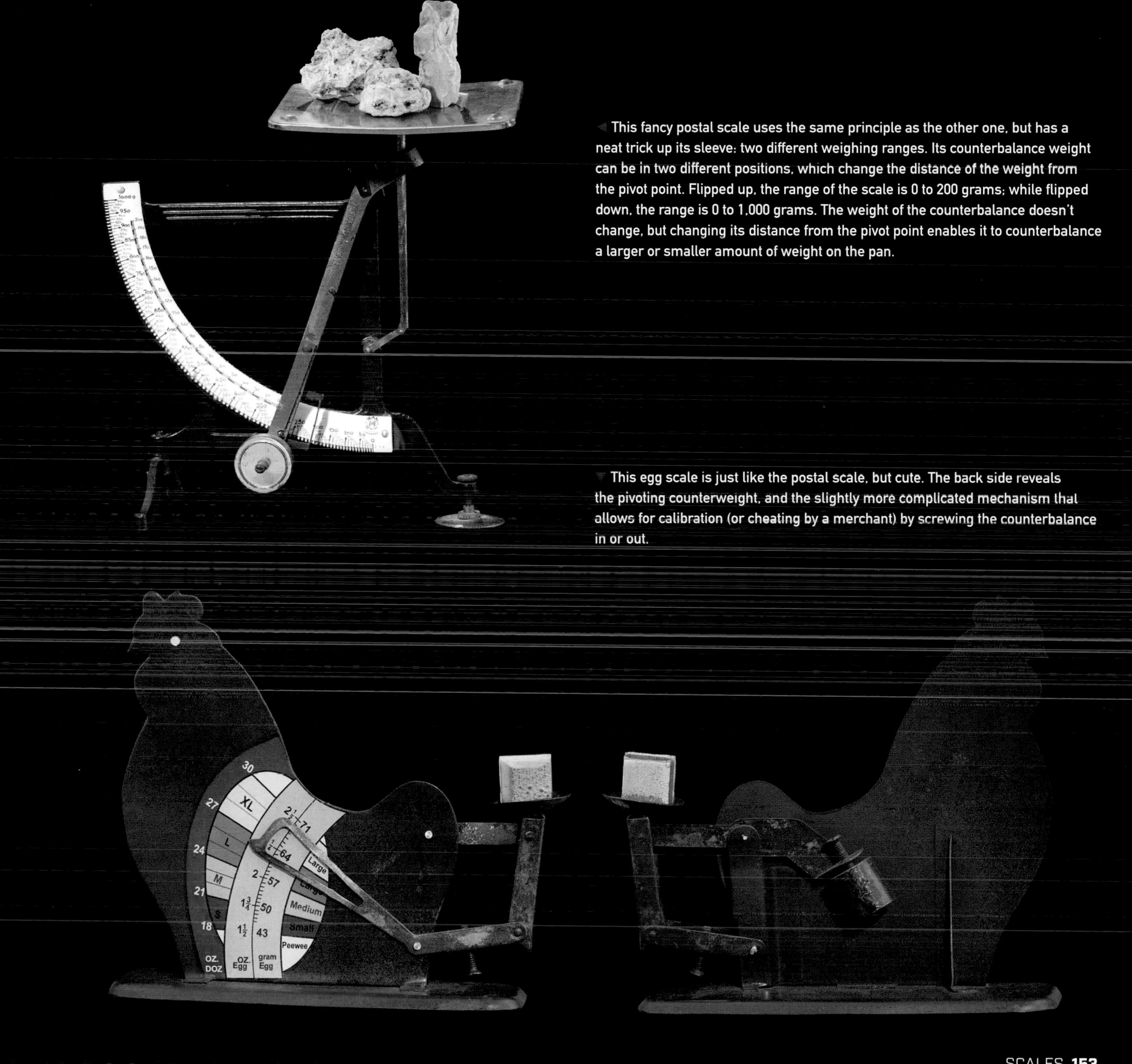

This fancy postal scale uses the same principle as the other one, but has a neat trick up its sleeve: two different weighing ranges. Its counterbalance weight can be in two different positions, which change the distance of the weight from the pivot point. Flipped up, the range of the scale is 0 to 200 grams; while flipped down, the range is 0 to 1,000 grams. The weight of the counterbalance doesn't change, but changing its distance from the pivot point enables it to counterbalance a larger or smaller amount of weight on the pan.

This egg scale is just like the postal scale, but cute. The back side reveals the pivoting counterweight, and the slightly more complicated mechanism that allows for calibration (or cheating by a merchant) by screwing the counterbalance in or out.

Toledo Balance

THIS BIG OLD Toledo brand scale looks (and is) complicated, but it uses the same principle as the postal scales we just looked at: counterweights that swing out away from the pivot point to increase their leverage. This type of "platform scale" is very convenient. You can put the thing to be weighed anywhere you like on the large open platform, and its weight will be read out on a big dial. This is pretty much as convenient as you can get! The price you pay for this convenience is a big, heavy, complicated mechanism.

Changing the "tare" of a scale means adjusting it back to zero after something has been put on it. For example, you might put an empty bucket on the scale, then adjust it to read zero again. Then you can add something to the bucket and weigh just the stuff you're adding, without having to do the math of subtracting out the weight of the bucket. On some scales the tare adjustment is really just rotating the dial or pointer, but this scale has a substantial counterbalance that is moved back and forth to add or subtract weight from the balance path. (Changing the tare is not the same thing as calibrating the scale, which means adjusting it to accurately read both zero and some fixed test weights.)

The weight of the dial pointer itself could influence the reading, if it isn't perfectly balanced around its center of rotation. Little counterweights let you adjust its balance point in both dimensions to exactly cancel out this potential source of error.

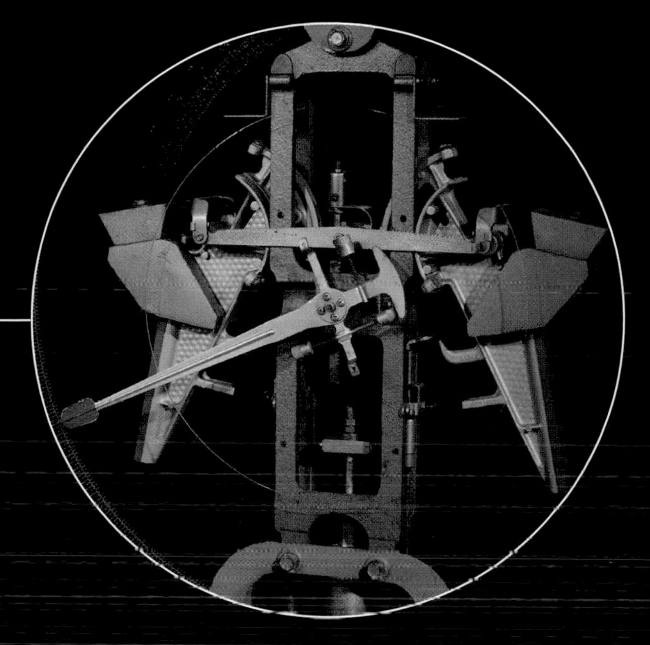

These counterweight arms hold a key reason this scale is built like a tank. You can see that the two counterweights are attached at nearly the very top of a pair of mounting rails that run down the top side of the "wings." If they were mounted farther down, they would have much more leverage, and it would take a lot more force to raise them up. That in turn would increase the maximum weighing capacity of the scale.

This scale can in fact be adjusted to have a *much* larger maximum capacity than its current 250 pounds. The same model, the Toledo 2181, came in varieties from 200 pounds (90 kg) to 2,000 pounds (900 kg) in maximum capacity. The only differences are the numbers printed on the dial face and the positions of the counterweights along the rails.

Obviously, a scale meant to handle something that literally weighs a ton needs to be strong!

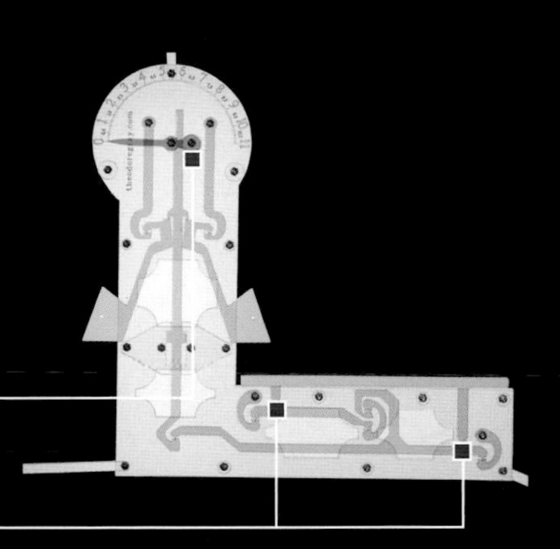

This simplified model shows how a Toledo-style platform scale uses multiple points of compound leverage and swinging weights.

A rack-and-pinion gear system magnifies the downward movement of the hanging rod and turns it into the circular movement of a dial pointer.

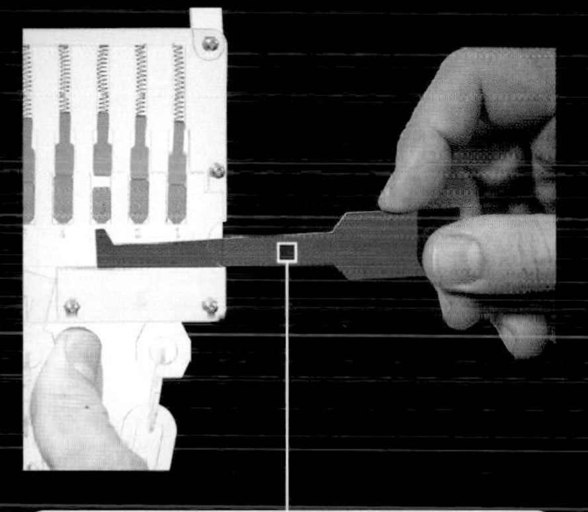

Leverage is used to reduce the force that reaches the measuring mechanism. In this two-dimensional model, the overall leverage is 15 to 1. For weight centered on the right side of the platform, there's a single 15:1 ratio between the long and short arms, while for weight centered on the left side, there's one 5:1 ratio, followed by a 3:1 ratio, which multiplies out to the same 15:1 ratio. No matter where you put the weight, there is an overall 15-to-1 ratio. This mechanical property is what allows you to place the weight anywhere on the platform without changing the reading.

At the top of the scale is what basically amounts to a fancy, heavy-duty version of the bent-arm postal scale we just saw. When the rod is pulled downward by objects on the platform, two reference weights pivot outward, increasing their leverage, until a balance of forces is reached.

WHY IS THIS THING SO BIG?

DID I MENTION that this thing is huge? It's built like a heavy machine tool, with quarter-inch-thick cast-iron case and massive cast-iron balance arms. The pivot points are hardened steel. But the maximum capacity of this scale is only 250 pounds, like a typical bathroom scale! So why build it so huge? Partly so that it can survive in a warehouse or factory environment where angry workers throw sides of beef at it day in and day out (this one supposedly came from a butcher shop). Maybe the occasional forklift backs into it. The insides are a delicate precision instrument able to weigh with a resolution of better than one part in a thousand (the gradations on the scale are in ¼-pound increments). It makes sense to protect this fragile and expensive investment with a case that's up to the demands of its environment.

This is the rough modern equivalent of the massive cast-iron Toledo scale. It's made of sheet metal and plastic, and weighs and costs about one-twentieth as much (how this is done is explained on page 160). Does that mean it's cheap modern crap? Sort of, but for good reason, and with many advantages. The design is responding to a different set of realities. The mechanism inside is not nearly as delicate, thus requires less protection. And if someone did drive a truck over it, a replacement would not be terribly expensive. In exchange for the weaker structure, you get a scale you can pick up and move around with one hand, instead of with a forklift.

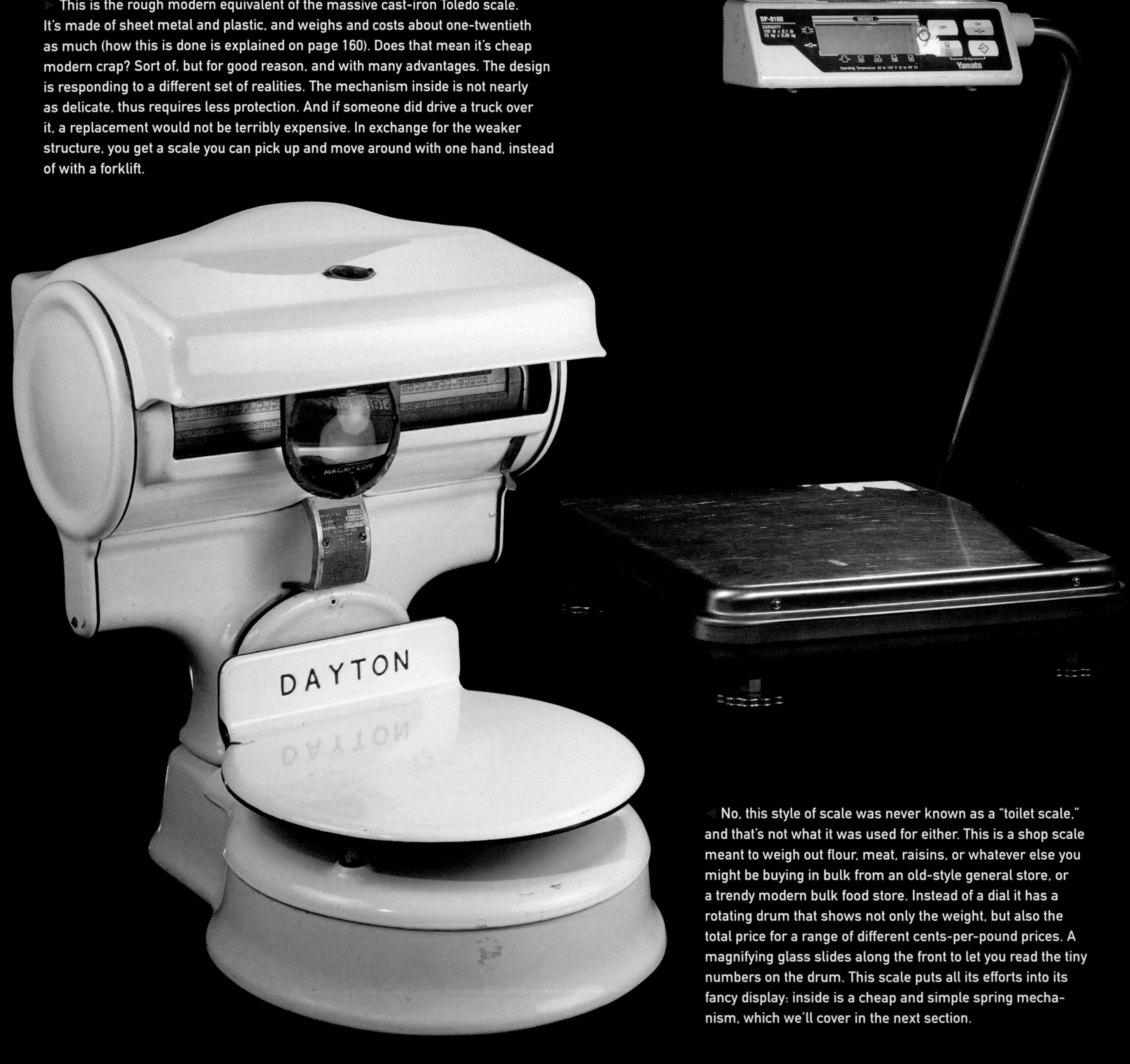

No, this style of scale was never known as a "toilet scale," and that's not what it was used for either. This is a shop scale meant to weigh out flour, meat, raisins, or whatever else you might be buying in bulk from an old-style general store, or a trendy modern bulk food store. Instead of a dial it has a rotating drum that shows not only the weight, but also the total price for a range of different cents-per-pound prices. A magnifying glass slides along the front to let you read the tiny numbers on the drum. This scale puts all its efforts into its fancy display: inside is a cheap and simple spring mechanism, which we'll cover in the next section.

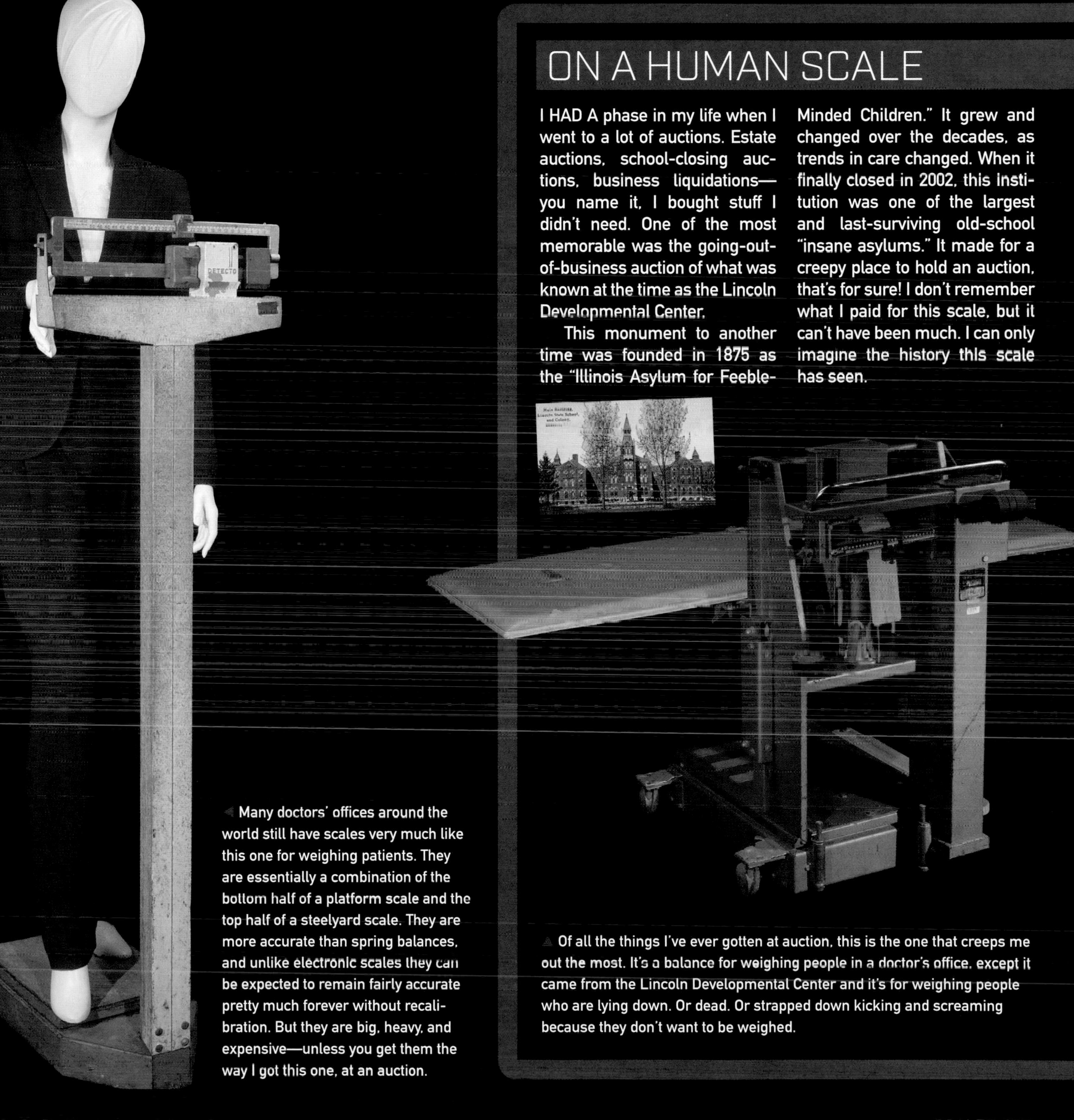

ON A HUMAN SCALE

I HAD A phase in my life when I went to a lot of auctions. Estate auctions, school-closing auctions, business liquidations—you name it, I bought stuff I didn't need. One of the most memorable was the going-out-of-business auction of what was known at the time as the Lincoln Developmental Center.

This monument to another time was founded in 1875 as the "Illinois Asylum for Feeble-Minded Children." It grew and changed over the decades, as trends in care changed. When it finally closed in 2002, this institution was one of the largest and last-surviving old-school "insane asylums." It made for a creepy place to hold an auction, that's for sure! I don't remember what I paid for this scale, but it can't have been much. I can only imagine the history this scale has seen.

◀ Many doctors' offices around the world still have scales very much like this one for weighing patients. They are essentially a combination of the bottom half of a platform scale and the top half of a steelyard scale. They are more accurate than spring balances, and unlike electronic scales they can be expected to remain fairly accurate pretty much forever without recalibration. But they are big, heavy, and expensive—unless you get them the way I got this one, at an auction.

▲ Of all the things I've ever gotten at auction, this is the one that creeps me out the most. It's a balance for weighing people in a doctor's office, except it came from the Lincoln Developmental Center and it's for weighing people who are lying down. Or dead. Or strapped down kicking and screaming because they don't want to be weighed.

Beyond Balance

THE BALANCES WE'VE seen so far work because of gravity. Objects are pulled to the earth by gravity, creating a downward force on each side of the balance beam. When those forces are exactly the same, the beam is in balance, and we say that the two things have the same weight.

Gravity can create a force, but so can a spring. In 1770 Richard Salter introduced the first really new idea in weighing since the invention of weighing: the spring scale. The weight pulls down, and a stretched spring pulls back up. When the forces are in balance, the stretch of the spring is a measure of the weight. This was a popular and enduring idea: to this very day you can buy scales of all sorts with the Salter name on them (and countless other spring scales made by people who've never heard of him).

The *problem* with spring scales is that they are not as accurate or trustworthy as balances. There is no way to look at one and know that it's reading correctly (as you can with an equal-beam balance), and they tend to drift out of calibration. Changes in temperature change the strength of the spring, age and overloading can warp the spring, and so on. For this reason (and because of intense lobbying by the Toledo company), spring scales are generally not permitted by law for measuring goods for sale. And companies that make non-spring balances are only too happy to point out how theirs are better. The "No Springs—Honest Weight" slogan of the Toledo scale company is truth in advertising. You can never entirely trust a spring scale.

A small spring scale

Heavy spring scale

The great thing about spring scales is that they don't require any sort of counterweight or delicate balancing point, so they can be much smaller and more robust. This one can measure up to 500 pounds: notice that the 50-pound weight hanging from it has barely moved the pointer! Yet the device itself weighs only a few pounds, and works in any direction. It can be banged around, dropped, or stepped on without suffering any harm.

THIS 100-POUND-CAPACITY scale looks superficially like a Toledo-style scale, but take the cover off and you find an imposter: inside, it's a cheap spring scale. In its defense, though, it came to me dead-on accurate according to my 50-pound standard weight—unlike the actual Toledo scale, which is currently about 5 pounds off at 50 pounds.

On the back side of the dial you get a clear view of the rack-and-pinion gear that magnifies the stretching of the springs. As you add weight and the springs get longer, the straight gear (the rack) is pulled down, rotating the small round gear (the pinion), which is connected to the pointer. Because the pointer is *much* longer than the diameter of the round gear, a small movement of the rack translates into a large movement of the end of the dial, making it easier to read small changes in weight.

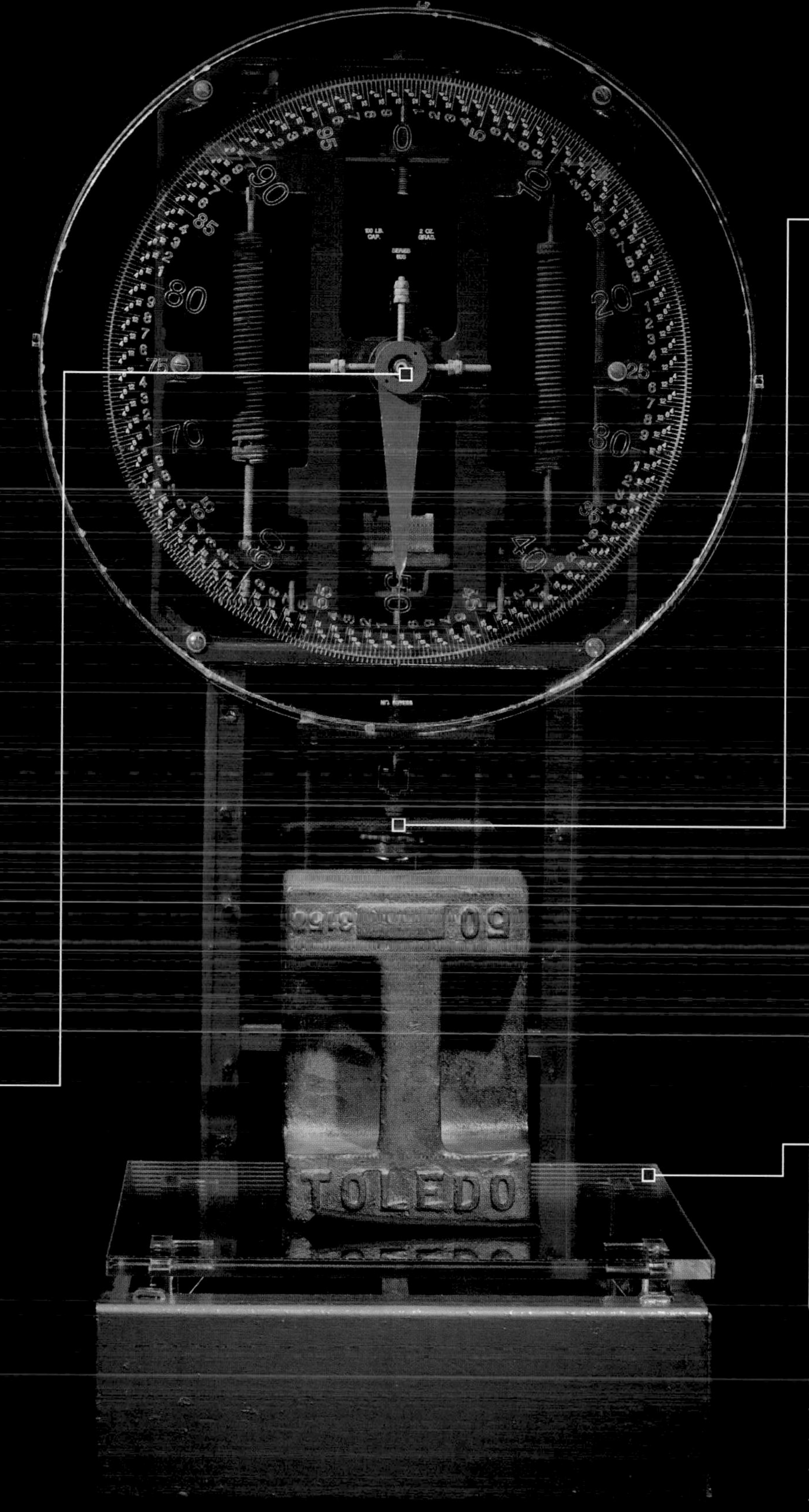

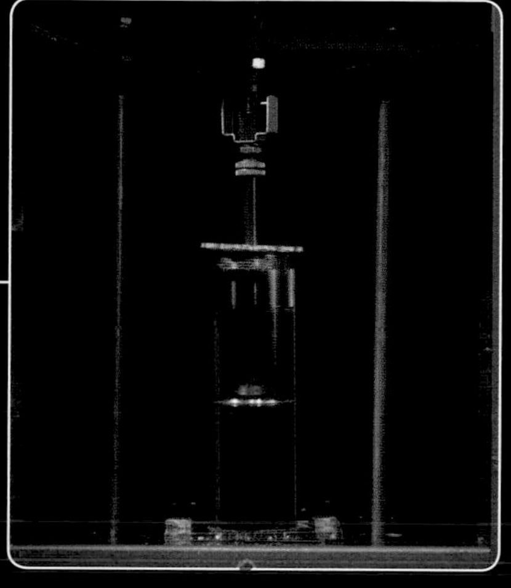

This is an oil-filled dashpot. Inside there is a loose-fitting piston that moves up and down through the oil when the dial turns. The viscosity of the oil slows down the piston, damping out vibrations so the scale comes to a final reading more quickly. Because liquid has no memory, there is no effect on the final reading, only on how long it takes to get there. If you ever need to transport a scale like this, remember the oil. It pours out if you tip the thing on its side to fit in your white Toyota Sienna minivan. Not that I would ever do anything as dumb as that.

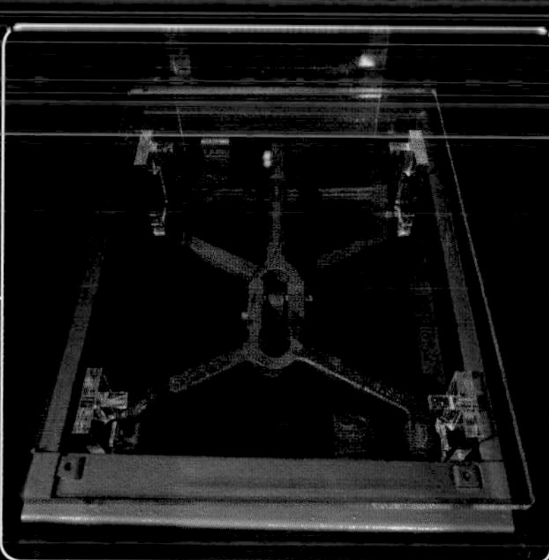

I've replaced the metal weighing platform with a clear acrylic one so you can see the set of lever arms that transfer weight from the platform to the spring. (See page 155 for a diagram that shows how these levers work.)

Despite being untrustworthy, spring scales are very popular. They are good enough for a lot of things. Do you really need to know to better than plus or minus a few pounds how much you weigh? A bit of plausible deniability might be a good thing. This bathroom scale shows how leverage can be used to allow for a smaller, weaker spring. The weight of the person standing on the scale presses down on the short arm of a teeter-totter beam, while the spring pulls back on the long arm. The spring has more leverage, so it doesn't have to pull nearly as hard to balance the force generated by the weight of the person

Simple non-electronic kitchen scales are nearly all spring scales. They are bigger and less accurate than electronic scales but have the huge advantage that they don't require batteries and tend to keep working, moderately accurately, for decades or centuries. I don't think I've ever had a cheap electronic scale that worked for more than a few years.

Strain Gauge Scales

MOST KITCHEN AND bathroom scales for home use today are fully electronic, with neither springs nor weights. They use strain gauge sensors, which can be fairly accurate, but often aren't. Although they seem fancy and modern, strain gauge scales are really no different than spring scales. They just use a solid block of metal as the spring, and an electronic way of measuring how far the spring has stretched. I'll explain how these things work by way of a super-sized model.

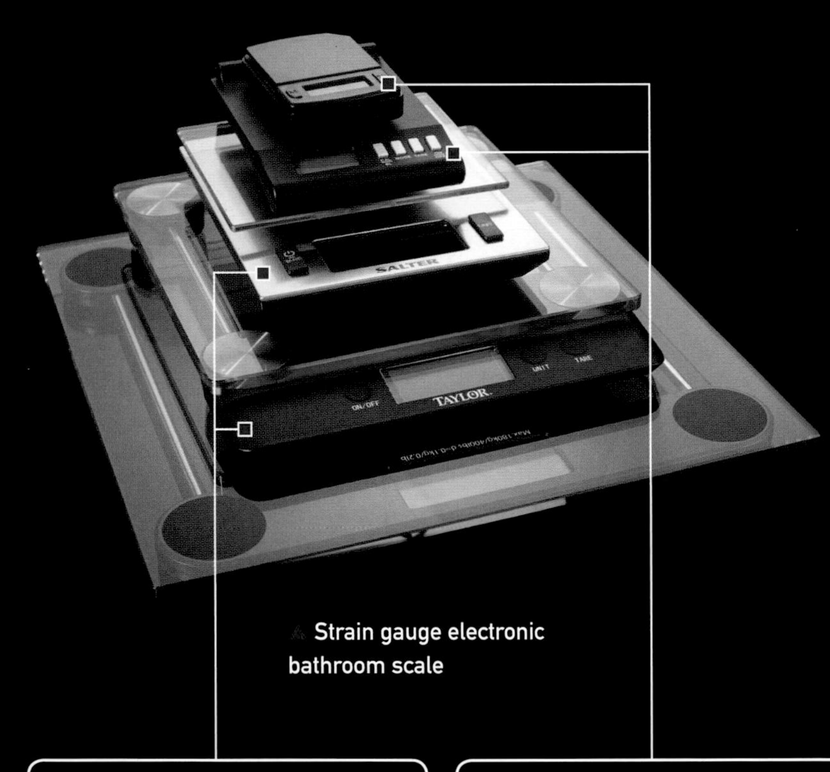

▲ Strain gauge electronic bathroom scale

Kitchen scales that take batteries all have strain gauge sensors these days. They work fine. Not super-accurate, but plenty good enough for cooking.

Strain gauge sensors allow accurate scales to be made very small—pocket-sized, even smaller than the old opium scale on page 143. And sometimes used for the same purpose.

Here we have a rubber tube filled with salt water connected to a meter that measures the electrical resistance through the water in the tube. With no weight on the hook, the tube is relatively short and thick: it has a resistance of 7.28 kΩ. When we put a weight on the hook and stretch out the spring, the tube gets longer and thinner. That makes it harder for the electricity to flow through the tube: it has to go farther (because the tube is longer) and it encounters more resistance along the way (because the tube is thinner). The resistance is now up to 10.49kΩ. If we calibrated the resistance measurement with some known weights, we could use this device to weigh things.

Of course this is a silly contraption to build. It would be easier to just put some labeled marks up next to the rubber tube to measure its length directly. That would simply be a spring scale. But what if the tube was so stiff that it stretched by only a microscopic amount? And what if we wanted to read out the weight not on a physical scale, but on an electronic display?

This real strain gauge works exactly the same way as the spring-and-hose contraption, except it uses a block of aluminum instead of a spring, and instead of water-filled hose, it uses a zigzag pattern of thin aluminum traces printed onto the surface of an insulating sheet of plastic. When force is applied to this assembly (for example, by stepping on one end of it), the aluminum bar bends very slightly. This slight bending is enough to slightly stretch and thin out the aluminum traces, which causes their electrical resistance to increase by just a bit. (There are many traces going back and forth in the same direction in order to increase the amount by which the resistance changes, to make it more accurate and easier to measure.) With suitable calibration, the resistance can be measured and translated into a digital display of weight. These devices are more accurate than spring scales, but as we will see in the next section, they have their problems.

Weighing with Sound

STRAIN GAUGES SEEM very modern and solid state, with no moving parts. But this is an illusion. In fact, they rely on the same kind of physical distortion of a piece of metal as a spring scale does. If the strain gauge didn't get measurably longer—actually changing shape—there would be no change in the reading on the display. That means strain gauge scales are subject to all the same problems as spring scales: they can wear out or drift over time, especially when under constant load. They are also quite temperature-sensitive. Bottom line: it's never a good thing when an instrument relies on bending a piece of metal for its performance.

There are other, very different and clever methods of weighing things, including the one shown on this page, which does it with sound.

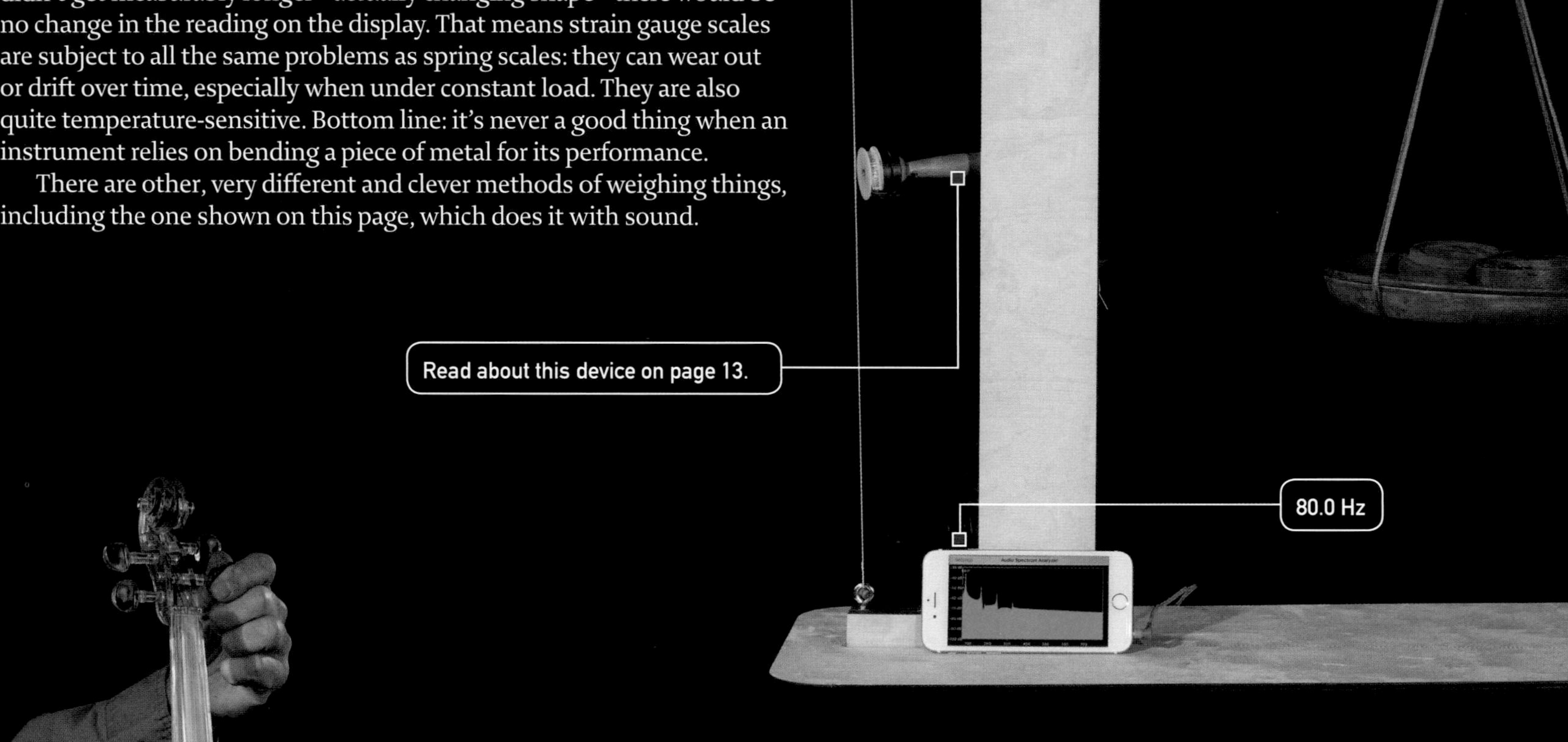

▲ Here we have a guitar string stretched between a fixed point and the short arm of a lever. On the long arm is a weighing pan. When the pan has only a small amount of weight in it, there is only a small amount of tension on the string, and it plays a very low note (80.0 Hz). When you add a heavier object to the pan, the tension increases and the string plays a higher note (116.3 Hz). The frequency of the note (how high or low it is) depends directly on the amount of weight in the pan, which causes the tension on the string to vary. If you measure the frequency after adding a range of known standard weights, you can calibrate this device and use it as a weighing scale—one that works by sound.

Nothing in this device actually has to move or bend for it to work (other than an insignificant deflection of the vibrating string). The lever arm transfers force to the string, but it barely moves at all. If the whole mechanism were made of perfectly stiff metal, it would still work, making this sound-based scale potentially more accurate than a spring scale. Nothing is being stretched or flexed to any meaningful degree, so there is much less chance of the measurement drifting over time. It's also not sensitive to temperature, because the vibration frequency, to a first approximation, depends only on the tension and the mass of the string.

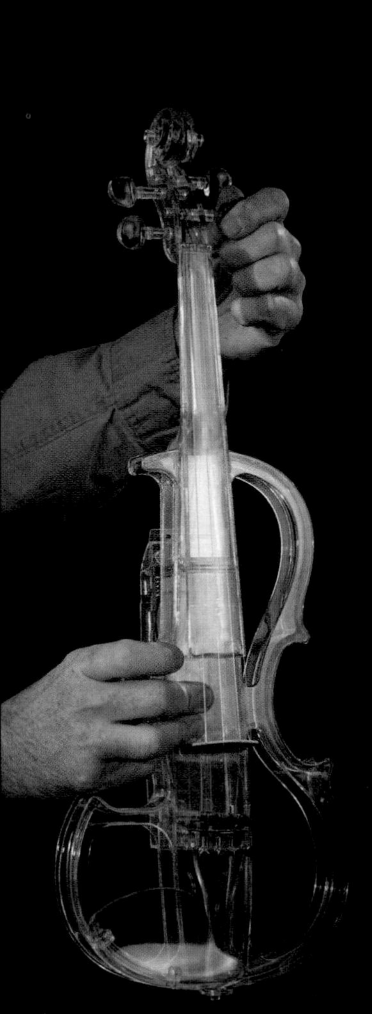

When you pluck a violin string, it vibrates. That vibration creates sound. If the string is stretched very tightly, it vibrates very fast (at a high frequency), which creates a high note. If the string is looser, it vibrates at a lower frequency, creating a lower tone. When you turn the tuning pegs on a violin or guitar, you are changing the pitch of the string by changing its tension.

The fact that the sound varies in direct relation to how tightly the string is stretched means you can use the frequency of the note as a *measurement* of how tightly the string is stretched. Therein lies a method of weighing things with sound.

▲ This is a very tough, practically indestructible object. It's made of thick, machined aluminum, watertight and strong enough to drive a truck over. That is, in fact, exactly what it's designed for: driving a truck over it. Or, to be more specific, it's designed to be mounted as part of the suspension system of a garbage truck in such a way that the weight of the truck rests on four of these units.

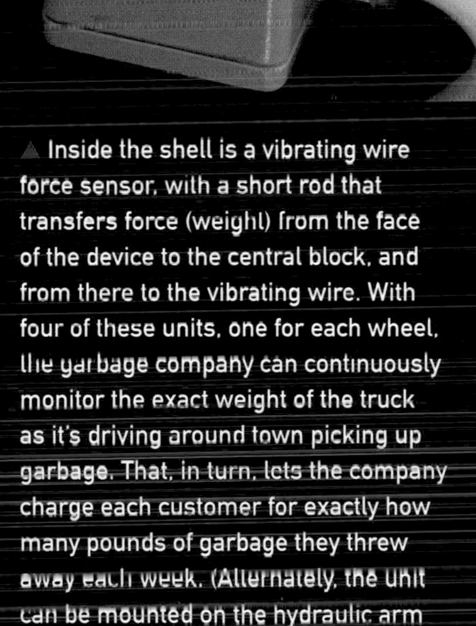

▲ Inside the shell is a vibrating wire force sensor, with a short rod that transfers force (weight) from the face of the device to the central block, and from there to the vibrating wire. With four of these units, one for each wheel, the garbage company can continuously monitor the exact weight of the truck as it's driving around town picking up garbage. That, in turn, lets the company charge each customer for exactly how many pounds of garbage they threw away each week. (Alternately, the unit can be mounted on the hydraulic arm that picks up garbage cans at each house.)

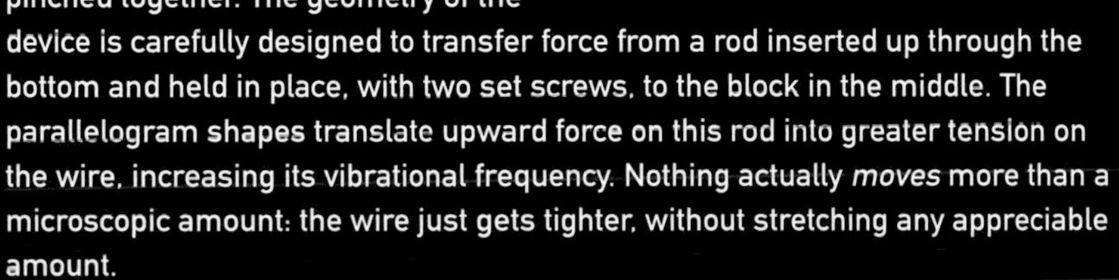

► This lovely, lovely device is called a vibrating wire force transducer. Using sound to weigh things, it's a commercial version of the model we just saw. Instead of a guitar string, it uses a short (less than ½ inch [1 cm] long) spring brass wire, which is held firmly on both ends by pairs of sapphire cylinders pinched together. The geometry of the device is carefully designed to transfer force from a rod inserted up through the bottom and held in place, with two set screws, to the block in the middle. The parallelogram shapes translate upward force on this rod into greater tension on the wire, increasing its vibrational frequency. Nothing actually *moves* more than a microscopic amount: the wire just gets tighter, without stretching any appreciable amount.

◄ I have a great fondness for this particular force transducer, both because it is a tremendously clever piece of engineering, and because it was invented by my Swiss grandfather, Armin Wirth, shown here in a photograph from 1923

The principle of a vibration frequency–based scale is taken to the extreme in the surface acoustic wave scale.

Just as in the case of the vibrating wire sensor, the surface acoustic wave scale has a rugged industrial case with a thick block of aluminum inside that transfers force, but no appreciable movement, to a tiny sensor. The difference is that the sensor in this scale is a slice of quartz crystal that has piezoelectric waves traveling across its surface. (See pages 120 to 124 for a discussion of how quartz crystals interact with electric fields to create resonant oscillators.) Stress in the crystal affects the waves, allowing that stress, and thus the weight on the scale, to be measured very accurately.

Weighing a Grain of Salt

SPRING SCALES, PLATFORM scales, Roberval balances, and electronic scales have completely replaced the ancient two-pan balance in virtually all common applications. The one area where the ancient design persisted longer than anywhere else was, ironically, in cases of very high-precision weighing. The main driver for the improvement of commercial scales was, and remains, convenience, and guarding against cheating. But for scientific work, accuracy is the obsession. Scientific scales have to be able to weigh things down to a part per million of error or less, and they have to be able to weigh even minute, almost invisibly small amounts of things. Today, high-precision electronic balances based on several technologies have replaced mechanical balances almost everywhere, but the old ones are still beautiful in their design and construction.

These high-precision balances are called "analytical balances," and they typically take advantage of a series of improvements on the original balance scale.

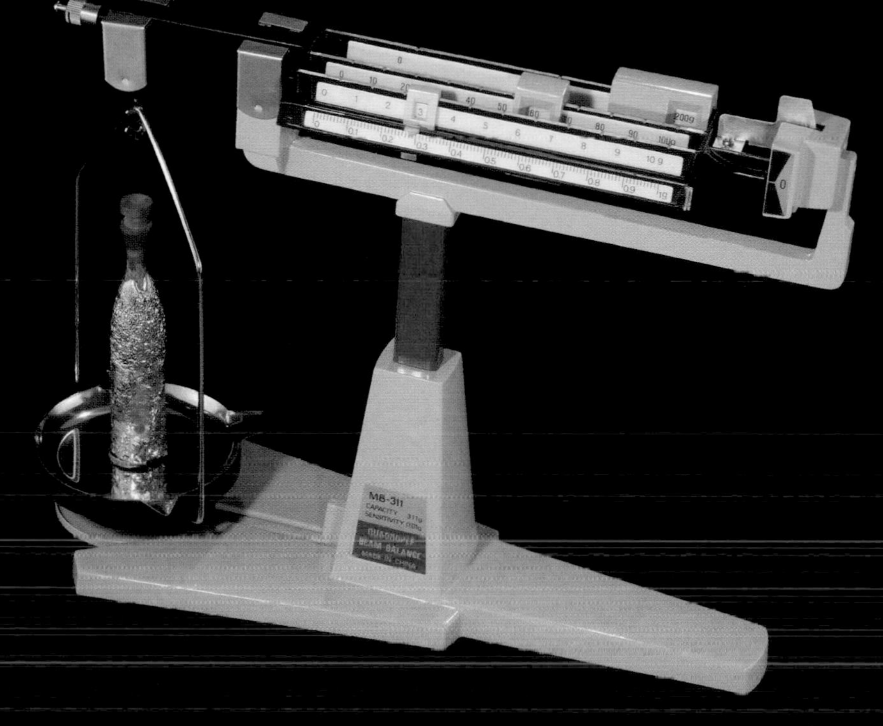

▲ Much like disposable razors (which used to have one blade but have now evolved to have as many as five), balances can be made fancier with more beams. The one pictured here has four separate beams, for 100-gram units, 10-gram units, 1-gram units, and 0.1-gram units. To justify the extra beam, the pivot point and general quality of the scale need to be higher.

▲ The triple-beam balance was used by generations of high school students all over the world. It's pretty convenient and pretty accurate, though not nearly as good as the balances we will meet later in this section. In essence it's a form of steelyard scale with a sample pan on the left and sliding weights on the right. Instead of one scale and one counterweight, this design has three beams with progressively lighter weights. On each scale there are detents (stops) that position the weight at exactly the right distance to balance the indicated amount of weight in the sample pan. The backmost scale goes in steps of 100 grams, the middle one in steps of 10 grams, and the front one in steps of 1 gram (this one can be slid between the marks to try to measure fractions of a gram).

▲ Going in the other direction, cruder balances sometimes have only two beams. (What appears to be a third beam in the back in the picture is actually a tare adjustment, for subtracting out the weight of an empty container placed on the scale.)

The Geometry of Balance

IN ORDER TO understand the beauty of the analytic balance, let's review the challenges of the two-pan balance.

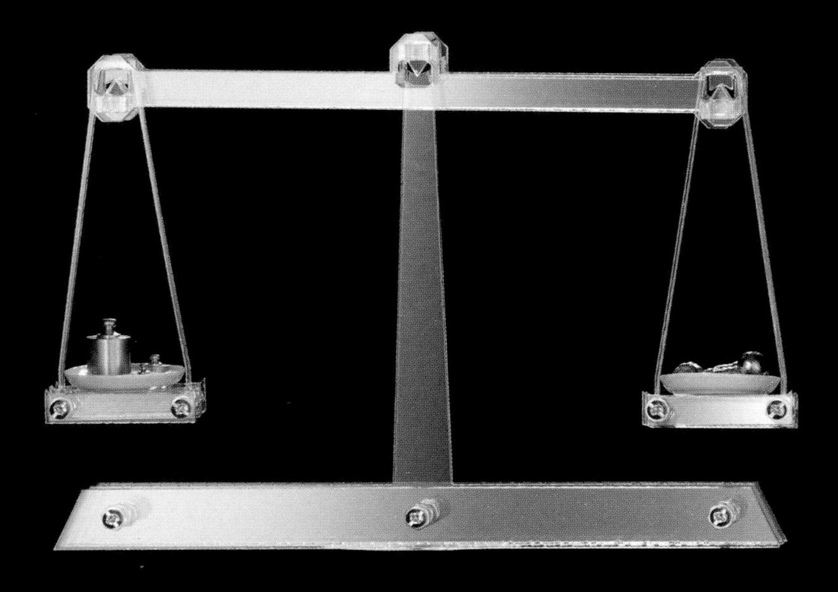

A perfectly ordinary two-pan balance has a straight, flat balance beam, with the three pivot points all in a straight line. With a straight beam, when the weights are perfectly equal, the balance will stay stationary in any position. This balance is similar to the Roberval design in that equal weight is indicated not by the balance moving to the level position, but by the fact that it's stable in any position.

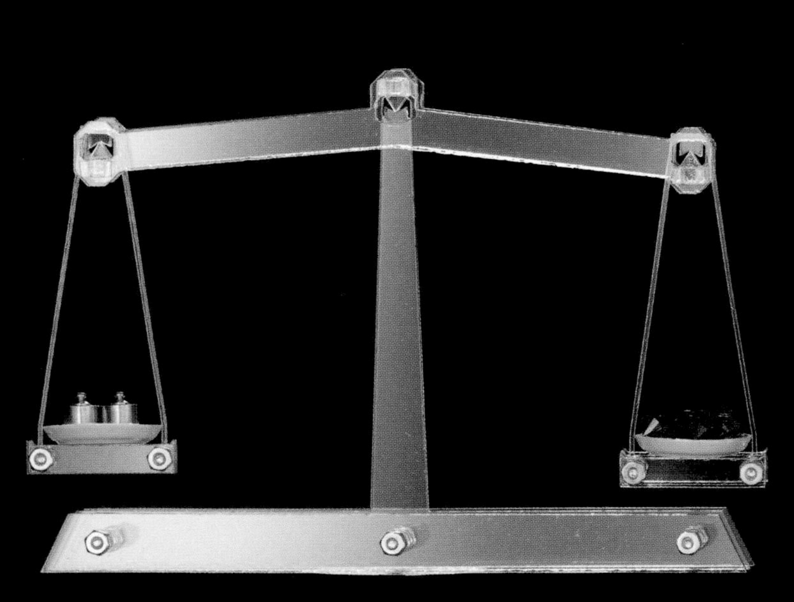

If you bend the balance beam so the central pivot point is slightly higher than the two pan-hanging pivot points, then the balance will act just a tiny bit like the postal scales we saw earlier. The bend means there is a small restoring force that brings the balance back to a level position when the weights are equal. If you bump it slightly out of level, it will come back to level, as long as the weight on both sides is equal.

If you instead bend the beam upward, you get something pretty useless: a balance that is never stable in the middle, no matter how perfectly equal the weights are on both sides. No matter what you do, it keeps trying to flip over to one side or another. No sensible scale is ever made this way.

When a downward-bent beam balance is perfectly in balance, it will be perfectly level. But when it's slightly out of balance—in this case with slightly more weight on the right—it will come to a stable position at a slightly tilted angle. This angle is predictable, and by adding a long pointer and a calibrated scale, you can read off the slight difference in weight between the two sides. This is very useful because it means you can get the last decimal place or two of a reading without needing teeny-tiny standard weights. You just need to put larger weights on the reference side until it's close enough to get the pointer within the range of the scale, then add or subtract the scale reading from the total of the standard weights you used. On the next page we'll see an example of a real balance that uses exactly this design.

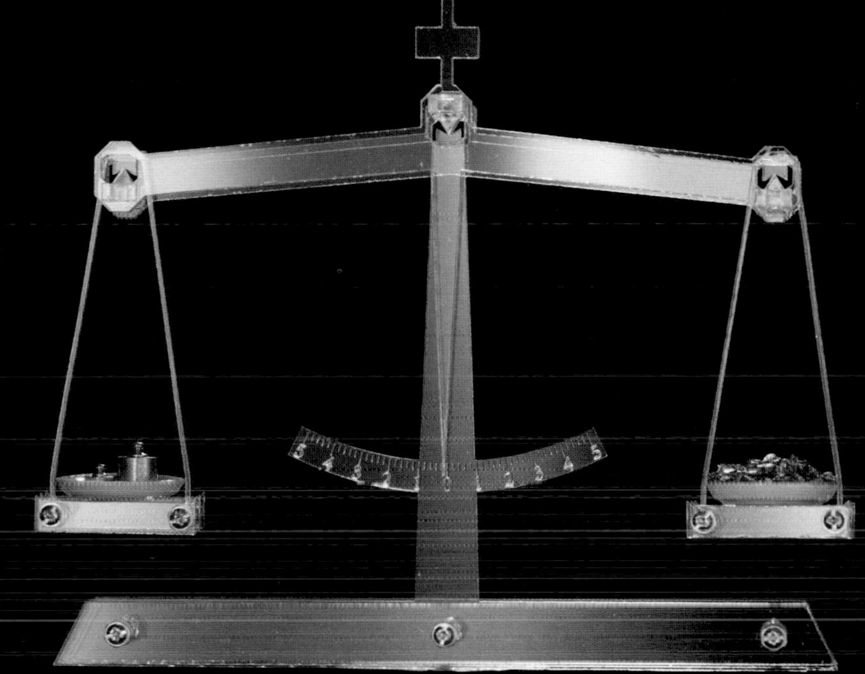

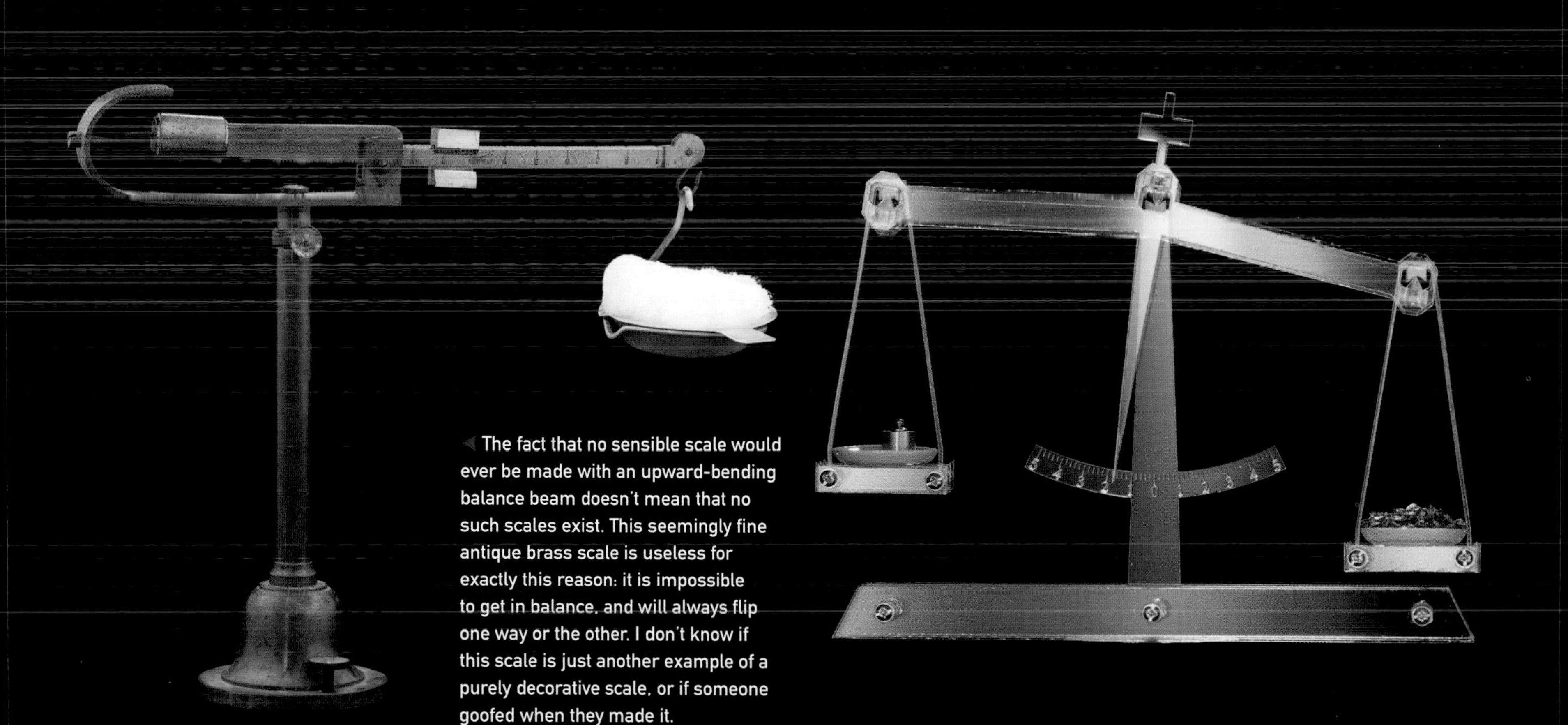

◀ The fact that no sensible scale would ever be made with an upward-bending balance beam doesn't mean that no such scales exist. This seemingly fine antique brass scale is useless for exactly this reason: it is impossible to get in balance, and will always flip one way or the other. I don't know if this scale is just another example of a purely decorative scale, or if someone goofed when they made it.

Anatomy of an Analytical Balance

ACHIEVING ACCURACY AND sensitivity in a balance is all about systematically eliminating sources of error, one after another. The biggest issue is the sharpness and alignment of the balancing point. Every good balance has "knives" and "anvils" made of something very hard. For industrial scales, hardened steel is used because it's tough. Precision balances ideally use hard, glassy crystal materials: agate was used first, then synthetic sapphire. This, being an older and not-so-expensive model, has steel knives.

This is a mid-range analytical balance made in the 1950s. It does not differ significantly from ones that were made hundreds of years earlier. You can see the very long pointer that reaches down to the calibrated scale at the base.

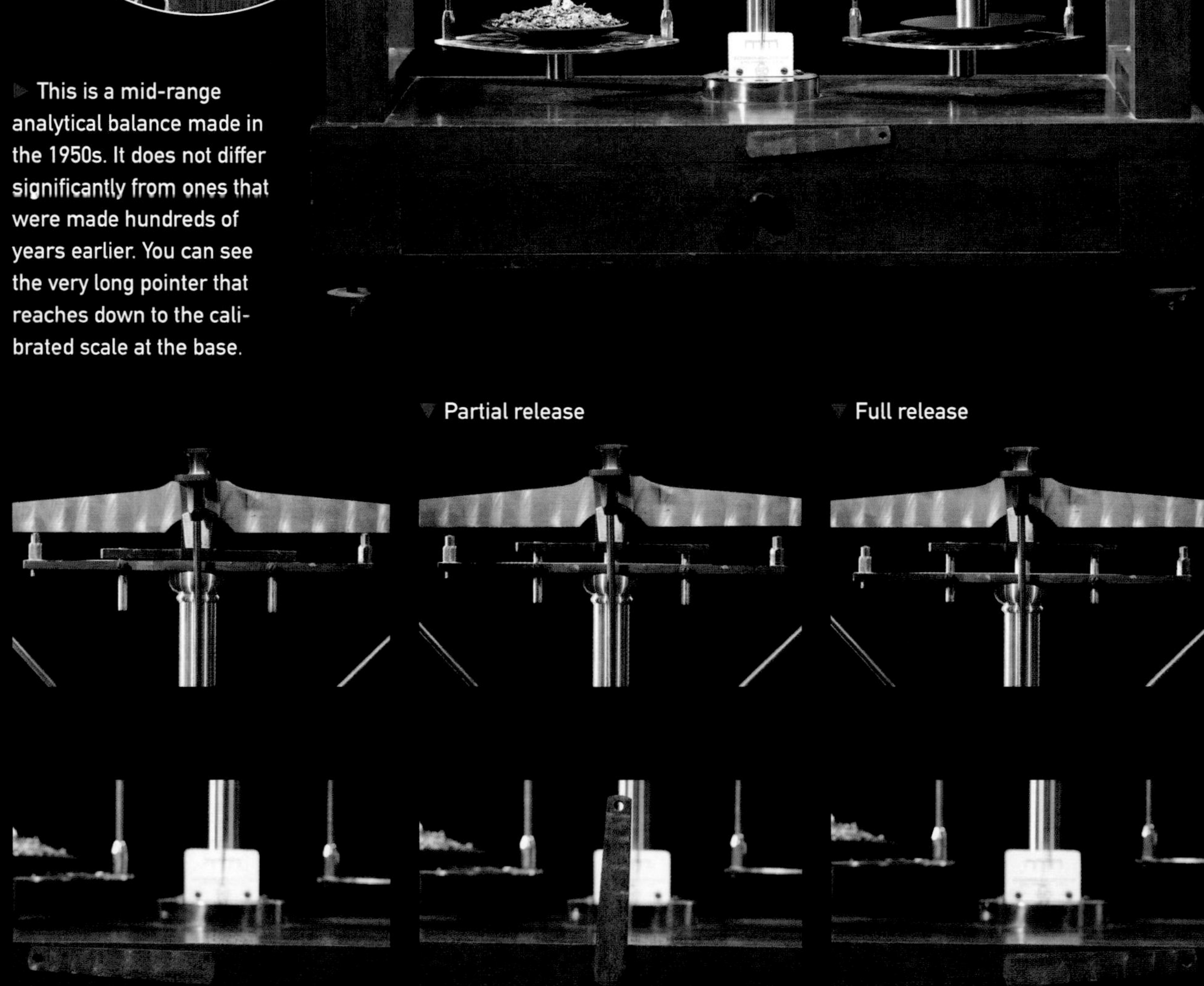

Knife edges are delicate and easily damaged, so precision balances always have a lever that lifts the balance beam up and off the anvil, protecting the edge while the balance is being moved, or while it's being loaded and unloaded (with samples and reference weights). This balance has a simple release mechanism that just lifts the balance beam completely up and off its knife edge. Fancier balances have more elaborate partial-release mechanisms.

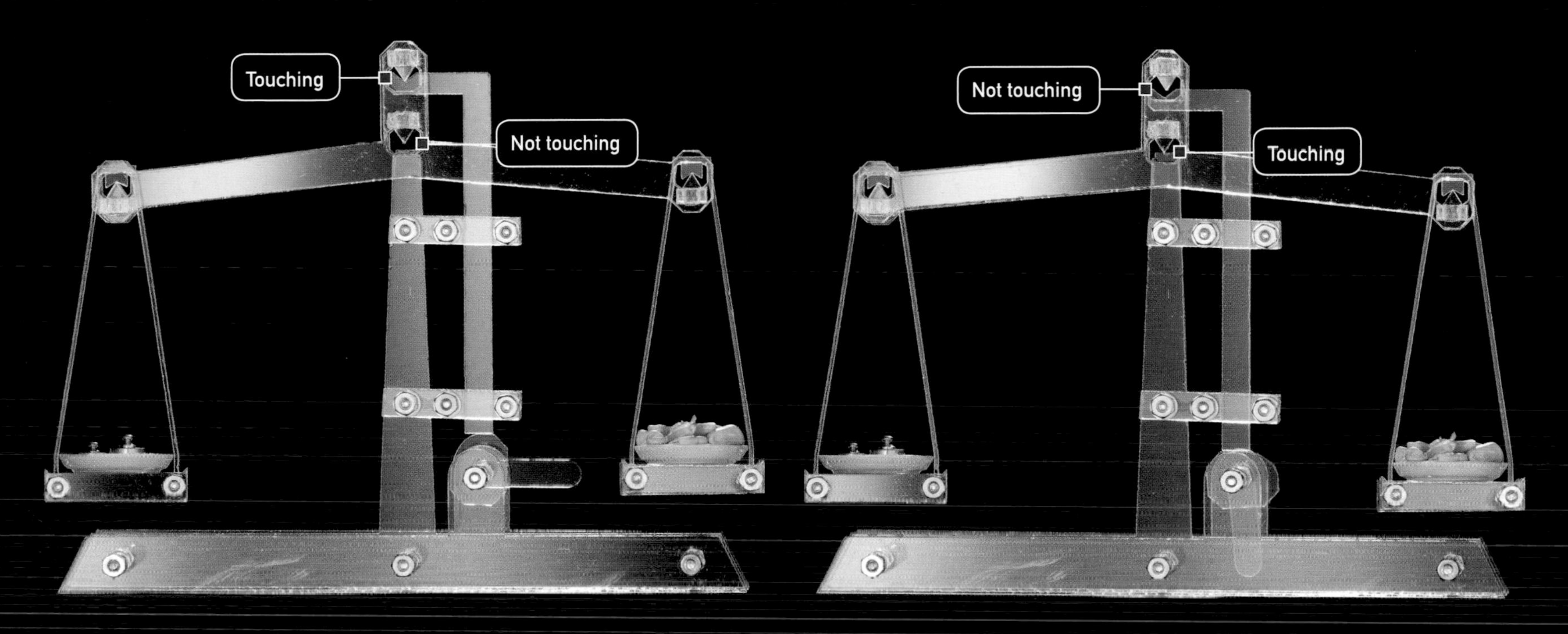

Complex release mechanisms are sometimes difficult to spot in real balances, so I've made a model to show how one clever type of partial-release mechanism works. With the release lever flipped up in the partial-release position, the balance beam is supported in the center by a robust, abuse-resistant, but not very accurate balance point. It's also suspended from a point well above its normal pivot point, which makes it tilt less out of level for a given difference in weight between the two sides. In this mode you can load up the pans without fear of damaging the delicate main knife and anvil. And you can quickly get the balance pretty close, using its less-sensitive movement to make a first approximation of the weight.

With the release lever flipped down, the upper, crude balance point is disengaged, and the balance beam is set down (gently, please!) on an expensive, super-sharp crystal balance point. If you did it right, you got the weights close enough so that the pointer (not shown in this model) stays in range of the calibrated scale and you can read off the total weight as described on page 167.

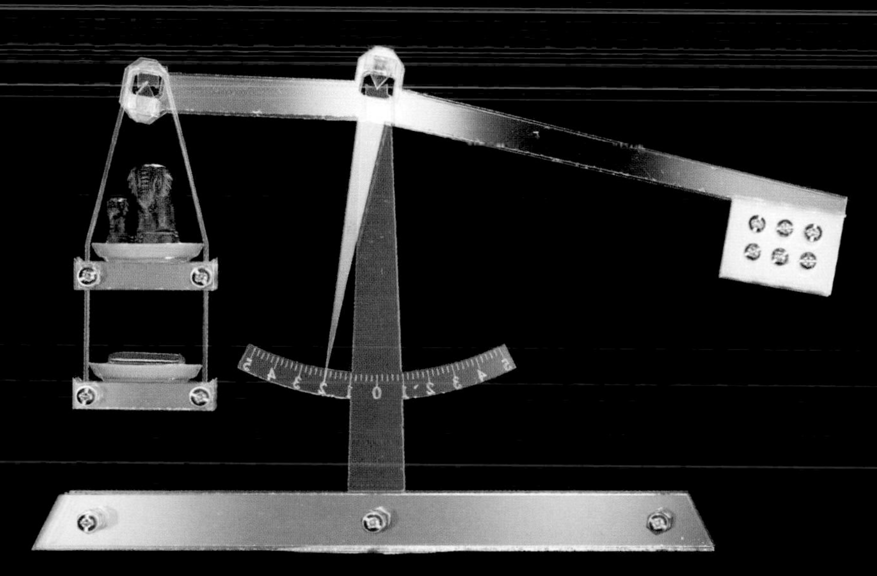

Two-pivot acrylic model

The final deviation from standard two-pan design that gets us to the modern (pre-electronic) analytical balance is the clever realization that you can put both the sample to be weighed and the reference weights on the *same side* of the balance beam, with a fixed standard weight on the other side. Instead of adding reference weights to balance the sample, you subtract reference weights until the total of sample + reference equals the fixed weight on the other side.

There are three advantages to this design. First, it means that the scale requires only two knife edges, not three (the fixed counterweight doesn't need to hang). Second, it means that when the beam is in balance, there is always exactly the same total amount of weight on it, whether you're weighing a heavy or a light sample. That means the inevitable tiny amount of flex in the beam will be consistent at all times. And finally, by moving the pivot point partway toward the sample pan, a lighter reference weight can be used without decreasing the maximum capacity of the scale. This reduces the total force on the central knife edge, increasing its accuracy.

THE ANALYTICAL BALANCE

THIS STUDENT-MODEL analytical balance from the 1970s shows off all the design features we've described: a slightly bent balance arm, two knives instead of three, reference weights on the same side as the sample, and an indicator that lets you read off the deviation from level balance to get the last two decimal places of the weight. And it has one more innovation: a clever system of cogs and levers that adds and subtracts reference weights automatically. We saw earlier that the most efficient set of reference weights would be successive multiples of 2, as in 1 g, 2 g, 4 g, 8 g, 16 g, 32 g, etc. That's exactly what this scale has. The user sees nice decimal numbers displayed on the front panel, but inside the cogs are picking from the multiple-of-two weights to equal the displayed amount, hiding the annoying math from the user.

The three big dials set the 10, 1, and 0.1 gram amounts. Each of these dials operates a cam mechanism as shown on the next page to add and remove reference weights above the sample pan.

The very last digit of the weight, the 0.0001 g position, is read by adjusting this knob until the gap indicator is exactly aligned with one of the marks in the projected image. This digit is reading tenths of a milligram (ten-thousandths of a gram). In realistic situations this digit is not likely to be fully accurate: you would have to work very hard to calibrate the scale and make sure it is perfectly level and free of dust to get it tuned up to the point where it could reliably measure tenths of a milligram. (The scale will easily indicate a *difference* of a tenth of a milligram or less, you just can't necessarily read an absolute weight that accurately.)

The next two digits of the weight, the 0.01 g and 0.001 g amounts, are read out on this optical scale. On the back of the balance beam there is a microscopic etched glass scale that moves up and down with the beam. A light and lens focus an image of this scale on the screen at the front of the scale. This is the equivalent of the calibrated out-of-balance pointer described earlier, just much more sensitive.

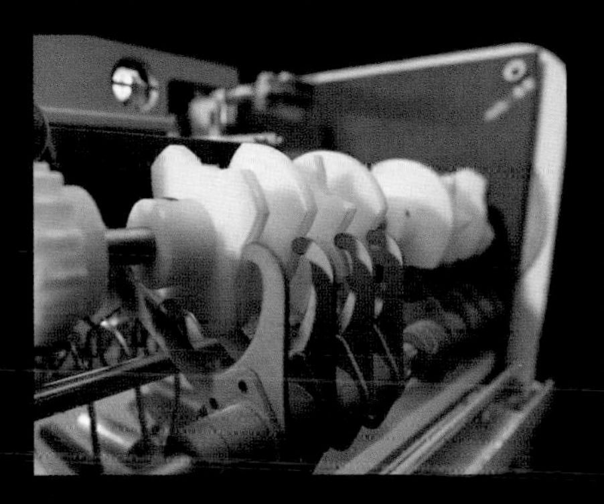

The lifter arms are controlled by cams (oddly shaped disks), all mounted on a common shaft, which is connected to a dial on the front of the scale. Each cam has a unique shape that causes its lifter arm to be up or down in a different set of positions. Working together, these cams count in binary.

These small weights let you calibrate exactly how much deflection of the beam there is when it is very slightly out of balance. That has to be calibrated because that slight deflection is what drives the optical scale that reads out the last three digits of the weight.

This scale has sapphire knives and anvils, the standard for analytical balances since the mid-1900s.

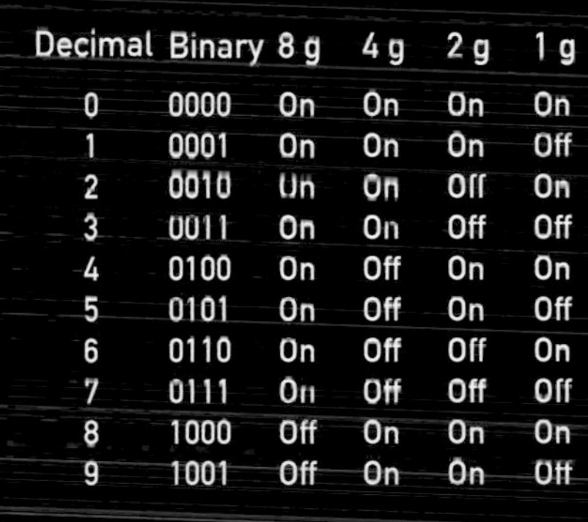

Decimal	Binary	8 g	4 g	2 g	1 g
0	0000	On	On	On	On
1	0001	On	On	On	Off
2	0010	On	On	Off	On
3	0011	On	On	Off	Off
4	0100	On	Off	On	On
5	0101	On	Off	On	Off
6	0110	On	Off	Off	On
7	0111	On	Off	Off	Off
8	1000	Off	On	On	On
9	1001	Off	On	On	Off

Decimal numbers, the kind we normally use, have digits from 0 to 9 in each position. Binary numbers have only 0 or 1 in each position. Instead of having 1s, 10s, 100s, and 1000s positions, in binary we have 1s, 2s, 4s, and 8s positions. Each set of cams lets you dial in a four-digit binary number using a dial that reads in decimal. (Note that the cams are out of order, to keep the frame more evenly loaded overall.)

The reference weights are all hanging from a complicated frame above the weighing pan. It looks like a mess, and none of it is very precise, but that actually doesn't matter. As long as nothing touches the frame except those weights that are meant to be hanging from it, the accuracy of the scale is maintained. Each weight has a lifter arm that is either down, allowing the weight to rest on the frame without interference, or up, keeping the weight completely away from the frame.

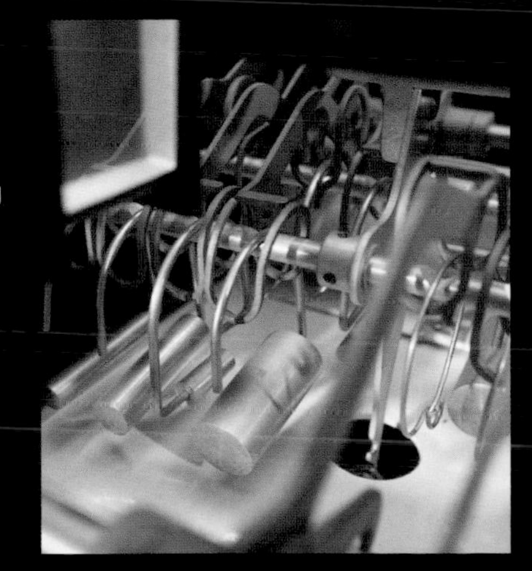

Once again, modern technology deprives us of all the beautiful guts in our machinery. This modern strain gauge analytical balance has about the same accuracy, at about a tenth the price, of an old balance beam analytical balance. You still have to be careful about leveling it and keeping the pan clean, but there's no messing around with clunky dials and tricky optical scales. You just put the thing on the pan and read its weight. Chemistry students today are unlikely to ever see a mechanical analytical balance, and there are even disturbing reports of university faculty who have never used one. Well, back when *I* was a student we still used mechanical balances, and *we liked it*.

This modern strain gauge analytical balance is mechanically uninteresting.

The Most Accurate Scale in the World

THIS IS THE most accurate scale in the world. It lives in the basement of NIST (the National Institute of Standards and Testing) in Washington, DC, and it just, during the writing of this chapter, contributed to the redefinition of the kilogram, from which all other definitions of mass derive.

Back on pages 140 to 141 we saw a selection of standard reference weights used to calibrate accurate balances. They come in various degrees of accuracy, with the best being "traceable" to an appropriate national standard of weight. The ultimate standard against which all such weights are calibrated is—or was—the "International Prototype Kilogram," a cylinder of platinum-iridium alloy kept carefully locked up in a vault in Paris. Until November 2018, the weight of that cylinder defined one kilogram, period.

On pages 132 and 133 we learned about the moment when our clocks became more accurate than the earth. When the measured time of noon appears to change from one day to the next, we believe our clocks over the rotation of the earth—because we know them to be more accurate.

Now we have a scale that is so accurate that if we weigh the International Prototype Kilogram—the most carefully preserved and cared for lump of metal in the world—and find that it appears to have changed weight, we believe the scale over the supposedly unchanging lump of metal.

How can a carefully guarded lump of metal change weight? It's not entirely understood, but it's long been suspected that it has been getting lighter, by about 50 millionths of a gram or one part in 20 million, over the last hundred years. No one knows why, but now at least we have an instrument that can remove any doubt as to whether it's the lump or the scale that is drifting.

The new balance works by comparing the downward force generated by the weight of a physical object with the upward force generated by an electric current. It thus relates the fundamental units of mass with those of time and distance. This is a big deal. The meter used to be defined with a standard meter bar, also kept in Paris. Now it's defined in terms of the speed of light. The second used to be defined in terms of the revolution of the earth; now it's defined by properties of the cesium atom. One by one, all the fundamental units of measurement have been made universal and unchangeable by relating them to things that exist and can be measured anywhere in the universe, not just in Paris. The kilogram was the last holdout, the last standard that resisted being made universal. But no more.

The moment in 2018 when the international committee on weights and measures voted to replace the standard kilogram with this instrument is arguably the most important moment in the history of measurement. It completes the dream of a complete system of measurement standards that has no standard objects.

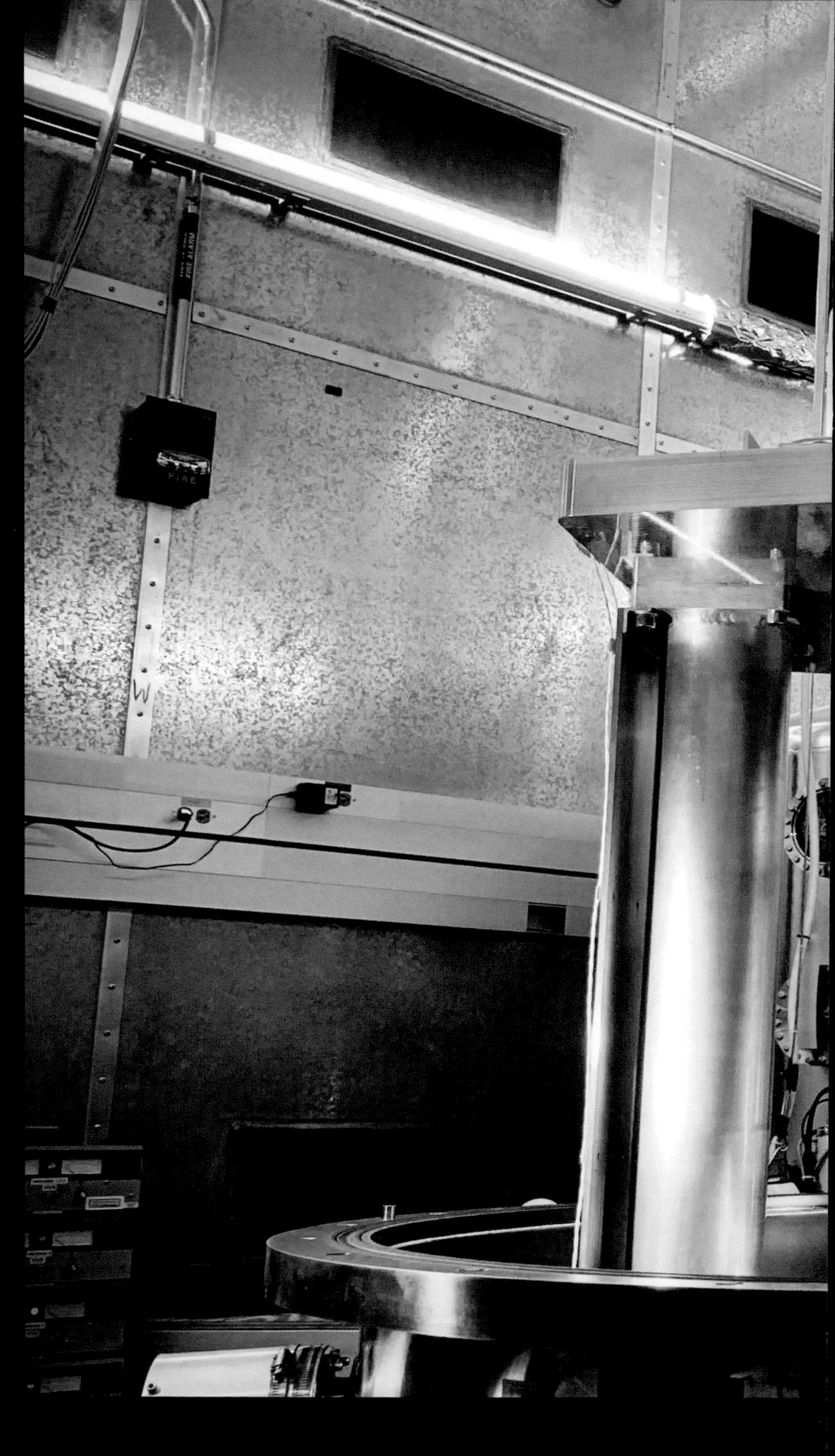

NIST

Heavy Weight Scales

AT THE OPPOSITE end of the spectrum from delicate analytical balances we find the workhorses of weighing: scales meant to weigh cars, trucks, trailers, trains, and airplanes. Some of these scales use exactly the same methods as smaller platform and electronic scales, just with more leverage applied to magnify the range of weights they cover. But there are also entirely different weighing methods that are only suitable for very heavy jobs.

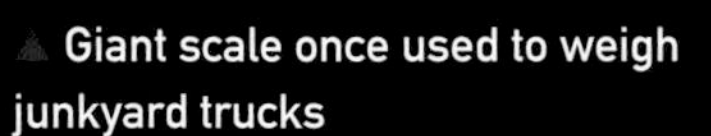

▲ Giant scale once used to weigh junkyard trucks

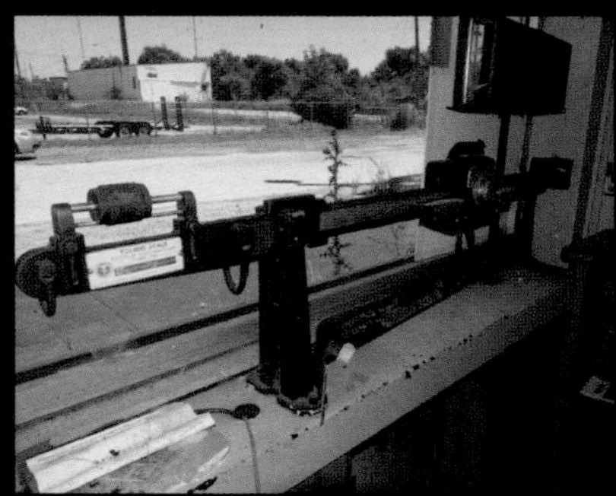

▲ Balance beam in the junkyard office

This scale was used to weigh trucks hauling scrap metal to a recycling yard. The truck was weighed when it arrived and again after it had been unloaded: the difference was the amount of material the driver needed to be paid for. I'm using the past tense here because, although this same system is still used at all junkyards today, this particular one closed down years ago and its huge old scale has rusted to ruin.

Underneath this truck-sized platform is a system of beams and pivot points very much like those in the platform scale seen earlier—except much bigger, underground, and currently submerged in water. The end point of the levers is underneath the office window (behind the random bush growing up against the building). Leading up from the last lever is a vertical rod that reaches up inside the office to the balance beam.

In the junkyard office is a balance beam much like you would find on an old doctor's scale, except instead of reading off fractions of a pound, it reads off fractions of a ton. It was a bit of a surreal experience being in this abandoned office for the first time in probably fifteen years. When I was a kid I used to buy zinc roof flashing here to make castings with. Later I bought steel plates for my friend Jim to make his plasma cut art. But that life is gone, Jim is some years dead, and so is this scale in its sad office.

Above is the junkyard balance beam cleaned up a bit. It was made, probably in the 1960s, by the Toledo Scale Company, and has a maximum rated capacity of 100,000 pounds (45,000 kg) otherwise known as 50 tons (45 metric tons).

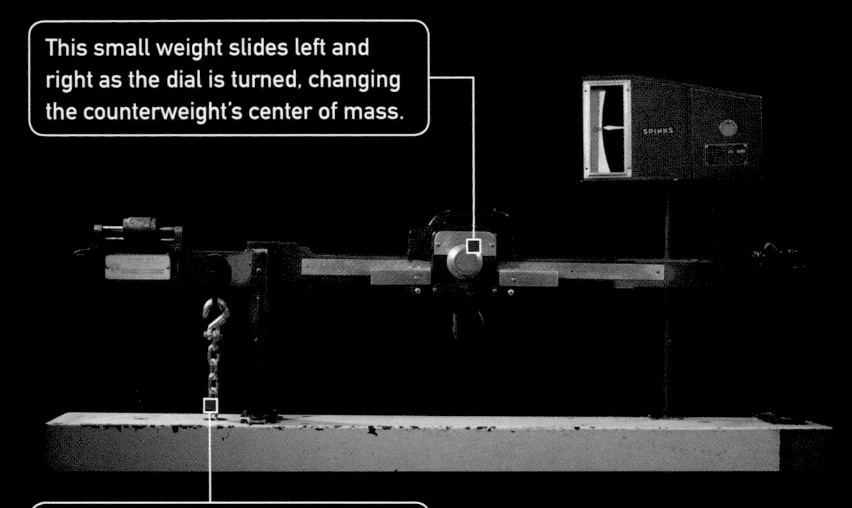

▼ The major tick marks on the beam scale are in units of 10,000 pounds (5 tons), and the minor tick marks are 1,000 pounds (½ ton).

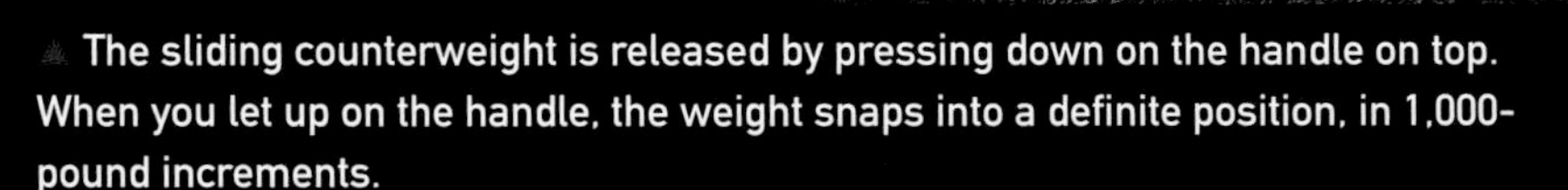

▲ The sliding counterweight is released by pressing down on the handle on top. When you let up on the handle, the weight snaps into a definite position, in 1,000-pound increments.

The counterweight isn't just a simple mass: it has a calibrated dial that moves a second, smaller counterweight (the bar with gear teeth on top) sideways as the dial is turned. This secondary weight acts a bit like one of the smaller beams on a multi-beam laboratory balance. A full turn of the dial moves the smaller weight by the equivalent of 1,000 pounds, or one minor tick mark on the main beam. The dial clicks firmly at its 10-pound minor tick mark increments, so the scale is not meant to be read to anything finer than these 10-pound increments. (10 pounds on a scale of 100,000 is one part in ten thousand, which is pretty good for any scale, and in practical use this scale was probably not that accurate most of the time. The scrap yard had other, smaller scales for weighing valuable metals such as copper.)

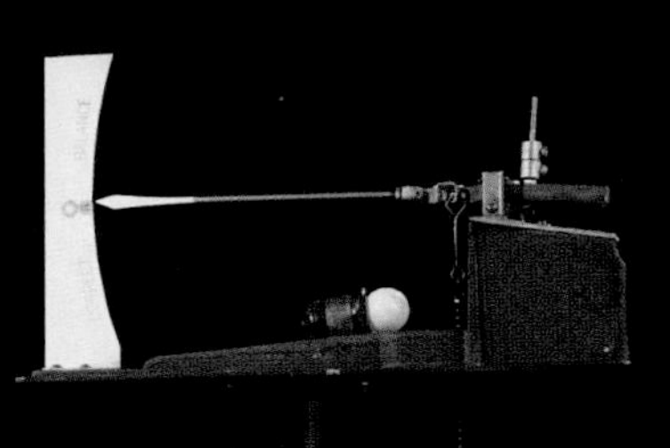

▲ This is an aftermarket add-on, not made by Toledo, which adds an oil dashpot to the scale, to help it settle down to a stable value more quickly. It also adds a pointer with a high degree of mechanical amplification to indicate when the scale is exactly level. That makes it more accurate, or at least easier to read. (The whole scale, including this pointer, is visible both from inside the office and through the office window from the platform outside. This dial would make it easy for the driver to see that the scale was in balance, perhaps helping increase trust in the transaction.)

▲ These slots on the front of the counterweight lead to a very clever feature of his scale: a built-in printer! There are raised numbers on the bottom of the main eam and the dial. You put a slip of paper with a sheet of carbon paper on top of it nto one of the slots, then squeeze the brass handle underneath. This presses the arbon paper and receipt up onto the raised letters, leaving a print of the current etting of the scale on the receipt.

SCRAPPY MCSCRAPFACE SCRAP YARD
SERVING YOUR IMAGINARY RECYCLING NEEDS SINCE 2019
"HIGHEST PRICES PAID"
FOR FIGMENTS OF YOUR IMAGINATION

		Price	Total
Gross	68 790		
Tare	37 520		
Net	31 270	10¢	$3,127

When a truck first arrived at the yard with a load of scrap metal, it would drive onto the scale, and the receipt would be placed in the "Gross" slot and imprinted with the total weight of the truck and scrap. (Gross, in this context, means the total weight of a container and its contents.)

After unloading, the truck would drive back onto the scale and the same receipt would be inserted again, but this time in the "Tare" slot. This slot leads to the same place inside, but with the paper shifted over a bit, so the tare amount is printed just below the gross amount. (Tare, in this context, means the weight of an empty container.) By subtracting the tare weight from the gross weight you can calculate the weight of the metal that was unloaded. Multiply that by the price per pound of the particular type of metal delivered, and you have the amount the driver needs to be paid.

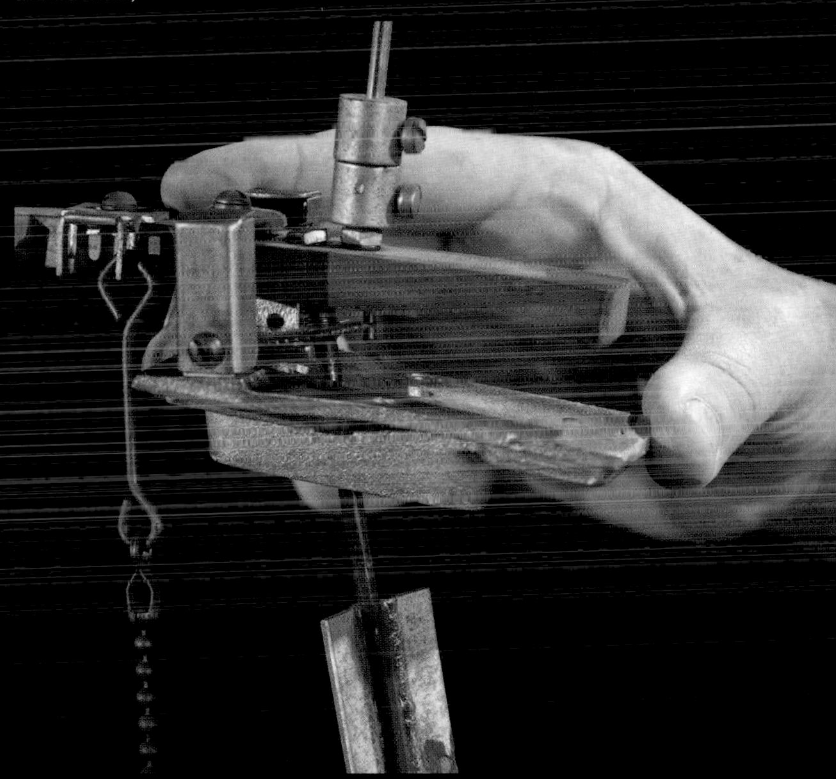

▲ A paddle pushing through viscous oil in this pot slows down the movement of the pointer. You can even adjust how much resistance you get from the oil by rotating the vane: edge-on there will be very little effect, but side-on it will take much more force to push through the oil.

All this is still exactly how the scrap metal business works (or any business where goods are received by weight), except the scales are electronic and the receipts are computer printouts. The metal itself hasn't changed.

ELECTRONIC SCALES HAVE pretty much taken over all ranges of weighing, but there are still a few applications where other technologies survive. For example, this trailer hitch scale is basically a nearly solid lump of steel that can weigh things up to 2,000 pounds (900 kg), with no electronics and no batteries required. It operates on a completely different principle from any of the other scales in this chapter: it is a hydraulic scale. Inside there are no levers, pivot points, springs, strain gauges, or any of the other parts of a traditional scale. There is only a short hydraulic cylinder and a pressure gauge.

Coming out the top of the scale is a peg on which you are meant to rest the tongue of a trailer (the part that is normally connected to a trailer hitch on a car). The goal isn't to weigh everything on the trailer; rather, it is to determine how much weight is resting on the trailer hitch.

It's good to measure the "tongue weight" of a loaded trailer because it's important when loading a trailer that at least about one-third of the total weight is pressing down on the hitch. Otherwise the load can become unstable when you're slowing down, leading to "fishtailing" and loss of control. More weight on the tongue both makes fishtailing less likely, and gives more traction to the rear wheels of the car, making them better able to resist sideways forces should it occur anyway.

▲ The trailer hitch scale is an all-in-one design where you can't really see what's going on.

▼ Here is a spread-out model I've made to illustrate the mechanism.

This is a hydraulic cylinder. It's filled with glycerin (hydraulic oil would work as well, but glycerin is customary for measuring instruments). When you press down on it, the liquid inside is forced out through a hole on the side. Rubber seals around the cylinder prevent leaks.

This high-pressure hose transfers hydraulic pressure from the cylinder to the gauge.

This is the same hydraulic cylinder cut in half so you can see the working parts and the hole that connects the chamber to the hose connector.

▲ As usual, modern technology ruins everything. This is also a truck scale, similar in size and capacity to the lovely old machine we just explored. But there's nothing interesting to see. There are no levers or beams, nothing underground, no balance beams or sliding weights. There's just what you see here, and a small electronic control box in the office.

The actual weighing is done with four strain gauges, one at each corner, fastened to metal beams that deflect a predictable amount when trucks drive on them. The readings from all of them are added up to give the total weight on the platform, regardless of how that weight is distributed. It's like having four entirely separate scales, one for each corner. The total weight is just the sum of the weights measured by each of the four scales.

This dial reads off the pressure reaching it through the hose. The pressure is a measurement of the weight on the cylinder: more weight gives more pressure. The ratio between weight applied and pressure generated is determined by the diameter of the cylinder. A larger cylinder can support more weight with a given amount of pressure. Even very small cylinders can easily support very heavy weights: this one is rated for ten tons. And pressure gauges are happy to read out very high pressures: several thousand PSI (pounds per square inch) is a perfectly normal operating pressure. As a result, small hydraulic scales can weigh tremendously heavy things—just not very accurately.

◄ Pressure gauges work by way of a "Bourdon" tube. When you increase the pressure inside a flattened, coiled tube, the tube inflates slightly, which causes it to straighten itself out.

▲ The movement of the tube is turned into rotation of the dial by a rack-and-pinion gear in the center.

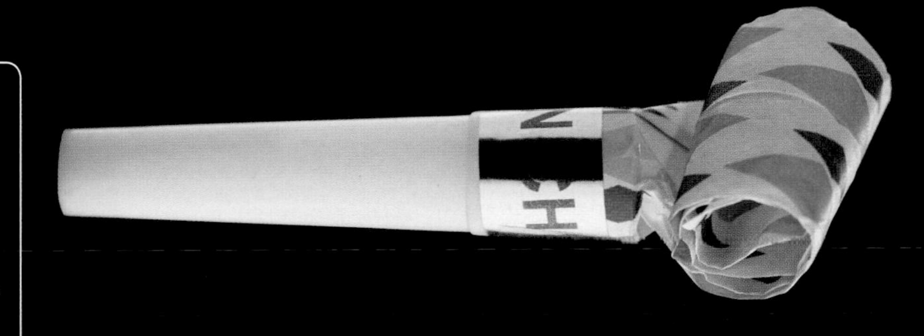

▲ These common party favors are a form of Bourdon tube. When you blow in the end, you increase the pressure in the flat paper tube, causing it to straighten out.

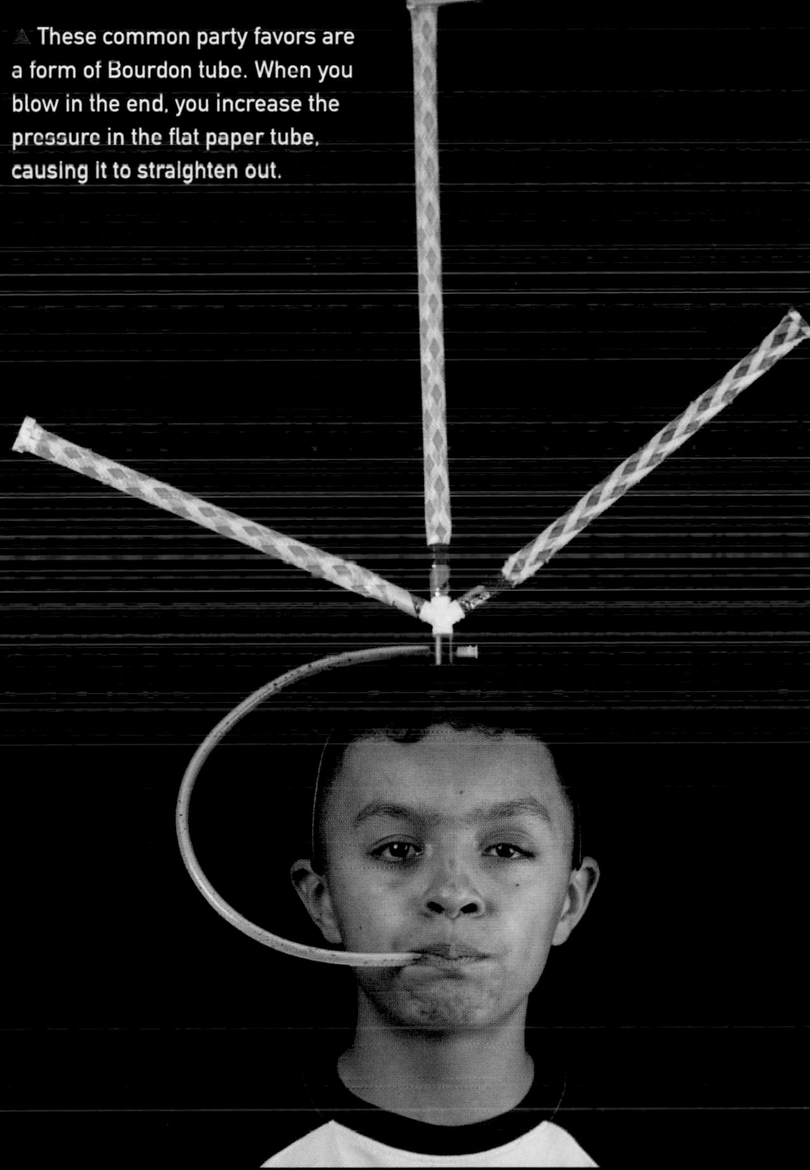

▲ If you put three paper Bourdon tubes on someone's head the result is very silly, totally worth it, and not good for weighing anything.

Weighing Without Weight

BALANCES AND SPRING scales work for weighing things because heavy objects are pulled toward the earth by gravity. Duh. That's what being heavy means, right? Well, yes and no. As you probably know, in space everything is weightless. Yet things are surely still "heavy" in some sense of the word.

Say you have a lump of iron metal that weighs 50 pounds here on earth. If you're buying iron by the pound, you'd have to pay for 50 pounds' worth in order to purchase this particular lump.

If you take the same 50-pound iron weight into space it will "weigh" nothing. If you try to put it on a scale, it will just float away. But it's still the same piece of iron. No one is going to give it to you for free just because it "weighs nothing." They are going to want to charge you the same amount as on earth, because it's still the same amount of iron.

If you're no longer paying for the downward force of the object, what are you paying for? In the theoretical marketplace of outer space, how do you determine how much this lump of iron should cost?

The name for the thing you'd be paying for is *mass*. Mass is an invariable property of a substance that doesn't change whether it's measured on earth, on the moon where it weighs one-sixth as much, or in space where it doesn't weigh anything. Mass, in other words, is not the same thing as weight.

Here on earth we almost always (except in physics class) ignore this issue and use the words *mass* and *weight* interchangeably. Gravity is close enough to the same anywhere on earth that you can use the downward force an object creates as a very good measurement of its mass. But when you're building rockets, or planning to live on another planet, it's something you need to worry about.

We have an intuitive sense of the difference between mass and weight. Mass is the thing that makes heavy objects difficult to get in motion—or difficult to stop if already in motion—even if they are supported in a way that makes their downward force (their weight) irrelevant. For example, a big boat floating in water is really hard to get moving, and once moving, is really hard to stop.

When the pilot of a big freighter screws up, there's nothing to be done but sit back and wait for the video to show up on YouTube. The boat is floating weightlessly on water, but it still has mass, and it's this mass that makes it hard to stop.

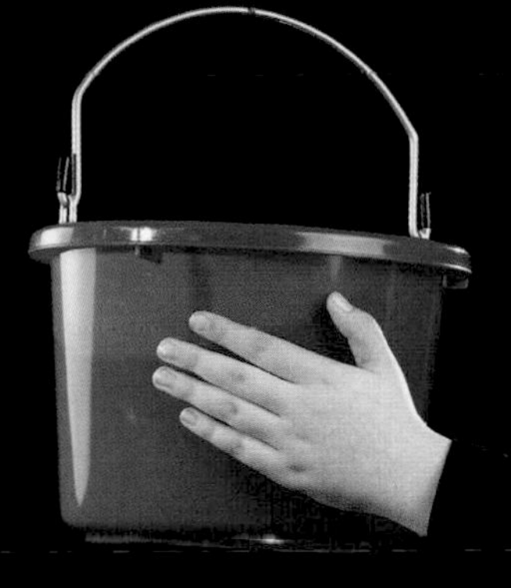

Imagine you have a hanging bucket that you're not to allowed lift up, or look inside of. Can you tell whether it's empty or filled with sand? Yes, of course! Just use your arms to swing it side to side. A light, empty bucket will move easily, and you can make it swing back and forth quickly. A bucket filled with sand has much more mass, and will be much more resistant to movement. Applying the same effort as you did to move the empty bucket, you'll only be able to move the high-mass bucket back and forth very slowly by comparison. This fact can be used as the basis of a weighing device that works even in zero gravity.

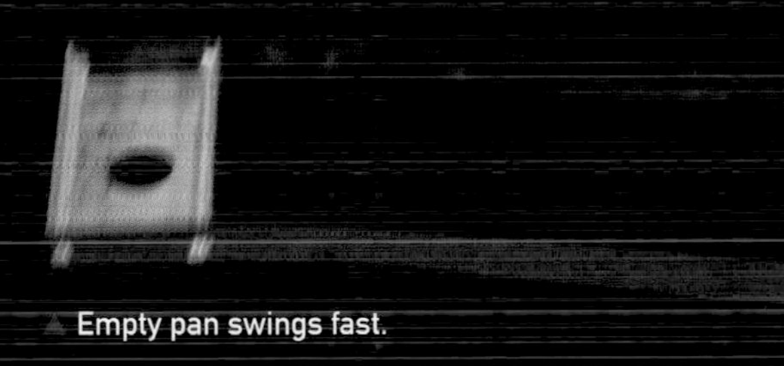

Empty pan swings fast.

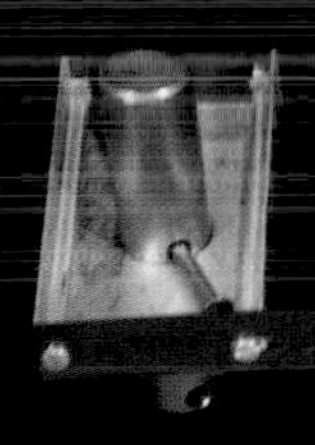

Full pan swings more slowly.

This "inertial balance" works equally well regardless of how much, if any, gravity is present. It has two spring-steel bands that like to stay straight and essentially act like your arms in the example above. If you set them swinging with no weight in the pan, they move back and forth quickly—several times per second. But if you put an object (i.e., a mass) in the pan, they swing more slowly. The heavier (more massive) the object, the slower it swings. If you count the number of swings per minute, you can calculate the mass of the object in the pan. This device is called an "inertial balance" because it uses the inertia (the unwillingness to be put into motion) of the object to measure its mass.

ON SPACE STATIONS where astronauts spend a long time away from gravity, it's important to keep track of their weight (er, mass) for health reasons. So these space stations have an inertial balance on board for weighing (er, measuring the mass of) the astronauts. This picture shows NASA Science Officer Bill McArthur using the inertial balance on the International Space Station. To use the balance, an astronaut grabs on and tries to make their body as rigid as possible while they are pushed and pulled back and forth by a big spring. A sensor measures the rate of their oscillation back and forth, and from this calculates quite an accurate weight. No, wait, I mean quite an accurate *mass*. In space you only have mass, not weight. On earth you have both.

There are very few examples of inertial balances used for practical purposes on earth. It's nearly always more convenient to use our handy gravitational field instead. This is an ad for an exception, and it has an interesting connection to another chapter in this book. It is a scale based on a clock.

Quartz crystal oscillators vibrate at an extremely stable frequency—so stable that at one point they were the most accurate clocks in the world. The only thing that will change the resonant frequency of a crystal oscillator is if you change its dimensions…or add weight to it. More weight slows it down, just like the examples we saw, only for smaller amounts of mass. How small? Because of how precise and regular the oscillations are, these devices can detect as little as one nanogram of added weight. That's one-billionth of a gram, or 0.000000001 g. This makes them useful for things like measuring the growing thickness of a microscopically thin layer of metal being deposited on a surface, or even tiny particles of smoke being blown at the sensor.

Weighing the Earth

WE'VE LEARNED ABOUT weighing things on earth, but what about weighing the earth itself? Here we need to be very careful about the difference between weight and mass.

How much does the earth weigh? The cheap answer is to say that the earth weighs nothing. It's floating in space, as weightless as an astronaut floating in a space station. But obviously that's not the answer we're looking for. We should instead be asking, what is the *mass* of the earth? Or, to put it another way, how much would the earth weigh if it were sitting on the earth?

▲ If you put the earth on a scale resting on a second copy of the earth, the scale would read the weight of the earth. But that's not very practical, right? To find a better solution, we need to think more deeply about how gravity works.

On the surface of the earth, what we experience as weight—the downward force—comes from the gravitational attraction created by the mass of the earth. But all objects create a gravitational attraction, not just the earth. The moon, Mars, a large mountain, even just a block of iron, all create their own gravitational attraction. The bigger the object, the more gravity it creates. We don't normally think of anything smaller than a planet as creating gravity, but that's only because the amount of gravity produced even by a large mountain is incredibly small.

▲ Consider an iron block sitting on the earth. We normally think the earth is pulling on the iron block, but in fact the iron block is also pulling on the earth with its own bit of gravity. To be precise, the earth and the block are pulling on each other with a force proportional to their masses multiplied together.

If we place the iron block on a scale we expect the scale to tell us the weight of the iron block in the gravitational field of the earth. But the reverse is also true. The scale is also reading out the weight of the earth in the gravitational field of the iron block. We are quite literally weighing the earth every time we weigh anything sitting on the earth. We're just not finding the weight of the earth *on earth*, we're finding the weight of the earth "on" some other, much smaller, object.

We know how to translate a weight measurement (the downward force) on earth into a mass (the amount of a substance), because we've calibrated our scales to the strength of gravity on earth. If we put the same object and the same scale on the moon, the mass of the object is still the same, but the downward force is less (about one-sixth as much) because gravity is weaker on the moon. If we want to translate weight measurements on the moon into mass measurements, we have to calibrate our scale to the strength of gravity on the moon.

How can we measure the strength of gravity on the moon? By taking a scale and our iron block to the moon and weighing it. Since we know the mass of the iron block, its weight (downward force) on the moon gives us the necessary calibration factor to translate moon weights into masses.

Having measured the strength of gravity on the moon, we could determine the mass of the earth by putting it on the moon instead of on a second copy of the earth. While this might be a *tiny bit* more practical, it's still just not going to work.

But suppose we take the iron block and instead of putting it on the moon, we put it *on another iron block* and see how much it weighs under the influence of the gravitational field of that block? That would let us calculate the strength of the gravitational field created by the iron block, giving us the calibration factor we need to turn weight measurements on the iron block into mass measurements. Then all we need to do is put the earth on the iron block and weigh it.

But putting the earth on the iron block is *the same thing as putting the iron block on the earth!* Turn this picture upside down, and you have exactly the same picture we had earlier of the block on the earth. In other words, as I said before, the weight of the iron block on earth is the same thing as the weight of the earth on the iron block.

The only hard part here is measuring the gravitational attraction of two iron blocks to each other (represented by the picture of the iron block on a scale on another iron block). Have you ever felt the gravitational attraction of two iron blocks toward each other? No, neither have I, because it is an almost infinitesimally tiny amount. The force is so tiny that it seems almost crazy to even try to measure it, but this is exactly what Henry Cavendish did in 1798, with an error of only about 1 percent.

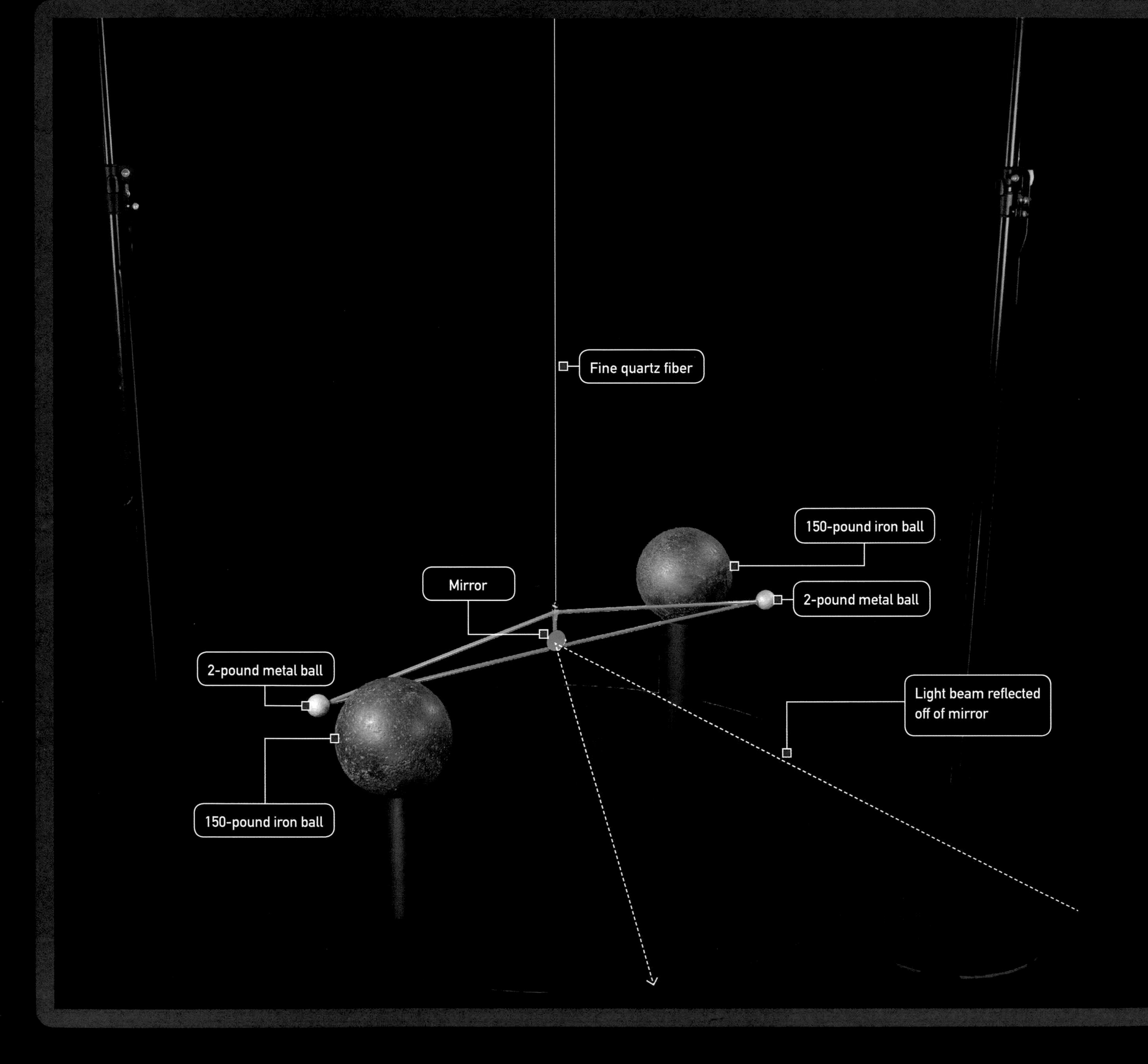
Fine quartz fiber
150-pound iron ball
Mirror
2-pound metal ball
2-pound metal ball
Light beam reflected off of mirror
150-pound iron ball

THE CAVENDISH EXPERIMENT

HENRY CAVENDISH, ONE of the most important scientists of the 1700s, was so shy he almost never talked to people. Which is fine, because instead he used his time to weigh the earth properly. He used an excruciatingly sensitive quartz fiber torsion balance to directly measure the force of gravity between two pairs of heavy balls. That is to say, he measured the weight of a small ball, not in the gravity of the earth, but in the gravity of a larger ball. Then, having measured the strength of the gravity created by the two balls, he could use the weight of the large ball on the earth (which is the same as the weight of the earth on the ball) to calculate the total mass of the earth. In what is without a doubt one of the single most remarkable achievements in precision measurement of a difficult-to-measure thing, he got the answer right to within about 1 percent of the best modern value: 5,972,200,000,000,000,000,000,000 kilograms (6,583,000,000,000,000,000,000 tons).

Everything about this experiment is ridiculously sensitive, yet it has to be carried out with the heaviest balls you can handle. For this picture I was able to cheat and use strong nylon fishing line to hold up the two small balls. Cavendish had to do it with a delicate quartz fiber. I don't know if history records how many fibers he broke trying to get one mounted and hanging properly, but I'm going to guess at least twenty.

It's fitting to end our discussion of scales with this astonishing device. A scale so delicate it had to be kept in a sealed room, observed only from a distance with a telescope—lest the gravitational attraction of the observer's body disturb the result. Cavendish built an instrument so perfect in conception, and so delicate in form, that it had to be kept as isolated from people as he kept himself.

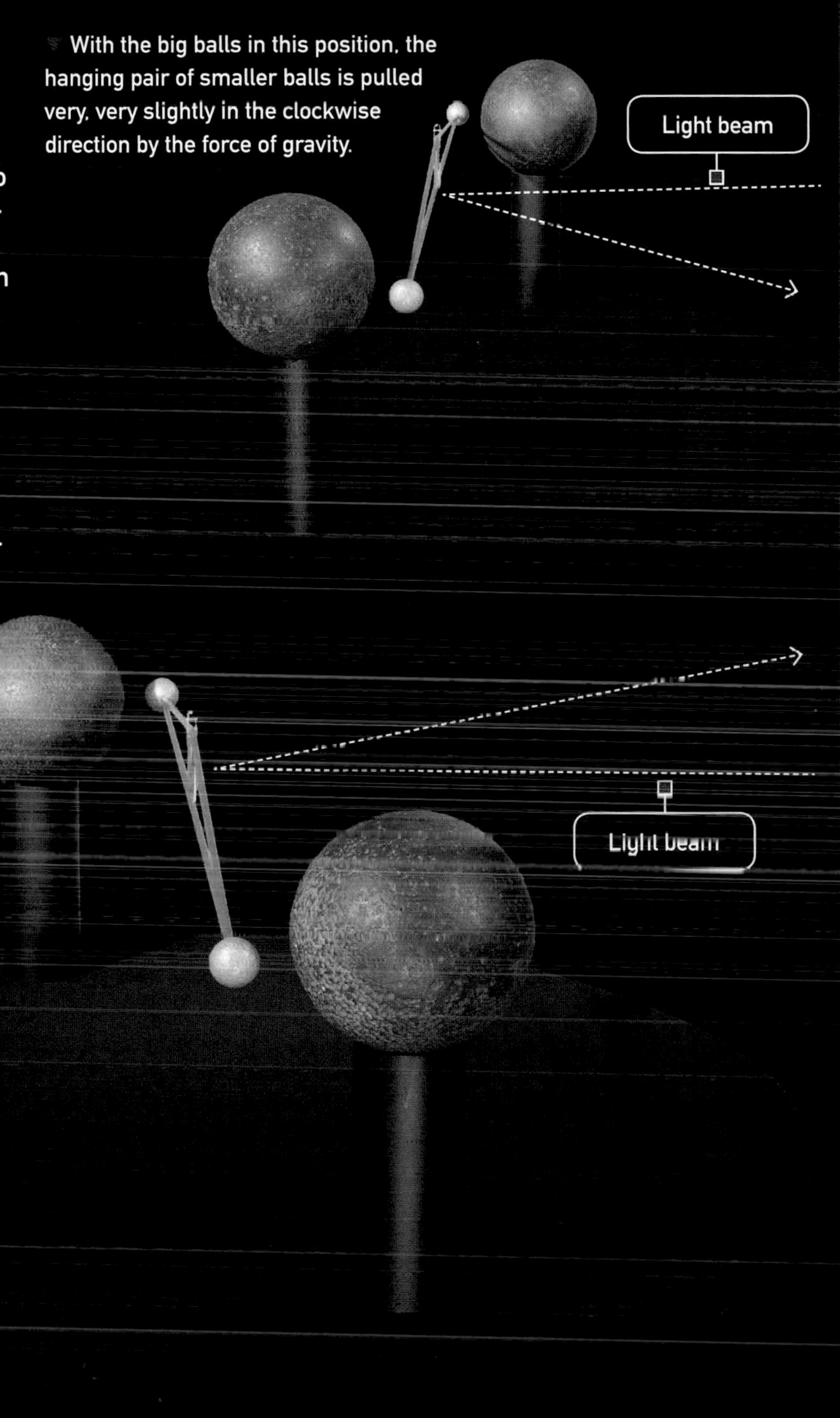

With the big balls in this position, the hanging pair of smaller balls is pulled very, very slightly in the clockwise direction by the force of gravity.

With the big balls switched to the other side, the small balls are now pulled very slightly counterclockwise. A beam of light bounced off the mirror mounted in the center of the hanging frame will move a little bit along a scale mounted to the wall far away (greatly magnifying the small movement of the frame).

Bearing a Burden

WHEN I WAS little, maybe ten or twelve, we had a pair of marvelous ball bearings in our basement workshop. *So satisfying*. Heavy and free-spinning, like bearings should be. Shiny and well-oiled. At first. Over time they started to rust, as my constant handling wore off the oil. I knew it was happening, but I did nothing. They started to stick. I grew up and moved away.

I never forgot those bearings. Not every day—but some days—I would think of them and feel sad that they were rusted, and ashamed that I was to blame.

I took apart a lot of things when I was young, and very few of them made it back together again in working condition. Those that did were usually not improved by my attentions, even if the goal had been to fix them. Sure, I had some successes, but those are not what I remember. Mostly I remember a feeling of defeat, of not being worthy of these fine things or the tools I mangled them with.

Later, over the course of many years, more and more of the things I worked on started ending up in better shape than I found them. It turns out I hadn't been an incompetent failure, I had just been a kid. Failure is how you learn. It defines you, not as a failure, but as experienced. Failure isn't who you are, it's how you become.

Old ruined bearing

FIX IT!

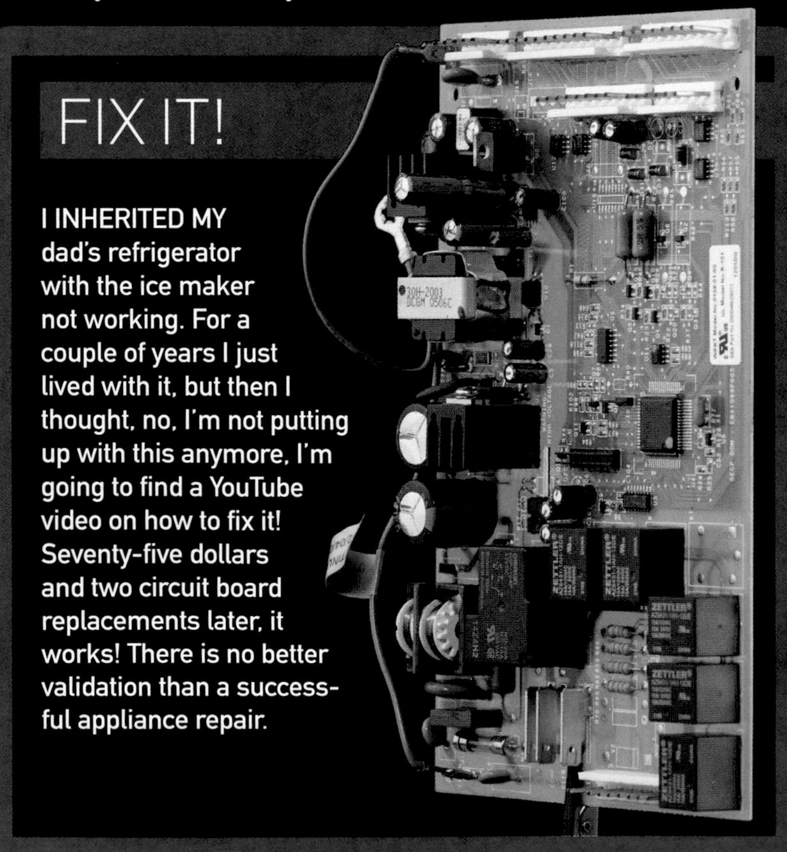

I INHERITED MY dad's refrigerator with the ice maker not working. For a couple of years I just lived with it, but then I thought, no, I'm not putting up with this anymore, I'm going to find a YouTube video on how to fix it! Seventy-five dollars and two circuit board replacements later, it works! There is no better validation than a successful appliance repair.

When my father grew too old to live at home, I moved back to the house I had grown up in—the house with the bearings in the basement. I was happy, and more than a little surprised, to find them again, just where I had left them. And a lot more rusted. One barely turned, the other not at all. For several more years that is where they stayed, until the writing of this book reminded me that I had new agency in the world. I was grown up now. I knew where to buy rust remover.

The locked-up one was so bad I decided to mill out the rivets in the spacer and completely disassemble it so all the parts could be given a thorough soaking in rust remover. I left them overnight, having had good success with that method on previous projects. In the morning I would reassemble the bearing with tiny #1 machine screws in place of the rivets.

I came down in the morning to find the solution bubbling. Please no. The sense of horror, loss, and disgust with myself was palpable. I had used the wrong kind of rust remover! Not the kind for making metal

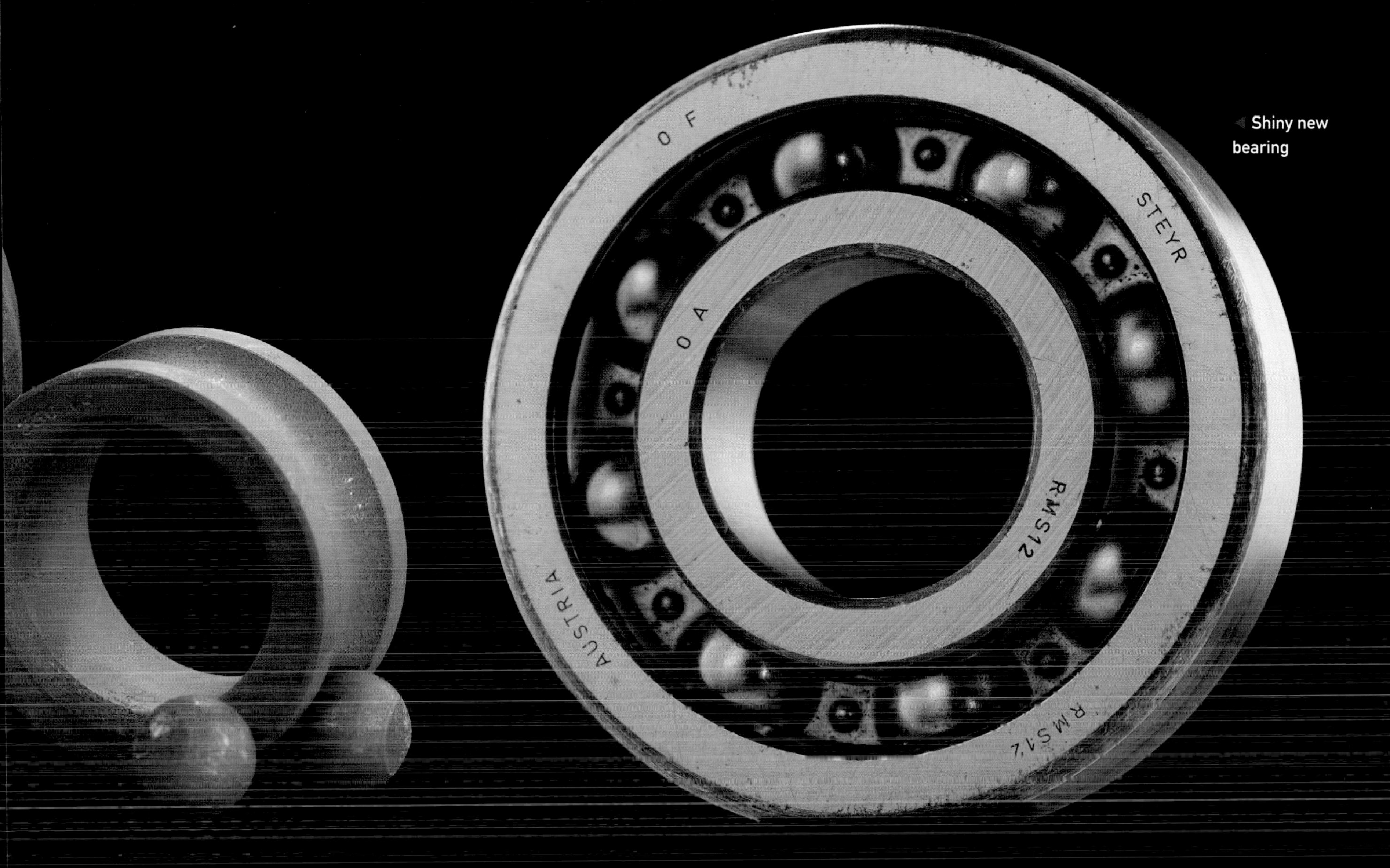

Shiny new bearing

shiny again, but the kind meant to etch the surface for painting. How could I have been so stupid! This bearing had waited half a lifetime for me to come home to it. For me to return and give it the love it deserved, and the oil it needed so that it could spin once again with a shiny humming sound. And I killed it.

How do you recover from something like that? Sure, I could buy a new bearing, just as good, but that's *not the same*. I left it on the bench in pieces. I could do nothing for it. It rests now on the shelf of remembrance, with the ashes of my family beside it.

Two weeks later, while engaged in a Craigslist transaction to acquire the Toledo scale, I found myself surrounded by the most marvelous collection of machine tool parts I have seen in a long time. You know, drill bits the size of your arm, straight reamers the size of your leg, lathe chucks to die for, and a jolly Russian eBay dealer at home in his chaos. We bonded immediately. I asked him, with great anticipation, if there were any large bearings to be found in his fine establishment. As is often the case if you ask for something "large" in this sort of place, I got more than I asked for. The *smallest* bearings he had for sale would nearly fit over my head, and they cost a fortune. Tempting, but too much, and in any case, *not the same*.

And then I saw it. In a box on the floor, mixed in with who knows what else, was the bearing that needed me as much as I needed it. A bit bigger, and a bit shinier, than the one I had twice ruined. Alone and forgotten. The Russian said I should just take it. He knew, somehow, that I needed it, and that this was not a commercial transaction. It is my bearing now, and I will not let it rust.

Ruining that bearing from my childhood was a sin that could be absolved only by an act of grace. It was fate that led me to the Russian, and it was through his grace that I have a new bearing to care for. Taking it home felt good. It felt like…redemption.

THE MAKING OF CLOTH

THREAD AND THE tools used to make cloth are as ancient as any human contrivance. Every culture, every continent, every people, have used spun and woven fiber for so long that we can only guess at the origins of the idea. Our very language is woven through and through with threads of metaphor—or should I say metaphors of threads. From the storytellers spinning their yarns, to the tales that knit together our shared experiences, allusions to thread form the warp and weft of our way of speaking, and our way of life. We are clothed in threads, we sleep on fabric, we walk on spun, woven, or knotted rugs, and sometimes our very existence hangs by a thread.

Even the word "text," the thing you're reading, has the same root as the word "textile," meaning cloth. As Robert Bringhurst writes, "An ancient metaphor: thought is a thread, and the raconteur is a spinner of yarns—but the true storyteller, the poet, is a weaver. The scribes made this old and audible abstraction into a new and visible fact. After long practice, their work took on such an even, flexible texture that they called the written page a textus, which means cloth."

ELDREDGE
TWO SPOOL

BE IT EVER so humble, this is the goal. In this chapter we are going to make a potholder. No cheating. We will begin at the very beginning.

When you talk about making something, there's always the question of what you're starting with. When people say they made a tie-dyed T-shirt, they usually mean that they colored an already-existing T-shirt. And frankly, the hard part of that process is making the T-shirt, not soaking it in some dye (which they also didn't make). Nothing wrong with "making" tie-dye T-shirts as a fun activity, but I wanted to go a little deeper. I wanted to make something out of cloth, starting at the very, very beginning, with the absolute bare minimum: a handful of cotton seeds.

Potholder the hard way

Fiber

FUNDAMENTAL TO THE whole enterprise of cloth is fiber: individual strands of something—anything—flexible. It could be animal hair, plant stems, insect cocoons, rocks, or something synthetic from a factory—if it's long and thin, reasonably flexible, and reasonably strong, someone has tried weaving it. Yes, rocks. Keep reading.

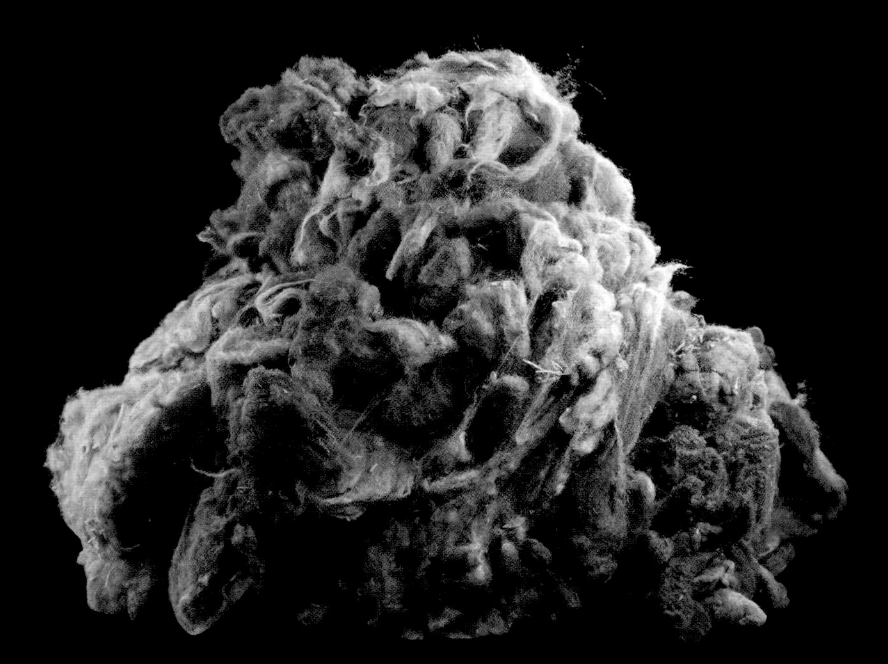

Wool, which comes from sheep, was for perhaps eight millennia the fiber of choice for clothing. It's a great fiber, warm, easy to spin, and durable. With plenty of sheep and fewer humans, it was an endless, renewable source of fiber. But as the human population expanded faster than the sheep population, wool became more and more scarce and, therefore, expensive. Wool can be made from similar fiber from other animals—angora rabbits, alpacas, and so on—but no matter the source, when there's a limit to the supply of animal fiber there will be precious few clothes to go around.

You can weave metal into cloth, but would you really want to wear a wire shirt? I mean, people do, but not so much since knights in shining armor went out of fashion as a career choice. Let's just say it's not a viable alternative for everyday wear.

Fiber made by insects (silkworms or, very rarely, spiders) is super-soft—a real luxury. But it's in even shorter supply than wool, difficult to clean, and if you ask me, a bit clammy. Again, insect fiber is just not practical for clothing the masses.

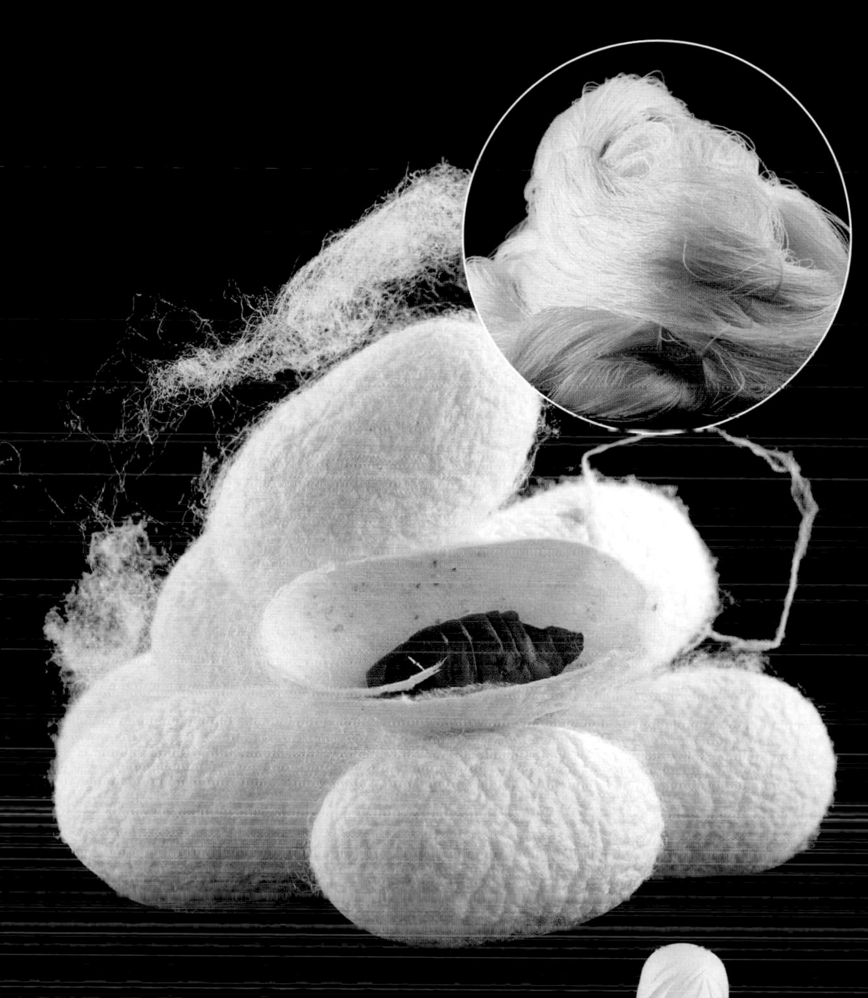

This is a heat-resistant furnace glove made from rocks, specifically from the fibrous and infamous rock called asbestos. It's more comfortable than a wire glove, but not particularly durable, and as the fibers break down they float into the air, get into your lungs, and quite reliably give you lung cancer. We don't use this kind of fiber anymore, for much of anything. Like wire, asbestos is also not a good alternative for clothes.

Now we're getting somewhere. This synthetic blanket, made in a factory from oil-derived chemicals, is as soft as silk (literally) and as cheap as rocks (almost literally). Man-made fibers produce the cheapest thread practical for making clothes. Huge amounts of synthetic fiber clothing are made and sold worldwide every year. Polyester/polyamide blended microfiber, as used in this blanket, really is wonderful stuff for the price. And, yet, synthetic fibers have a justifiably poor reputation when used for most kinds of clothing.

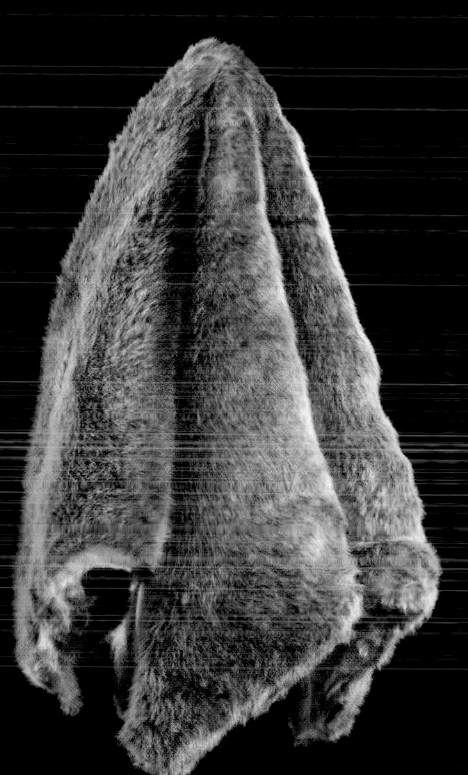

Behold the infamous polyester pantsuit, international symbol of tragic 1970s fashion, harbinger of poor taste to come. Simply put, polyester, in common with all synthetic fibers, *looks cheap*. It also tends not to breathe, which means sweat and odor build up on the person wearing the suit and even in the fibers of the suit itself. Modern synthetics have managed to overcome some of the limitations of breathability and cheap appearance. Synthetics are also great for certain kinds of performance clothing (waterproof clothes or quick-dry athletic wear, for example, are best made of synthetics).

But for everyday wear next to your skin, there is no synthetic as comfortable as the one great hero fiber we've yet to mention: cotton.

Planting

COTTON IS THE world champion of fiber, hands down, no competition. Cotton is ***the*** fiber of choice for clothes. It's soft, warm, breathable, comfortable, never itchy, durable, and strong. And thanks to modern farming and industrial practices, it's incredibly cheap to produce and universally available. Cotton, more than any other fiber, is what clothes the world. So cotton is what we're going to talk about in the rest of this chapter.

The first step in making anything out of cotton is to grow some of it. Cotton comes from cotton plants, which come from cotton seeds, which come from cotton plants. It's been going around this way for quite some time. In order to grow cotton, I had to break in and start somewhere in this cycle. Clearly seeds are the most basic form of cotton, so that's where I started: with a bag of cotton seeds purchased from a seed dealer.

▲ Cotton seeds are fuzzy when they come out of the cotton gin (see page 200), but for commercial-scale planting the seeds are de-fuzzed, then coated with an insecticide and graphite powder to help them run smoothly through industrial planters.

I planted about one-fortieth of an acre (1,000 square feet) of cotton at my compound in Central Illinois (plus a few backup pots in town in case I screwed up the field). I'm not a farmer and I have a history of failing to grow plants, so I worked hard to do everything by the book this time.

Planting is, in principle, a simple thing: put cotton seeds about 1 inch (2.5 cm) underground every 6 inches (15 cm), in rows 24 inches (60 cm) apart. My helpers and I planted by hand about two thousand seeds this way in two hours. Sounds OK, right? But that was to plant one-fortieth of an acre. Planting even one whole acre would have taken us *ten days* of eight-hour workdays. There are around seventy *million* acres of cotton grown each year. To plant that much cotton in a one-month planting window with a workforce as pathetic as us would require about forty million people to drop everything and plant cotton for a month. Needless to say, this isn't how it's done.

INDUSTRIAL SEED PLANTER

A MODERN INDUSTRIAL-SCALE seed planter (seen here folded up for the winter) can plant a thousand acres in a day, driven by one person. This machine plants twelve rows at a time and works twenty thousand times faster than we did! Even the smaller two-row mechanical planters used on less advanced farms are a hundred times or more faster than doing it by hand.

▲ Each row of the machine has its own self-contained "row unit," connected to a common drive shaft (and in this fancy version also to a vacuum hose that helps reliably select seeds for planting).

Cotton seeds fill the hopper.

A spray nozzle supplies a bit of fertilizer to get the seeds off to a good start.

Hidden under the seed hopper is the part of the planter that is hardest to get right: the seed selecting mechanism.

Angled, toothed wheels press the soil back together, closing the furrow after the seeds have been placed.

A pair of disks hidden between those two wheels cuts a shallow groove (furrow) for the seeds to fall into.

A final wheel presses the soil gently down over the seeds.

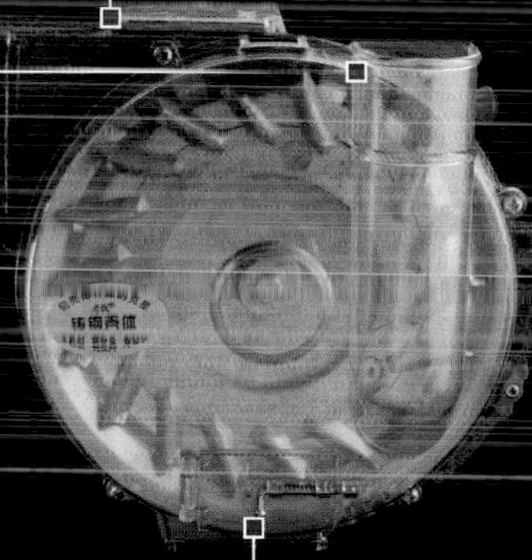

Seeds from a hopper enter here.

This is the heart of the planter. Seeds flow from the hopper down to the side of the disk pictured here, which rotates counter-clockwise. The disk picks up seeds one at a time in its slanted teeth, which each have a little seed-sized cup on the end. At the top, a small brush nudges one seed at a time to fall down behind the disk and from there out the bottom of the machine and into the dirt.

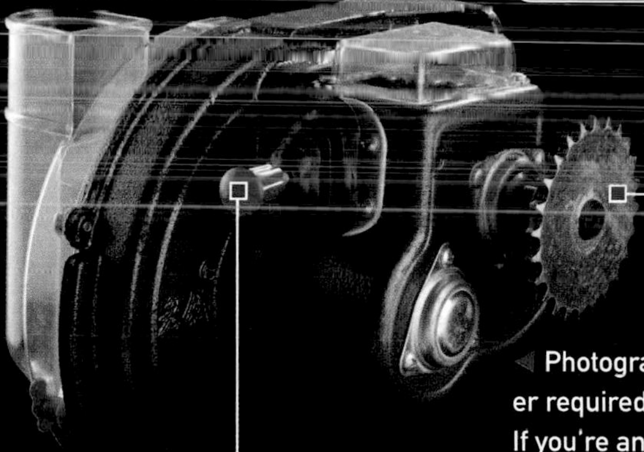

This sprocket gear is connected by a chain (like a heavy-duty bicycle chain) to the drive shaft running the planter. Turning the drive shaft turns the seed disk inside the machine.

Seeds fall out from here, through a tube, and into the slot cut in the ground.

This lever is like the gear shifter in a manual transmission car. It changes the speed of the disk relative to the drive gear. That lets you change the spacing of the seeds, or adapt to different drive shaft speeds. This same mechanism can be used to plant many different kinds of seeds, which may require very different spacings.

Photographing the guts of the planter required getting one into our studio. If you're an experienced farmer you'll notice that, while similar, this is not the exact mechanism used in the planter shown. If you're an experienced farmer you'll also know why I didn't want to pay for the actual John Deere part. (I picked this one up at a farm supply store in rural China for one-sixth the list price of the pretty green version.)

Growing

AFTER PLANTING THERE is a period of five to six months of waiting while the plants grow, interspersed with worrying and occasionally watering and/or spraying chemicals.

One thing I've noticed listening to my local radio station, which does agricultural reports several times an hour, is that farmers are never happy with the weather. Now I know why. For the whole summer that I was a micro-farmer, I was unhappy with the weather every single day. It's either too cold or too dry or too wet, and if it's perfect then we focus on the forecast, so we can be unhappy about the future instead. Real farmers can even be unhappy about the weather in entirely different countries! For example, around here if the weather is particularly good in Brazil the local farmers are unhappy because it means there will be a good harvest down there, meaning lower crop prices on the world market, meaning lower profits for them.

Exciting! My potholder emerged from the dirt about a week after planting.

Cotton has been grown by so many people in so many places around the world that every step of the process has been examined in excruciating detail. The best possible way to grow it has been worked out for each region. The plant has been engineered to maximize production, and chemicals are used to fine-tune its growth each season. There is even a plant hormone that can be sprayed on at a particular stage to convince the plant to stop growing taller and instead start putting effort into growing more and larger cotton bolls. Another chemical, applied at the end of the growing season, gets all the bolls to stop growing and split open before the first frost. Yet another makes the leaves drop off so the cotton can be mechanically harvested more efficiently.

As a dedicated micro-farmer I tried to follow all these best practices in miniature, and it paid off.

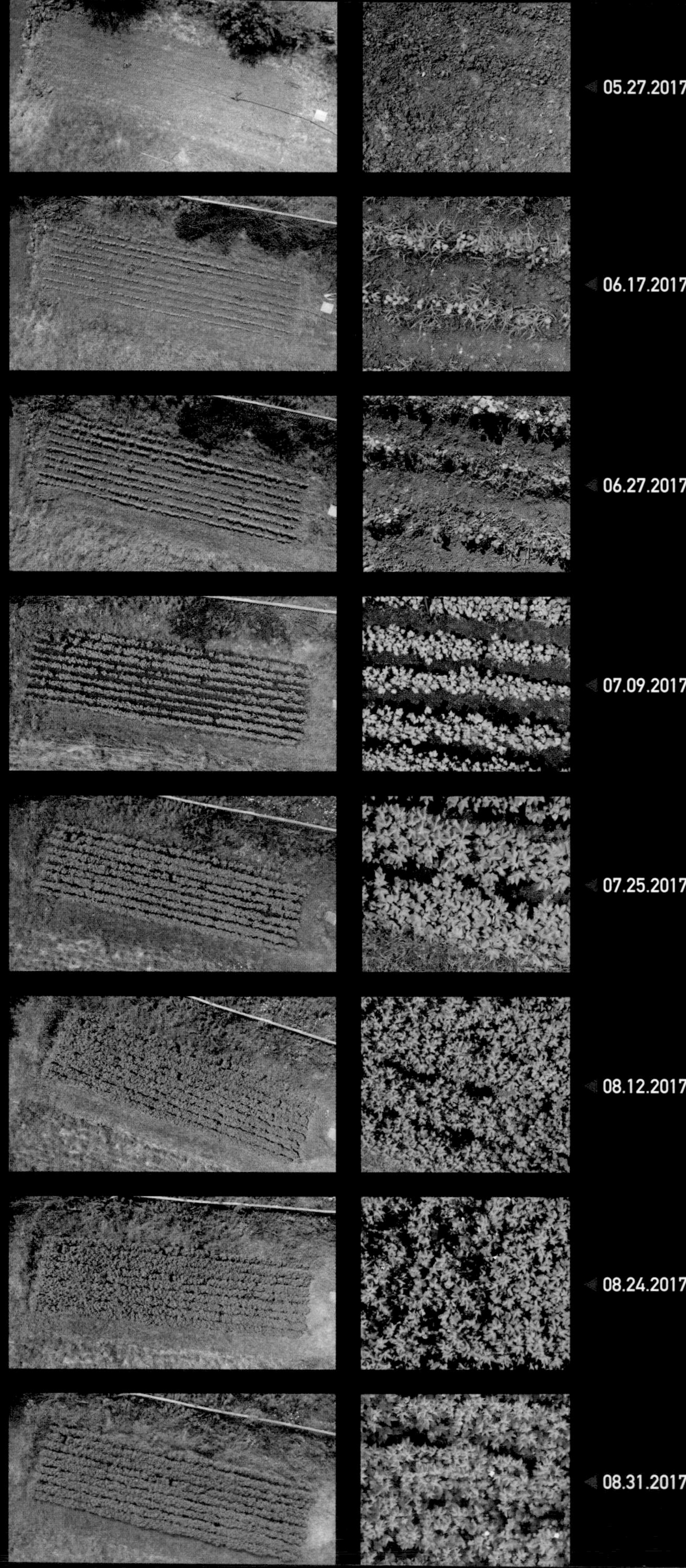

Cotton takes about 160 to 170 days to reach maturity, and I'm at the very far northern end of where it's possible to grow cotton in the United States. Cotton fields are associated in the popular imagination with the warm South because that's where it grows best. Fortunately, even in Central Illinois, with reasonably favorable weather (though I was still unhappy about it), and with the help of a few well-timed chemical applications, my cotton field grew and matured in time for a decent harvest before the first frost.

▲ My field was tiny, but cotton is grown on a vast scale. Worldwide, something like 100,000 square miles (260,000 square kilometers) are planted in cotton every year, most of it in China and India, with the United States coming in a close third by total production volume. That's half the size of France just in cotton fields! This particular field is outside the village of 南塘疃村 (Nan Tang Tuan Cun), about two hours by train south of Beijing, China. Many families in the village grow, pick, and profit from their own small section of the field.

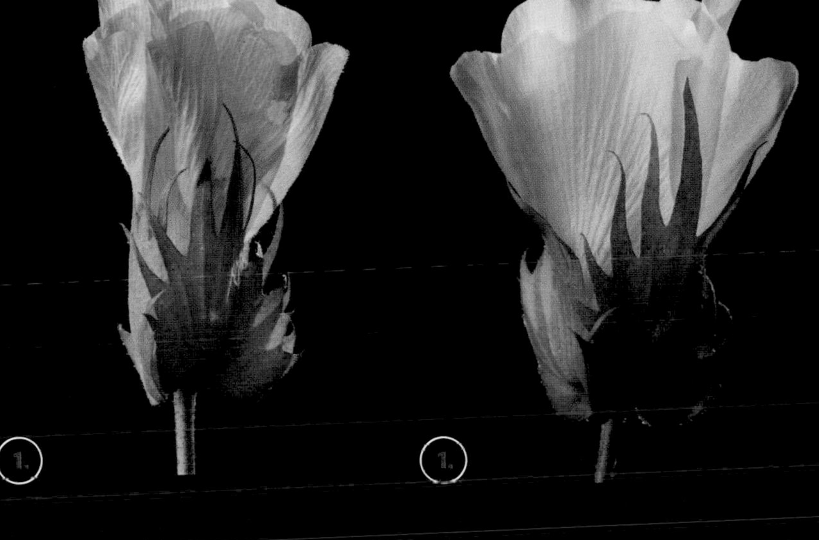

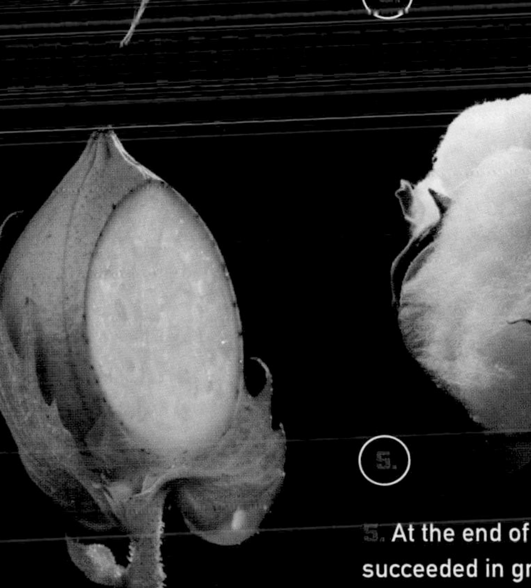

1. After the plants have been growing for a couple of months they start to produce these delicate pink and white flowers, the first sign that your cotton crop might be a success!

2. After a while the flowers dry up and fall off, leaving behind a small bulb called a boll.

3. The bolls begin to grow, eventually reaching about the size of a golf ball (1.5 inches or 3–4 cm in diameter).

4. Midway through the growing season the bolls are filled with a sort of jelly that is slowly transforming itself into a compact mass of seeds and fibers.

▲ Cotton plants are not soft-stemmed annuals like corn, wheat, or soybeans. If you let cotton plants grow for several years, they will turn into tall bushes with hard woody stems, which you can see starting to form here on these six-month-old cotton plants. But, even in climates where the winter won't kill the plant, people cut them down every year because it turns out you get a better yield of cotton by starting fresh each year.

5. At the end of the season comes the glorious moment when you know you've succeeded in growing real cotton! The bolls dry up and split open while still on the plant. The jelly has resolved itself into beautiful puffy soft cotton fibers ready to be made into cloth. Well, not really, while lovely and soft, these bolls are still a long

Picking

PICKING COTTON BY hand is pretty much the definitive example of backbreaking farm labor that machines have relieved us of. But with only one-fortieth of an acre to harvest, doing it by hand was the only sensible alternative for me. My back was angry with me for weeks.

▲ Worker picking cotton in China

In most parts of the world, including many parts of China, cotton is picked by machine. But there are still places where it's done the old way, by hand, one boll at a time. By harvest time the plants are so tall that this woman almost disappears among them! When cotton is picked by hand, it's done over the course of a month or more, as the individual bolls mature and split open according to their natural timetable. The plants remain green and healthy while pickers walk through them multiple times over the course of the picking season. Once the picking has concluded, they are all cut down in preparation for next year's planting. They would not survive the winter anyway.

▲ Cotton ready for harvest in a cotton field in Caddo Parish, Louisiana

The situation is very different when cotton is going to be harvested by machine. In that case, chemicals (plant growth regulators sprayed from high-clearance sprayers and crop-dusting airplanes) are used to force the plants to bring all their bolls to maturity simultaneously, and then drop all their leaves. The result is a sea of sticks and white puffs with little or no green in sight.

▼ Aerial view of the same Louisiana cotton field

From the air it almost looks like snow, but when this picture was taken at Logan Farms in the deep south of Caddo Parish, Louisiana, it was so hot that my drone batteries almost failed from overheating. That's not snow, it's cotton.

The other difference between this field and the one in China is that the whole thing, and several others like it, are owned and farmed by one family, with a crew of no more than a dozen part-time seasonal employees. In all, they plant and pick about the same area of cotton every year as the entire village of Nan Tang Tuan Cun.

Mechanical picker at work in the Louisiana field

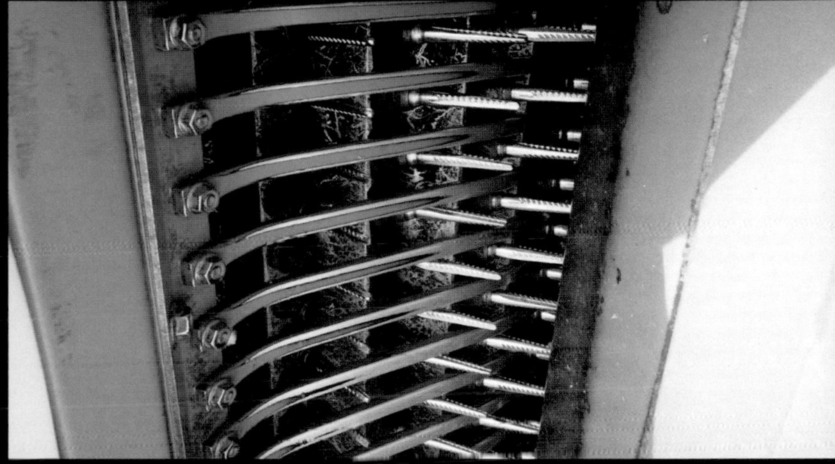

Inside the picking head

It may take a village to raise a child, or pick a field of cotton, but with this machine you don't need one. A modern cotton picker (by which is meant a machine, not a person) can pick about 100 acres (40 hectares) a day, thousands of times more than can be done by hand. Remarkably, mechanical cotton pickers did not become commercially successful until the 1950s. Given that the other labor-intensive phases of cotton production had been mechanized far earlier, we can only speculate how the history of the American South, including the history of slavery, might have been different had a mechanical cotton picker been developed around the same time as the cotton gin (see page 198)

Inside the picking head we see the solution to the hard part of mechanical cotton picking: how to get the bolls off the stems quickly and reliably. Each of the horizontal spikes spins rapidly, and the frames holding them swing around from left to right, while simultaneously rotating around a vertical axis, dragging the bolls into the machine on the right side. It's a complicated dance that points to what must be hundreds and hundreds of gears inside. You can see from the picture on the left just how big this picking head is: the people are basically standing inside its arms. You can also see a tank of water in the picture. That's because all the spikes have to be kept wet, so the cotton fibers will stick to them as they are spinning. The picker keeps a supply of water on board, which must be refilled periodically. Using water to make the spikes sticky was one of many key ideas that made mechanical cotton picking possible.

This is a seriously big machine.

While it's working, the picker is automatically rolling the cotton up into a giant cylinder in a bin at the back. When a roll is finished (every few minutes), it is automatically shrink-wrapped in plastic and birthed out the back end of the machine.

The most recently finished roll is carried on the back of the picker while it keeps working, starting to form a new roll in the now-closed bin. When the machine reaches the edge of the field, the finished roll is dropped where it can be conveniently picked up later. The machine never stops moving.

Finished rolls are lined up in groups of four, because the trucks that transport them to the next phase of processing can hold four at a time. The trucks have a belt-driven scooping ramp; they just back up to the row of four rolls and suck them all up in one gulp.

▼ A pile of cotton bolls fresh off the plant. That's *bolls*, which contain seeds, not *balls*, which are pure cotton fiber, no seeds.

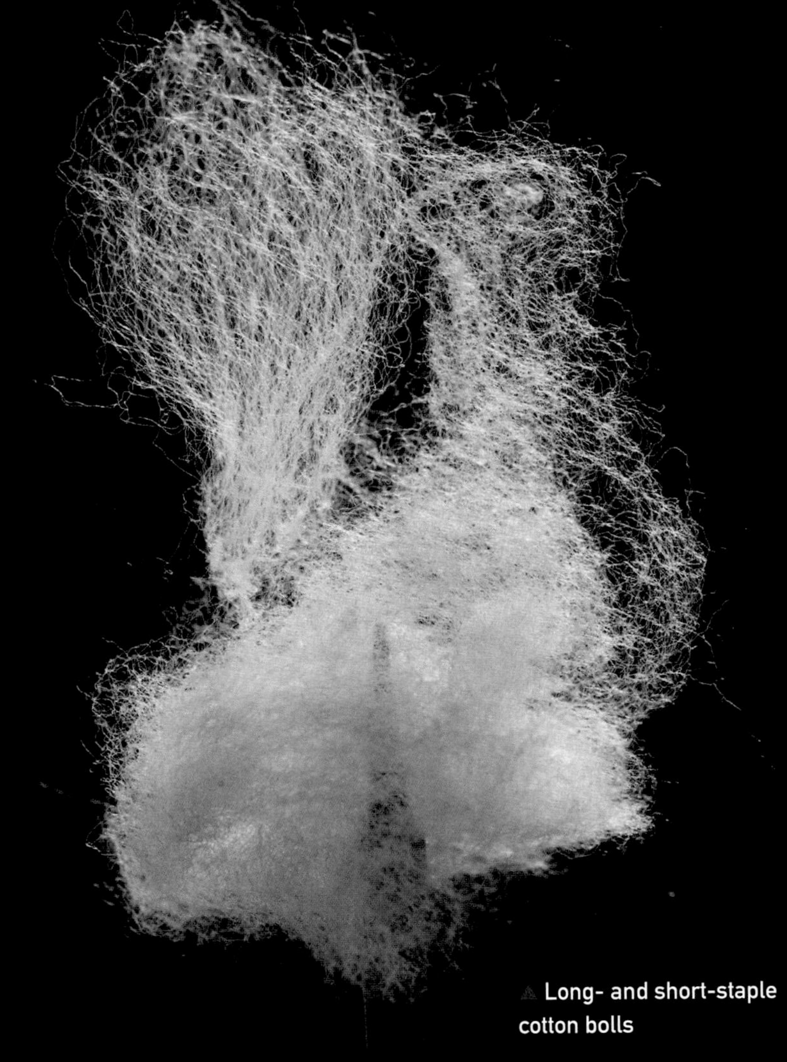

▲ Long- and short-staple cotton bolls

After picking and separating the husks, what you get is a pile of cotton bolls (or a compacted roll of them if you're using a mechanical picker). That's "bolls," not the more familiar "balls" like you get at the drugstore. Cotton balls are pure cotton fiber, while bolls also contain seeds. And therein lies the main problem with cotton.

We care a lot about the fiber, because it's the part of the plant that we turn into fabric, but what about the plant? For the plant, it's the seeds that matter for its own survival and reproduction, not the fibers. So, why does the plant even bother to make the fibers? There is no definitively known answer to this question. The best theory is that the fibers absorb water and provide a consistent, protected environment for the seed to germinate. They also make the seeds harder for animals to eat—have you ever tried eating a cotton ball?

Because the purpose of the fibers (from the point of view of the plant) is to protect the seeds, the fibers are, not surprisingly, *very firmly* attached to the seeds. Separating them is such a hideous chore that an entire phase of American history turned on the invention of the machine that finally made this job easier.

There are several varieties of cotton grown in the world, which can broadly be divided into long-staple and short-staple varieties. (In this context "staple" means fiber, so long-staple cotton has long fibers and short-staple cotton has shorter fibers.) All cotton plants grow their fibers at about the same rate; the difference between long- and short-staple cotton is in how long the bolls grow before opening up and drying out. Long-staple cotton can only be grown in parts of the world with a very long growing season (i.e., near the equator).

Egyptian cotton, sea-island cotton, and Pima cotton have long fibers and produce the highest-quality fabric, but are more expensive because there isn't that much tropical land available to grow these varieties. The long fibers are relatively easy to pull off the seeds.

The far more common short-staple variety of cotton, also known as upland cotton, can be grown nearly anywhere in the southern half of the United States and in many other parts of the world. It constitutes 95 percent of the American cotton harvest and 90 percent of all the world's cotton. Unfortunately it's much harder to grab onto and pull these shorter fibers off their seeds.

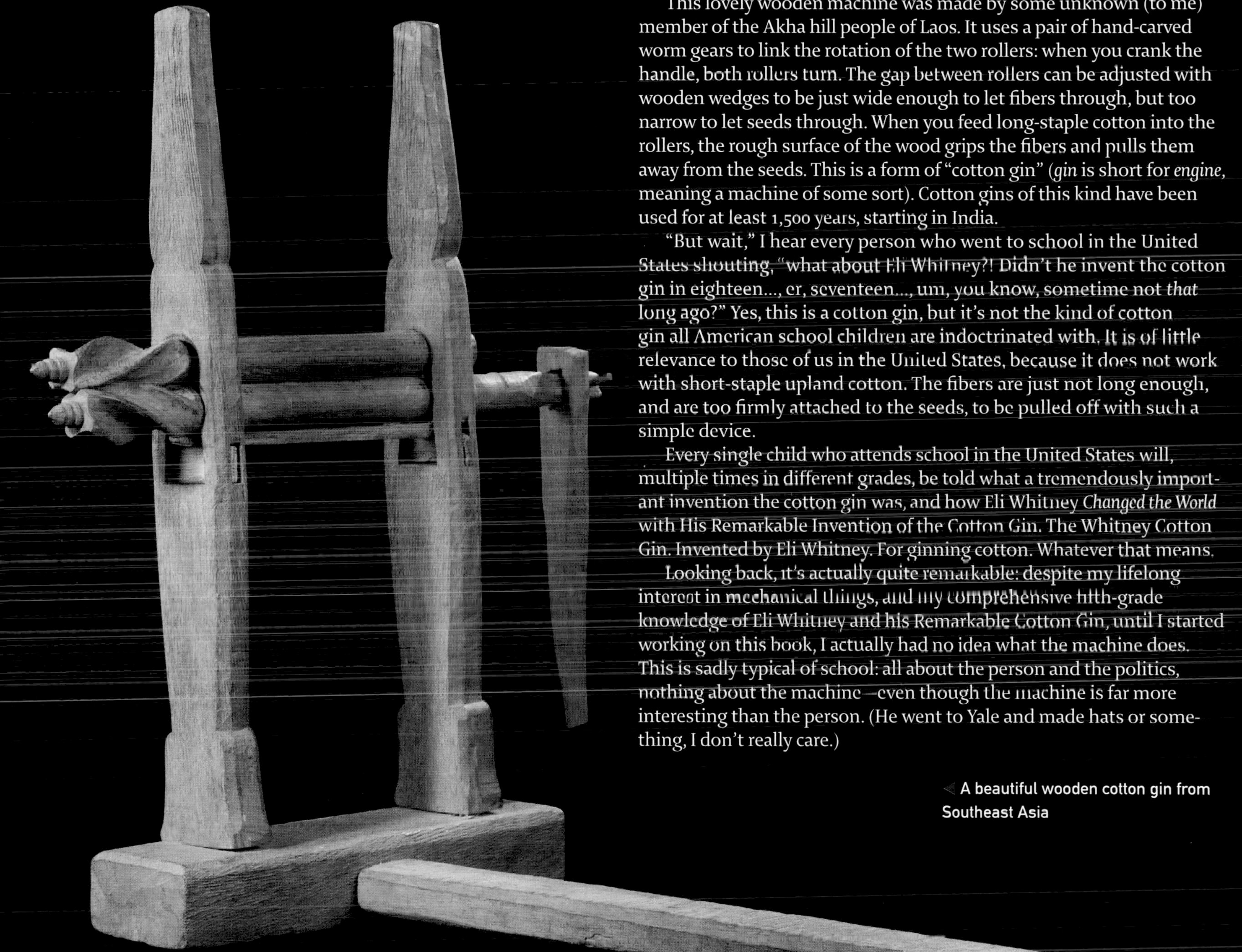

This lovely wooden machine was made by some unknown (to me) member of the Akha hill people of Laos. It uses a pair of hand-carved worm gears to link the rotation of the two rollers: when you crank the handle, both rollers turn. The gap between rollers can be adjusted with wooden wedges to be just wide enough to let fibers through, but too narrow to let seeds through. When you feed long-staple cotton into the rollers, the rough surface of the wood grips the fibers and pulls them away from the seeds. This is a form of "cotton gin" (*gin* is short for *engine*, meaning a machine of some sort). Cotton gins of this kind have been used for at least 1,500 years, starting in India.

"But wait," I hear every person who went to school in the United States shouting, "what about Eli Whitney?! Didn't he invent the cotton gin in eighteen..., er, seventeen..., um, you know, sometime not *that* long ago?" Yes, this is a cotton gin, but it's not the kind of cotton gin all American school children are indoctrinated with. It is of little relevance to those of us in the United States, because it does not work with short-staple upland cotton. The fibers are just not long enough, and are too firmly attached to the seeds, to be pulled off with such a simple device.

Every single child who attends school in the United States will, multiple times in different grades, be told what a tremendously important invention the cotton gin was, and how Eli Whitney *Changed the World* with His Remarkable Invention of the Cotton Gin. The Whitney Cotton Gin. Invented by Eli Whitney. For ginning cotton. Whatever that means.

Looking back, it's actually quite remarkable: despite my lifelong interest in mechanical things, and my comprehensive fifth-grade knowledge of Eli Whitney and his Remarkable Cotton Gin, until I started working on this book, I actually had no idea what the machine does. This is sadly typical of school: all about the person and the politics, nothing about the machine—even though the machine is far more interesting than the person. (He went to Yale and made hats or something, I don't really care.)

◁ **A beautiful wooden cotton gin from Southeast Asia**

Eli Whitney's Cotton Gin

TO EXPLAIN THE function of Eli Whitney's invention, I've made this transparent acrylic model of his original design, generically known as a saw gin. The key parts are a set of what look a lot like circular saw blades, and a comb through which the saw blades rotate. The hooked teeth of the blades catch the cotton fibers, while spaces between the teeth of the comb are too narrow to allow the seeds to pass through.

▼ Whitney's first cotton gin was a small, hand-cranked machine, not that much more complicated than the older roller gin, but with a few crucial differences that allowed it to work with all kinds of cotton.

▼ This is the "saw" from an industrial cotton gin. The teeth are small and very sharp, but shaped in such a way that they do not catch the seeds, only the fibers.

A single manually operated cotton gin like this, properly made of metal and operated by motivated adults, could process 45 pounds (20 kg) of cotton a day, compared to a single pound a day for a person working by hand. Any time a machine provides an almost fifty-fold increase in productivity, that has consequences. But this is still nothing compared to the machines of today. One modern power gin can process those same 45 pounds of cotton in about half a second! The factories we'll meet in a few pages process over 100,000 pounds (45,000 kg) of cotton per shift using just a couple of gins operating together.

◀ Here we see my transparent cotton gin being operated by child labor, as is tradition. We actually used this gin to process the 20-odd pounds of cotton I grew.

The crank on the back is turned so that the drum holding the saw blades is turning clockwise when seen from this side.

Raw cotton bolls (seeds with attached fibers) go into this hopper.

The bottom of the hopper is made of a comb with widely spaced teeth. The bolls are much too big to fit through the slots, but seeds that have been picked mostly clean of fiber are small enough to fall through to the bottom of the machine, where they can be collected.

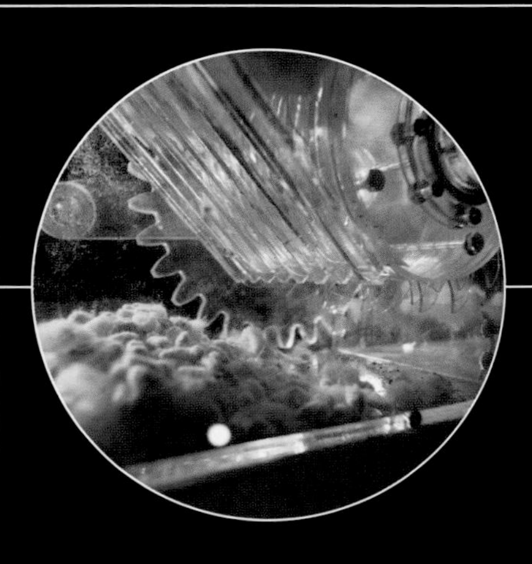

Cotton gins in operation are just about the dustiest things in the world. There's cotton fiber flying in all directions! (The industry term for loose cotton fiber is "lint" and that is a very good word for it.)

The gear driving the brushes is smaller in diameter than the gear it's connected to on the saw blade drum. That means the brushes are moving in the same direction as the saw teeth, but faster, so they can lift the cotton fibers forward off the ends of the teeth. The brushes are spinning fast enough that the fibers fly off and out the back of the machine.

This is the heart of Eli Whitney's invention: a saw tooth pulling a little tuft of cotton fiber through the narrow gap in the comb. The gap is too small for cotton seeds to fit through, so they remain trapped on the other side. The fibers are pulled through and then swept off the saw blades by the brushes on the other side. With each pass through the comb the tooth may pull through only a small amount of fiber, or none at all, but that's OK because the saws are spinning fast. They just keep picking away at the fiber until the seeds are clean enough to fall through the wider gabs in the comb at the bottom of the hopper.

▲ Here is the result of ginning all my harvested cotton. It looks pretty similar to the pile of cotton bolls on page 196, but there's an important difference: this is pure fiber spit out of the cotton gin, with no seeds inside. Only about one-third of the weight of an intact cotton boll is fiber, so this pile is also much lighter.

Cotton Gin

▲ This is the other output of the cotton gin: the seeds. On the left are about the number of seeds I started with, and on the right are the seeds I ended up with. As you can plainly see, I got a *lot* more than I started with! That's the point of farming. In fact, in many cases—corn, wheat, soybeans, and so on—the seeds are what we eat, so the *only* point of farming is to make more seeds. A few of them are saved to plant next year; the rest are eaten. In the case of cotton, the seeds are more of a by-product, but are still economically important, both for planting next year and for their oil content.

▲ Cotton seeds that don't get replanted are pressed into cottonseed oil, which is used for deep frying and as an ingredient in many processed foods. It's a common arrangement for cotton gin operators to make deals with farmers in which the service of ginning the cotton is paid for by letting the gin operator keep the seeds (which they sell to be made into oil).

▼ No discussion of cotton can escape the shadow of slavery, represented here by these slave shackles, which in the United States goes hand-in-hand with cotton plantations. You might think that the cotton gin, which saves such a huge amount of labor, would have reduced the need for slaves, and thus maybe helped eliminate the institution. But exactly the opposite was true. Prior to the invention of the cotton gin, growing short-staple upland cotton wasn't very profitable, even with slave labor. Once it became possible to clean the cotton mechanically, it suddenly became a lot more worthwhile to plant, tend, and harvest the cotton. Demand for slaves to do all the work *other* than ginning skyrocketed, and didn't let up until the Civil War put an end to it in 1864.

Gins on Opposite Sides of the World

WE'VE SEEN HOW I ginned a few pounds of cotton; now let's see how it's done on a bigger scale, in two countries on opposite sides of the world: China and the United States. Representing China is the village of Nan Tang Tuan Cun (南塘疃村), about two hours by high-speed train south of Beijing. Representing the United States is the Gilliam Gin Company of Caddo Parish, Louisiana.

After the cotton is hand-picked in China, it is bundled in plastic sheets, loaded onto trucks, and taken to the gin. (Gin is the name for both the machine and the whole factory that separates fibers from seeds.) The farmer is paid by weight, so the truck is weighed full and again empty. This one delivered about 10,000 pounds (4,500 kg) of cotton.

In the American system, one farmer supplies the gin with enough cotton to run for a week, so no weighing is necessary: each farmer just gets back the fiber from their own raw cotton, however much that may be. In China an individual delivers only enough cotton to keep the factory going for maybe an hour, so separating the output is impractical, and each incoming load needs to be accounted for.

Weigh-in

Unloading the incoming cotton in China

Weigh-out

Test gin at the Nan Tang Tuan Cun factory

You can guess a lot of things about one system by analogy with another. For example, I know that corn (grown all around where I live) is sold by weight, but it's the *dry* weight that counts. If the corn is wet, you don't get paid for the weight of the water. Every grain elevator that buys corn has a moisture meter to determine the water content, so I guessed—correctly—that every gin operator also has a moisture meter. But there's an additional complication with cotton that I didn't guess: cotton varies in the ratio of seeds to fiber, and the fiber is what you're getting paid for. So each batch of incoming cotton also has to be test-ginned using this small gin in a side room, to measure how much fiber you get from each kilogram of cotton bolls. (Again, in the American system this isn't necessary because each farmer gets back their own fiber.)

It takes about twenty minutes to unload the cotton from the truck at the Chinese gin. Then two guys rake the freshly unloaded cotton into the most powerful vacuum cleaner I've ever seen. A 12-inch (30 cm) pipe stretches several hundred feet (100 m or more) from here all the way to the far side of the factory, where the bolls are taken into the ginning process.

Unloading the incoming cotton in Louisiana

The system in America is a bit different. The rolls made by the mechanical picker are dumped, by built-in conveyor belt, off the back of the truck in a matter of seconds. They land on a "walking floor," which uses a set of alternately shifting aluminum slats to move the bales away from the truck.

Cleaning the Cotton

THE WALKING FLOOR moves the giant rolls slowly into an enormous, spinning, sucking death machine, which chops them up and sucks them over to the gins. The input stage is very different, but from that point on, the ginning process is quite similar in both countries.

The giant vacuum deposits the bolls at the top of a monster of a machine, which separates out twigs and leaves. Reasonably clean cotton bolls fall down the chute at the front of the separating machine and are sucked up into a second overhead vacuum duct, which delivers them to the gins.

▲ The walking floor moving the rolls into the giant vacuum

▶ Cleaning machines separate out debris. This is the Chinese version; the American one is different in details, but similar in scale.

▲ The feed hopper (top and bottom) delivers reasonably clean cotton to the gin.

Ginning the Cotton

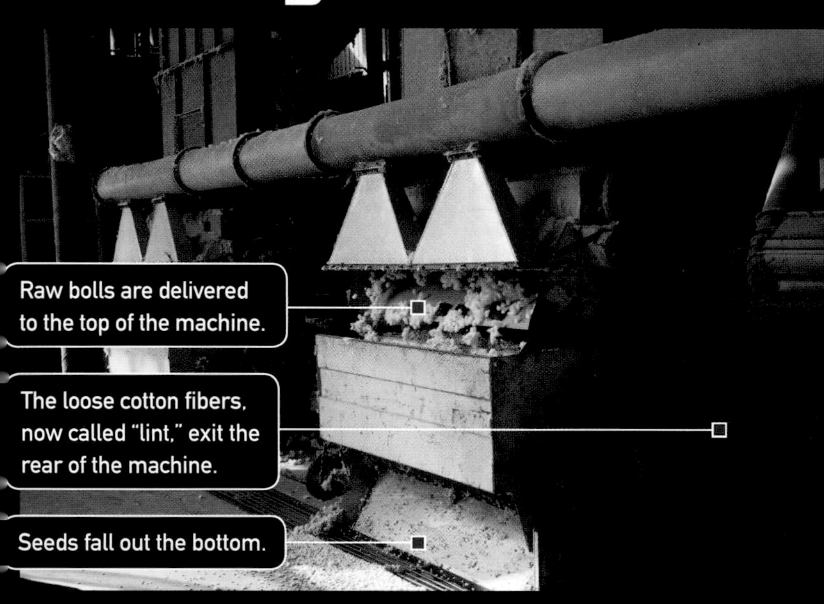

THE FIRST COTTON gins were small, hand-cranked machines, but they quickly grew into something much bigger. This is what the cotton gin had evolved into by the mid-1900s: a large, very dusty machine powered by electric motors. Others are newer and bigger, but all work in essentially the same way as Whitney's original design.

Inside the gins we find the saws and a comb just like in the model on page 199, except much bigger, and in much larger numbers. In operation these saws spin very fast, ripping apart the cotton bolls so fast that seeds pour out the front of the machine.

Twig separator

The fibers leave the back of the gin in a strong current of air flowing through this flat metal duct. Notice the strange back-and-forth S-curve: that's a twig separator. If you look closely you can see an open slit on the top right of the S shape, and another one on the bottom left. Cotton fibers are very light, and they easily follow the flow of air around the S-curve in the same way that soap bubbles or smoke goes with the flow of whatever air they're in. But heavier twigs, leaves, or bits of seeds can't turn the corner as quickly and end up flying out through the slits. The exact locations and widths of the two slits are adjusted so that almost no fiber, but most of the sticks and leaves, come flying out. (In both the Chinese and American gins, several more stages of cleaning follow, using basically the same technologies in both countries.)

American gins are bigger, but there are fewer of them. Overall I think these two factories can process about the same amount of cotton per hour. (What's not captured by either of these still photos is the mesmerizing river of cotton falling down the front of the machine. Imagine Niagara Falls, except it's cotton.)

Baling the Cotton

AFTER GINNING AND cleaning, nearly pure cotton lint is blown over into the next room, where it lands in a monster compacting machine, which packs the loose fibers down into dense bales ready for shipment.

Loose cotton lint enters here.

The lint is first collected and repeatedly compressed in the left-hand column: a ram comes down every few seconds to press down a new layer.

You can tell from the diameter of these cylinders that they are designed to create a serious amount of force!

After a few minutes, enough fiber has accumulated on the left side to make a bale, and the entire lower half of the machine rotates by 180 degrees, interchanging the position of the two columns. The right side is where the real compacting happens.

When the bale is fully compressed, the compacting chamber lifts up and workers tie off baling wires around the cotton before the pressure from above is let off. You can tell from the deep, thick ribbing that there must be a *huge* amount of pressure inside these massive steel compacting chambers.

Released from the compactor, the bale is wrapped in plastic, weighed, and sent to the warehouse.

The American baler is nearly identical, except built upside down. Instead of hydraulic rams pushing down, it has hydraulic rams pushing up, forming a bale near the top of the machine (which is mostly sunk down in a pit, making it harder to photograph). Amusingly, this means both machines are actually pushing the cotton in the same direction. (Get it? Since they are almost literally on opposite sides of the world, if one is pushing down and the other is pushing up, both rams are actually going in the same direction.)

▲ There is something satisfying about inventory. It's calming in a way. It is human effort, good fortune, wealth, and prosperity made visible. This warehouse can easily hold millions of dollars' worth of cotton, hard-won but safe now—in the bag, as it were. The room fills up during picking season, then empties slowly over the following year as the cotton is sold off to the factories that spin it into thread. The ebb and flow of its contents reflect the prosperity of the village whose livelihood it represents.

▲ The fibers are what pay the bills, but you have to deal with the seeds somehow! Remember the river of seeds pouring out the bottom of the gin? They land in this trough in the floor, which has a screw auger inside that pulls the seeds off to the side.

▼ While the fibers are on their way to the spinning mill, the seeds are unceremoniously dumped out in the yard by screw auger conveyor. They will be sold for processing into oil or animal feed.

This auger pulls the seeds out of the factory.

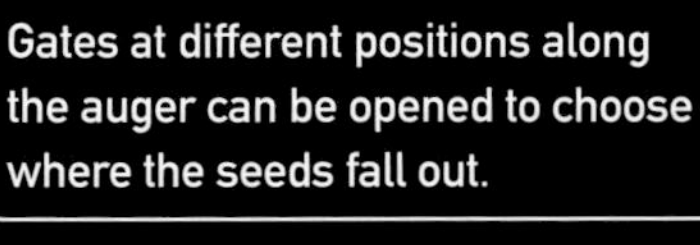

Gates at different positions along the auger can be opened to choose where the seeds fall out.

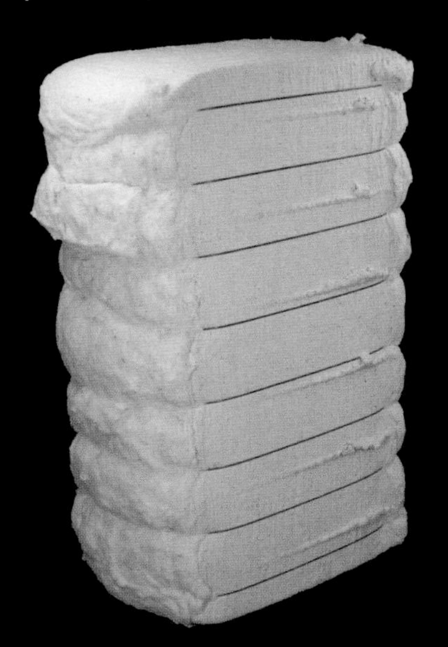

◀ A standard bale of cotton weighs 500 pounds (230 kg). As with large gold bars, there is natural variation from bale to bale, so each one is individually weighed when it's sold. The value of one bale of cotton varies every day with the market price of cotton: at the time of writing, mid-2018, a bale like this was worth about $600 on the world market—a bit less at the gin, a bit more once transported to its end user. (I picked this one up for $500 straight off the ginning floor in Caddo Parish.) About 120 million of these bales were produced worldwide in 2018.

◀ I couldn't take a bale of cotton home from China, but I could drive one back from Louisiana. Thus did I find myself going down the highway, sweating with my arm out the window of a rusty pickup with no air-conditioning—the only proper way of driving through the South. (OK, it's not a pickup truck. I don't own a pickup truck anymore, just this cargo van. But I promise it's old, starting to rust a bit, and was, for a time, on blocks out front.)

▲ A tale of two countries, two systems, and two factories, different in detail, but unified in the goal of making world-class bales of commercial cotton. The factory on the left is in China, the one on the right is in the United States.

The Industrial Revolution

> You got to jump down, turn around
> and pick a bale of cotton
> You got to jump down and turn around
> and pick a bale a day
>
> **—Lead Belly, "Pick a Bale of Cotton"**

THE AMERICAN FOLK song "Pick a Bale of Cotton" was most famously recorded by Lead Belly in 1940. Earlier versions—whose lyrics I cannot repeat here—were working songs, along the lines of "that [African-American man] from Shiloh can pick a bale a day, can you?" Having personally picked about one twenty-fifth of a bale of cotton in my life, I can tell you for darn sure that *I* could never pick a whole 500-pound bale in one day! The best hand-pickers, working under the whip, could pick as much as one-third of a bale in a day. So *maybe* there did exist a person who, in a go-for-broke effort, could pick a whole bale in twenty-four hours (after which, John Henry style, they would likely fall down dead).

A machine, on the other hand—well, that's another story. The mechanical pickers we saw earlier can easily pick 200 bales a day or more. At normal rates of work, that one machine can replace at least six hundred people. Repeat that experience over a thousand industries and you have the industrial revolution.

Machine hall of the manufacturer Richard Hartmann in Chemnitz, c. 1868

Prior to the invention of the cotton gin, cotton fabric was an expensive luxury. Many of the stages of production were entirely manual. But once huge amounts of cheap cotton fiber started becoming available, people started working on making the rest of the production process more efficient and mechanical. The whole of the northern half of the United States and large swaths of England were utterly transformed by the rush to mechanize the spinning and weaving of cotton. To a large degree the industrial revolution was based on, and fed by, cheap cotton fiber made possible by the cotton gin.

Despite the important economic and social reasons for considering the cotton gin a key invention, I think it deserves its place in history even more as a metaphor. People had been growing and processing cotton for literally thousands of years. Generation after generation they did this almost entirely by hand. Weeks were required to make a single piece of cloth, enough for a shirt or maybe a nice dress.

Think about it. People labored their whole lives, as had their ancestors going back fifty generations, doing *the same thing*, never once stopping long enough to figure out that a simple machine, something anyone could have built even in ancient times, would have collapsed their workload by a factor of fifty. The human race has a dedication to shortsightedness that is truly impressive.

The cotton gin serves as an example of the powerful idea that, hey, maybe we could make a *machine* to do that instead! It was this idea, more than any material change in the world, that created the industrial revolution. Once the idea took hold, clever new machines started popping up everywhere: machines to make chain links or pretzels, machines to cut gears or slice bread, even machines to make fancy bows.

Before you turn to the next page and see how it works, think about what the machine might look like that made this bow from a spool of ribbon. How complicated do you think it is? How do you suppose it forms the loops and fastens the ribbon together in the center? How does it measure out the correct length of ribbon for each loop?

THIS BOW-MAKING MACHINE dates from the 1950s, well after the glory years of industrialization, but I think it's a perfect example of the end result of the train of thought for which we honor the cotton gin. It's remarkably simple, yet delightfully perfect in its operation. Load up the ribbon, pop in a plastic spike, crank the handle, and suddenly: bow!

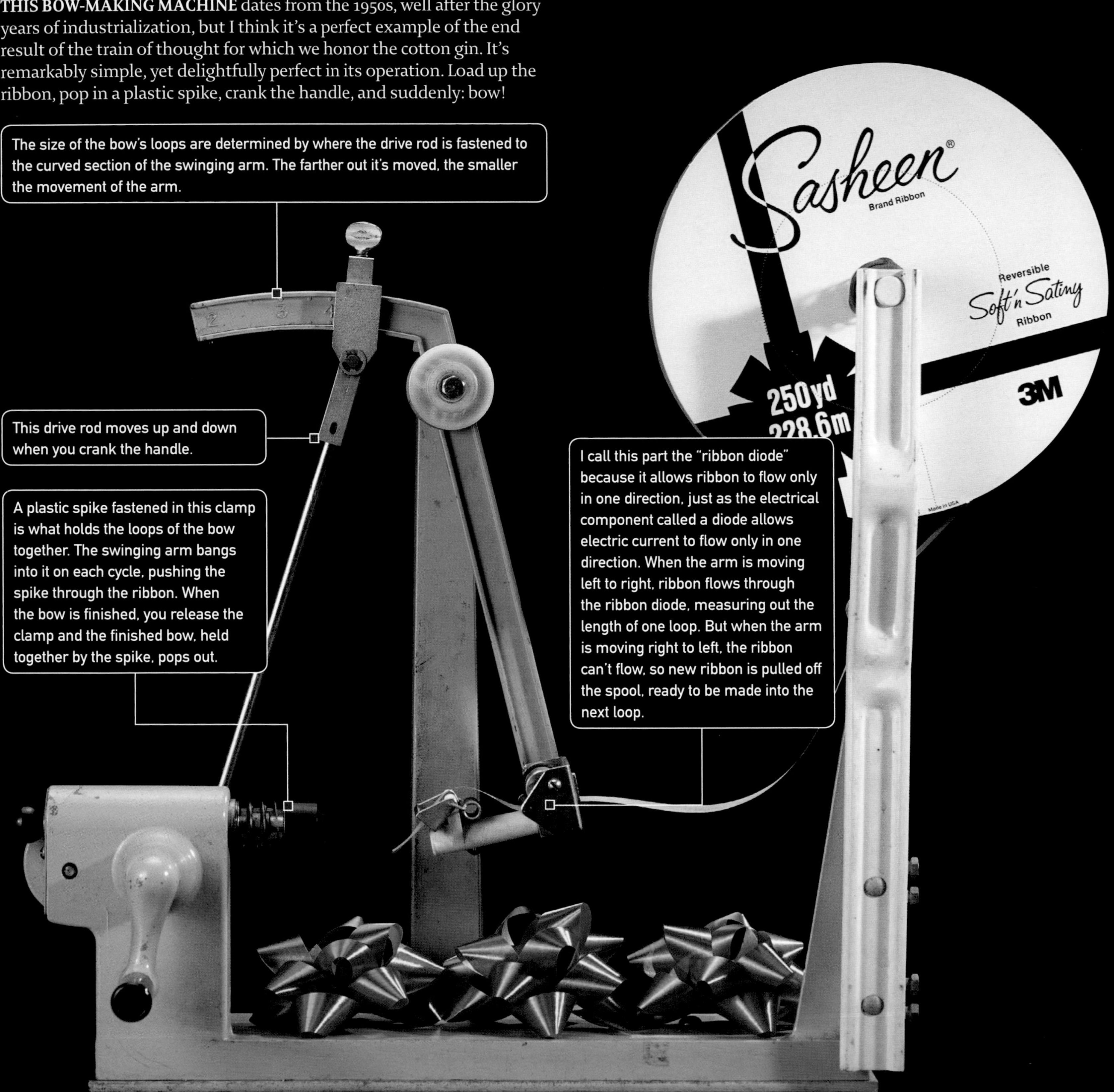

Carding

AFTER THAT BRIEF diversion into bow-making, let's get back to making our potholder. Now that we have separated cotton fibers, we need to start organizing them into thread or yarn.

Cotton fibers straight from the gin are called "lint," and that's exactly what they look like: random fibers going every which way, just like you might find in the lint filter of a clothes dryer. Remarkably, there is no difference between the fibers in this random pile and the fibers in finished woven cloth. The cloth is just a more *organized* form of the fibers, in which they have been aligned, twisted, and woven together.

This kind of blows my mind every time I think about it: a cotton shirt is nothing more than a pile of cotton lint that has been fancy-twisted into the shape of a shirt. With enough patience, you could unweave and untwist the shirt and turn it back into a pile of random lint without breaking a single fiber. The fact that it holds together for years is a testament to just how cleverly the twisting job was done.

The first step in organizing cotton lint into a shirt, or a potholder, is to get all the fibers lined up in the same direction, instead of going every which way. This is called "carding" the cotton. On a small scale it is done using carding paddles like these, which are basically very fine combs made with sharp, angled wire teeth. You brush the cotton repeatedly between two paddles until all the fibers are going in the same direction, then brush once in the opposite direction to release the mat of fibers from the angled wires.

The result of hand-carding a bit of cotton is called a "puni," a sort of sausage of aligned cotton fibers ready for hand-spinning. The process is simple, but painfully slow for anything more than a small bag of cotton. On an industrial scale everything is much more complicated. The simple one-step process of carding is replaced by five separate stages of re-forming, each with its own machine, before the cotton is ready for spinning.

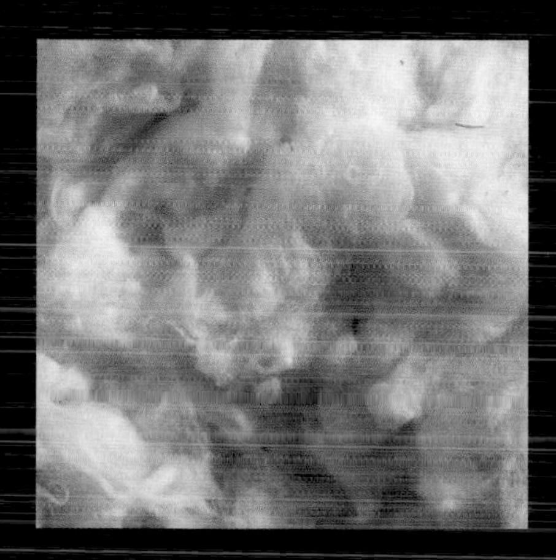

▲ Cotton lint

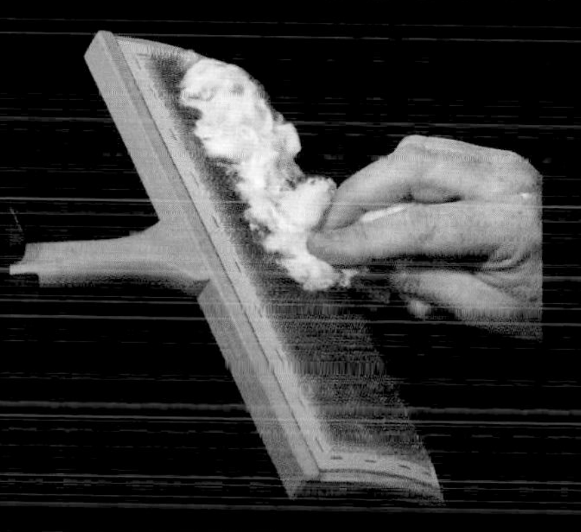

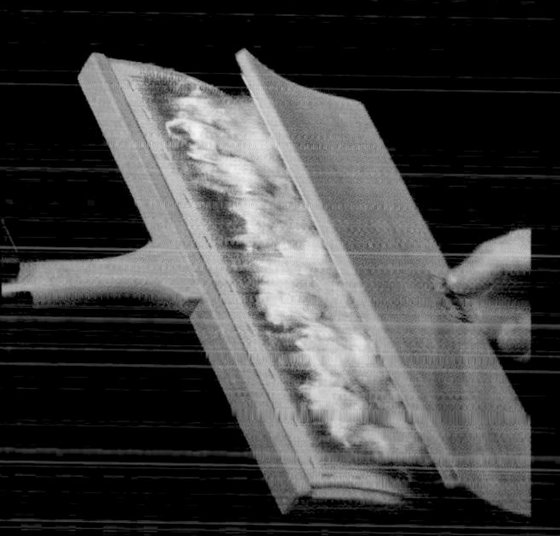

▲ Carding paddles

▼ A hand-carded "puni" ready for hand-spinning into thread

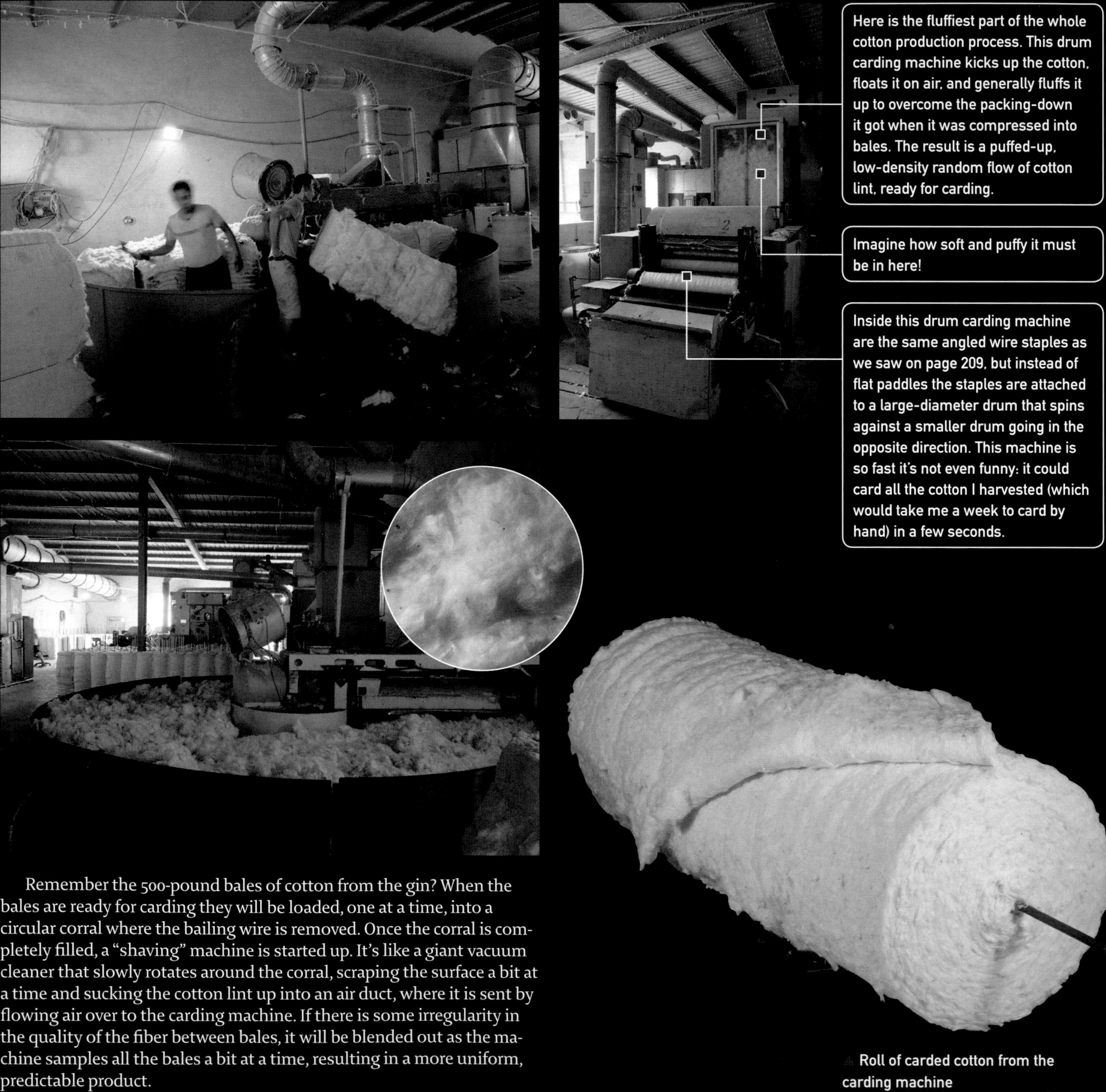

Here is the fluffiest part of the whole cotton production process. This drum carding machine kicks up the cotton, floats it on air, and generally fluffs it up to overcome the packing-down it got when it was compressed into bales. The result is a puffed-up, low-density random flow of cotton lint, ready for carding.

Imagine how soft and puffy it must be in here!

Inside this drum carding machine are the same angled wire staples as we saw on page 209, but instead of flat paddles the staples are attached to a large-diameter drum that spins against a smaller drum going in the opposite direction. This machine is so fast it's not even funny: it could card all the cotton I harvested (which would take me a week to card by hand) in a few seconds.

Remember the 500-pound bales of cotton from the gin? When the bales are ready for carding they will be loaded, one at a time, into a circular corral where the bailing wire is removed. Once the corral is completely filled, a "shaving" machine is started up. It's like a giant vacuum cleaner that slowly rotates around the corral, scraping the surface a bit at a time and sucking the cotton lint up into an air duct, where it is sent by flowing air over to the carding machine. If there is some irregularity in the quality of the fiber between bales, it will be blended out as the machine samples all the bales a bit at a time, resulting in a more uniform, predictable product.

▲ Roll of carded cotton from the carding machine

The carding machine turns out big rolls of carded cotton. They are the equivalent of the little cotton puni I made by hand, except each one is the size of a (small) person. While the cotton in this form could be spun by hand, it is not consistent enough for machine spinning. To make perfectly uniform thread day in and day out, the fibers need to be arranged more perfectly before spinning commences.

The rolls of carded cotton are stored temporarily in the factory, awaiting the next step of processing.

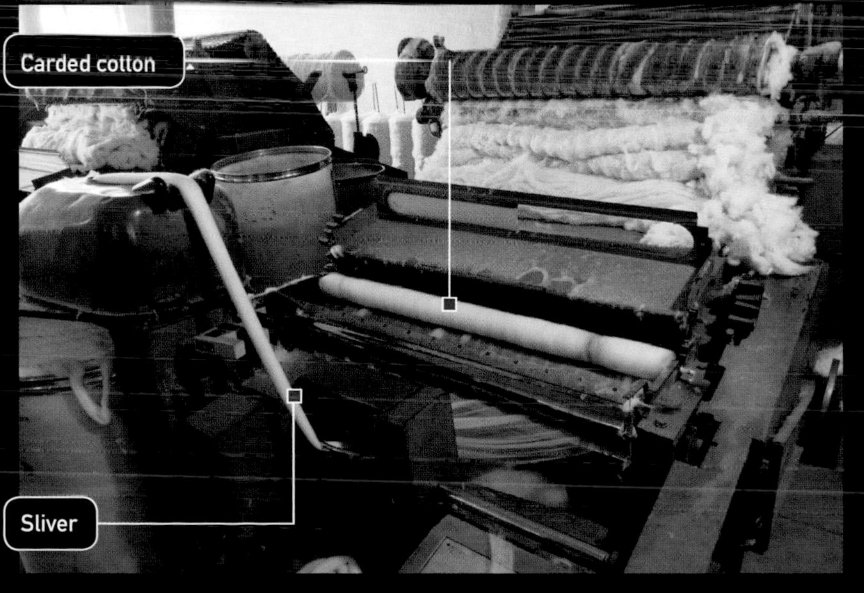

This machine unrolls the rolls of carded cotton and pulls them into a thick, loose rope called "sliver." It's the first of four stages of sliver, each thinner and more perfectly aligned than the one before.

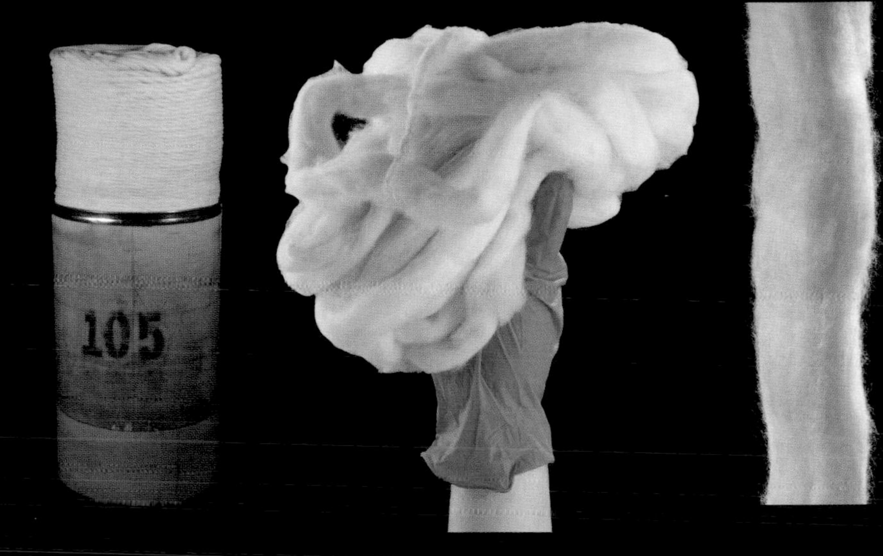

The stage-1 sliver, very slightly twisted, is temporarily coiled up in big drums before moving on to the next machine. It looks like inch-thick rope, but is actually very lightweight, loose, and easy to pull apart. The fibers are just aligned, not yet twisted around each other.

This machine unspools a dozen or so ropes of stage-1 sliver at the same time, blends them together, and stretches them out to form a new, thinner stage-2 sliver.

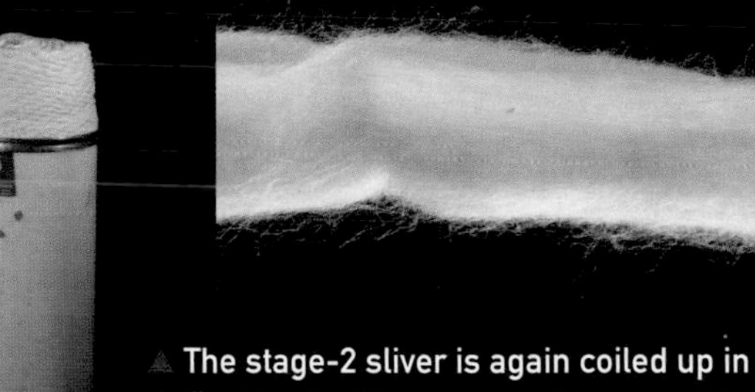

The stage-2 sliver is again coiled up in tubs and stored temporarily. It's a lot like stage-1 sliver, but thinner and more aligned. This process is like toffee pulling, except with cotton.

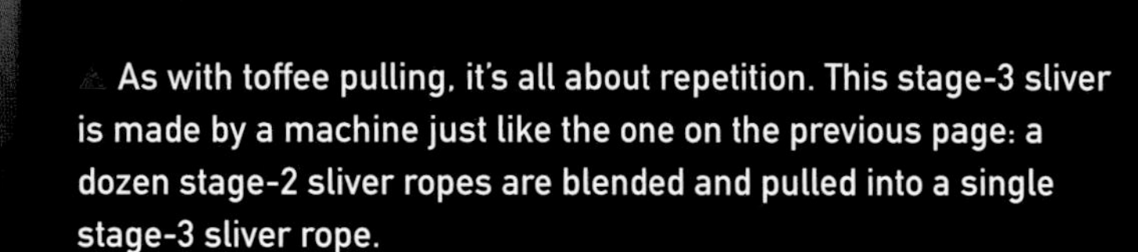

As with toffee pulling, it's all about repetition. This stage-3 sliver is made by a machine just like the one on the previous page: a dozen stage-2 sliver ropes are blended and pulled into a single stage-3 sliver rope.

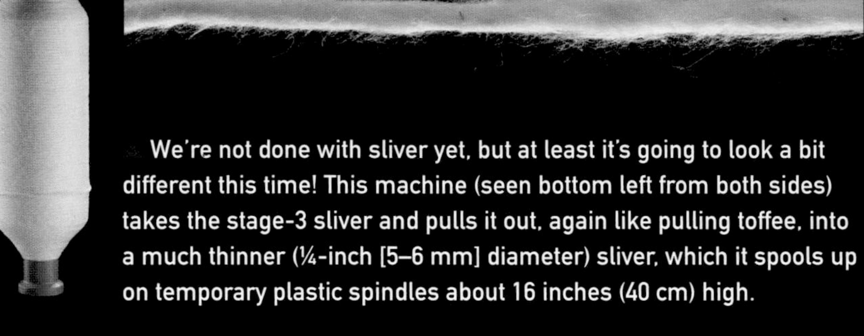

We're not done with sliver yet, but at least it's going to look a bit different this time! This machine (seen bottom left from both sides) takes the stage-3 sliver and pulls it out, again like pulling toffee, into a much thinner (¼-inch [5–6 mm] diameter) sliver, which it spools up on temporary plastic spindles about 16 inches (40 cm) high.

This is the part of the machine that replicates the human skill of feeding out fiber from a puni at just the right rate. Stage-3 sliver enters on the left and is pulled along by a series of four wheels, each turning a bit faster than the one before. The stage-4 sliver exiting on the right is going much faster, and thus contains much less fiber per unit of length (i.e., it's thinner).

Spools of sliver ready for spinning

The stage-4 sliver looks a lot like thick yarn, but it's still only very slightly twisted. If you pull on it even a little bit, it immediately comes apart. In this form the cotton is *finally* ready for the next stage, spinning. Spools of stage-4 sliver are stockpiled, ready for their big day in the spinning machines.

The whole factory is in continuous flow: each machine produces its output at, on average, the same rate as the next machine consumes it. In theory you could just take the different stages directly from one machine to the next. But in practice, having "buffers" like this, holding at least a few hours' worth of production, is good insurance against one or the other machine breaking down. In this factory it's the next stage, spinning, that is the rate-limiting step. All the other machines can catch up if they break down, so the ultimate rate of production is determined by how close to 100 percent of the time they can keep the spinning machines running.

Spinning

THE PROCESS USED to create thread or yarn from loose fiber is called spinning. This can be done slowly by hand, faster with a simple machine, or at tremendous speed with machines that spin a hundred yarns or more in parallel.

The ancient art of spinning is slow, meditative, and tremendously useful. Over the course of days, weeks, and months, one person working alone, or a village working together, can use the simplest of tools to create clothes, blankets, bags, and a hundred other useful items of spun, woven, and knitted fiber.

It takes quite a bit of practice and skill to use a spindle. One hand keeps the thing turning rapidly while the other hand feeds fiber from a puni at just the right rate to make thread of the desired thickness. Thin, fine thread is *much* harder to make than crude, thick yarn.

The word "spinning" describes exactly the entire process of making thread. Sliver can be pulled apart very easily because you don't have to break any individual fibers; they just slide apart. But once the fibers are twisted together they can't slide anymore. Breaking the thread requires breaking the strong individual fibers. There's nothing more to it than that: it's just twisting, which you get from spinning.

The problem with this sort of manual spindle is that making thread requires a *lot* of twisting. A simple machine can accelerate the process immensely.

The classic wooden spinning wheel of story and song, found in fairy tales from "Rumpelstiltskin" to "Sleeping Beauty," was not generally used to spin cotton. When it wasn't spinning straw into gold or putting people to sleep for a hundred years, this type of wheel was mostly used to spin wool. (The individual fibers in wool are much longer, two to three times as long as cotton fibers, so the equipment best suited to spinning it is different.)

Fairy-tale spinning wheel

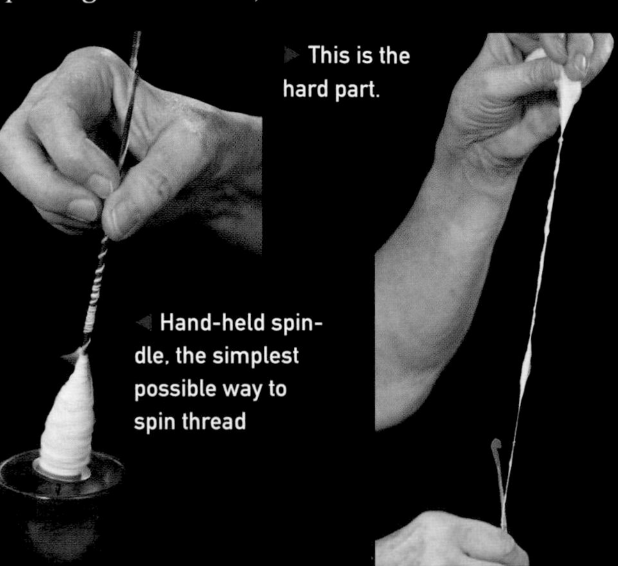
This is the hard part.

Hand-held spindle, the simplest possible way to spin thread

Mounted here is a spindle almost identical to the handheld spindle we just saw. The only difference is that it has a small pulley wheel on it so it can be driven with a belt.

The job of a spinning wheel is to spin the spindle really fast with minimal effort from the user. The bigger this wheel is, the more times around the spindle will go for each time the wheel turns.

Cotton, more than wool, played an important role not just in American history, but also during India's struggle for independence from Britain, led by Mahatma Gandhi. Control of the mechanized cotton industry gave Britain control of the population. Gandhi encouraged the people in the millions to take back economic power by making their own spinning wheels. The wheel typically suitable for spinning cotton is called a floor charkha. Proof of the historical and cultural importance of this machine is the fact that the image seen below isn't a real floor charkha, it's a small working model of one. People don't make commemorative models of machines they don't care about.

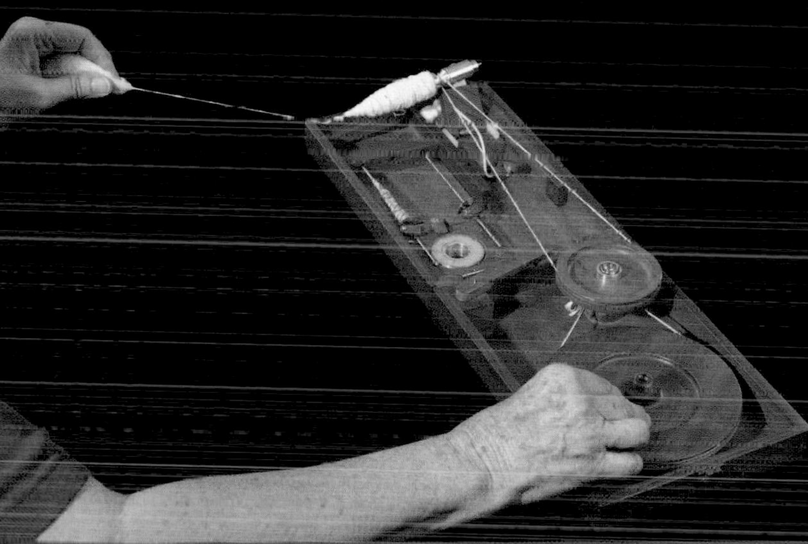

Indian charkhas come in a whole range of sizes, from large floor models down to the "book charkha," which is literally the size of a hardcover book. Since there is no room inside a book charkha for a large wheel (used in floor models to give a very high gear ratio to make the spindle go fast), it instead uses a compound pulley system. You crank a biggish wheel, which drives a small wheel, which is connected to a medium-sized wheel, which drives a tiny wheel on the spindle. This is exactly the same idea found in the compound gear chains in clocks.

FAIRY TALE MISTAKES

I ALWAYS USED to be confused by the "Sleeping Beauty" story, in which the unfortunate princess is pricked by a cursed spindle, falling into a deep sleep for a hundred years. The storybooks I remember seeing as a kid all had pictures of wool-spinning wheels, *which don't have anything sharp on them.* As is typical of fairy tales, the mechanical details were entirely ignored. For example, in Disney's *Sleeping Beauty*, the part she's shown touching is not even the spindle: it's the distaff! And it's not sharp! I'm sorry to say that the much-celebrated research department at Disney fell flat on their faces with this one.

The people who wrote the fairy tales weren't wrong, only the people who selected the pictures a hundred years later, after common knowledge of spinning had been lost. Cotton-spinning spindles are in fact needle-sharp. They just don't look like most of the pictures in today's books and movies.

The sharp part

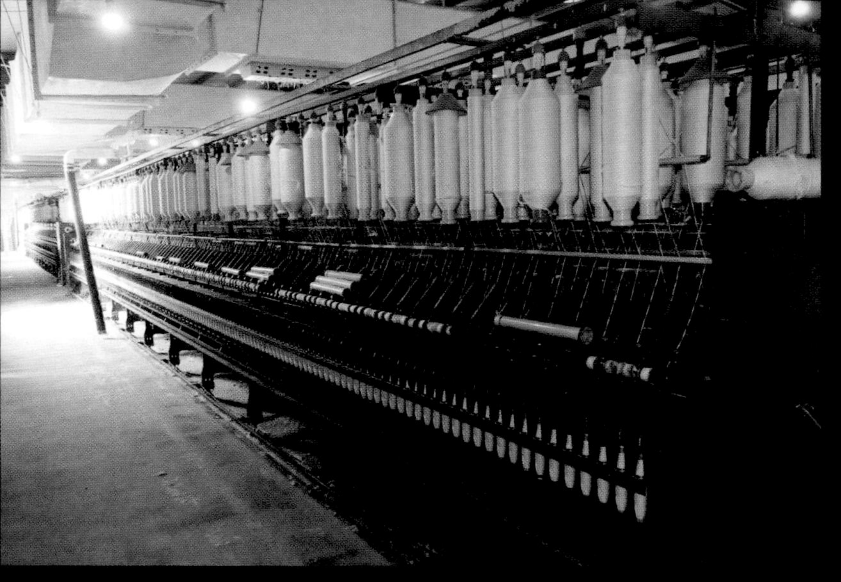

The industrial equivalent of a spinning wheel

In an industrial spinning machine, spools of stage-4 sliver hang at the top of the machine, and smaller spools of finished thread grow at the bottom. In between the sliver is gently elongated, then twisted at very high speed by the spinning of the spool at the bottom. A series of three progressively faster rollers (similar to the machine on page 212) stretch the stage-4 sliver to the right thickness to be spun. The actual spinning happens just beyond the last roller.

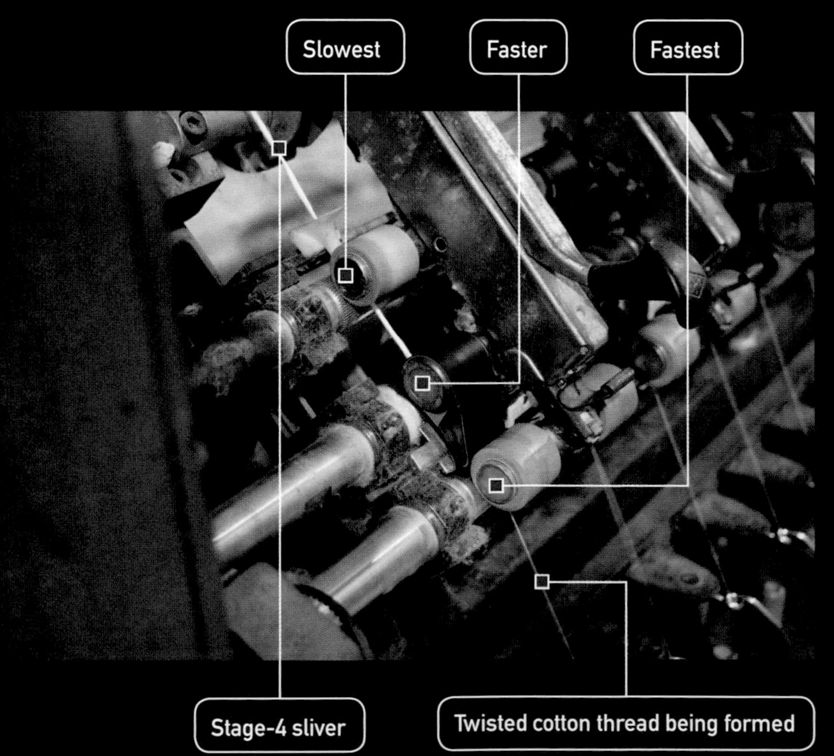

The output of the spinning machines is finished cotton thread, but it's not quite ready to leave the factory yet. The spinning machines spool the thread onto fairly small wooden spindles, each of which does not hold very much thread. (They need to be small so they can spin very fast as the thread is being created.)

Turning many small spools into one big spool means a lot of knots! The middle section of the respooling machine, shown at right, is an automatic knotting device. In about a second it creates a tiny knot and trims off the excess thread on each side, leaving a barely visible connection. Sensors in the machine detect thread breaks, and suction arms are able to recover both ends of thread and bring them together in the knotting device, restarting the machine automatically.

The air in the spinning room is hot and oppressively humid—even more so than the air outside. This is no accident, as you can see from these pipes delivering a steady fog of moisture to the room. The properties of cotton fiber are very dependent on humidity levels. To ensure consistent behavior as it passes through all the stages of stretching and spinning, the fiber needs to be kept at a constant, relatively high moisture level.

The thread from the small spools is rewound onto these much larger spools.

Small wooden spools from the spinning stage are loaded into this turret. They drop one by one to the bottom of the machine, where they are unspooled.

Rows of many dozens of respooling machines operate without human intervention. Workers only need to keep their turrets filled with fresh wooden spools, and return the emptied spools back to the spinning machines to be filled with new thread.

Automatic respooling and knotting machines combine the thread from many small wooden spools into one large spool with a cone-shaped cardboard core.

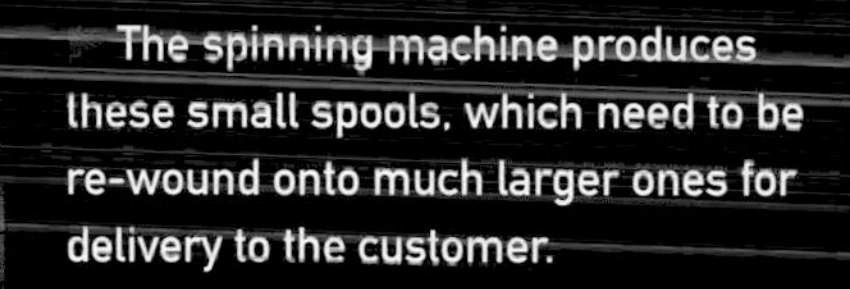

The spinning machine produces these small spools, which need to be re-wound onto much larger ones for delivery to the customer.

A cake of my cotton yarn!

Getting closer to potholder-land! We now have enough yarn to make a potholder, spun by hand with a book charkha, from the cotton I grew, ginned, and carded. I wish I could say I spun it myself, but sadly I just did not have the time to master the art of cotton spinning. Sue, a member of our local spinners' and weavers' club, took pity on me and spun my cotton for me. Cotton can be spun into much thinner thread than this, but I asked Sue to make me fairly thick yarn because it's much easier and faster to spin thick yarn than thin, and because it makes weaving go much faster.

Weaving

ONCE THE THREAD is spun and spooled, the next step is to turn it into cloth. There are two main ways to do this: weaving and knitting. Of the two, weaving is by far the oldest method, and by far the most commonly used. Knitted fabrics are nice, but woven fabric is one of the cornerstones of civilization.

Weaving is all about ups and downs. In the simplest pattern, called a plain weave, every yarn alternates from being on top to being on the bottom, in all directions. (This cloth is made of thick rope so you can easily see how it's put together, but of course normally cloth is made with much thinner yarn or thread.)

This close-up of rope cloth shows a plain weave pattern, the simplest of all.

THE SIMPLEST LOOMS

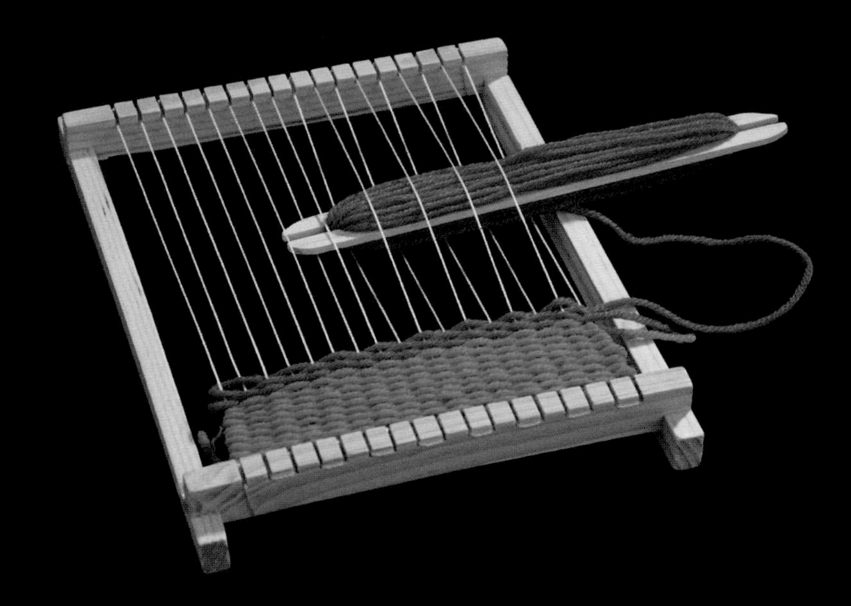

A lot of weaving can and has been done with almost no technology. In the simplest looms, threads called the warp yarns are stretched between two sticks or bars. To make cloth, you push a shuttle wrapped with more yarn (called the weft yarn) through the warp yarns, dipping the shuttle up and down, over and under, one yarn at a time. The weft yarn pulled behind the shuttle is left woven between the warp yarns when the shuttle comes out the other end.

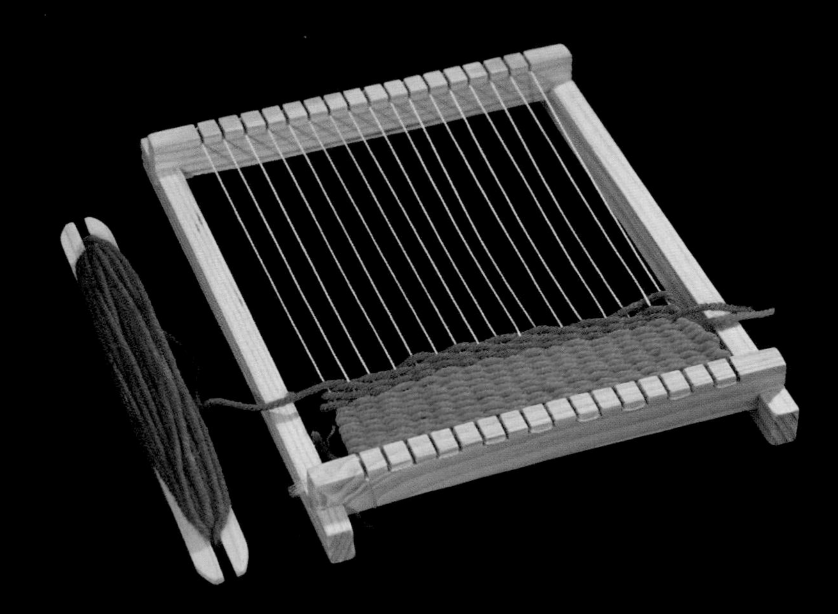

After the shuttle is pulled all the way through, the newly placed weft yarn is combed down to pack it in with the previously woven weft yarns. Do that a few hundred or a few thousand times, and you have a bit of cloth.

By far the most time-consuming part of weaving by hand is having to push the shuttle, or a sort of leader-stick with no yarn on it, alternately up and down through the stretched warp yarns. This gets *insanely* tedious after a while! Fortunately there is a better way: heddles. In this child's toy loom the heddles are made of blue plastic, but work exactly the same way heddles work in larger looms—by lifting every other yarn all at once.

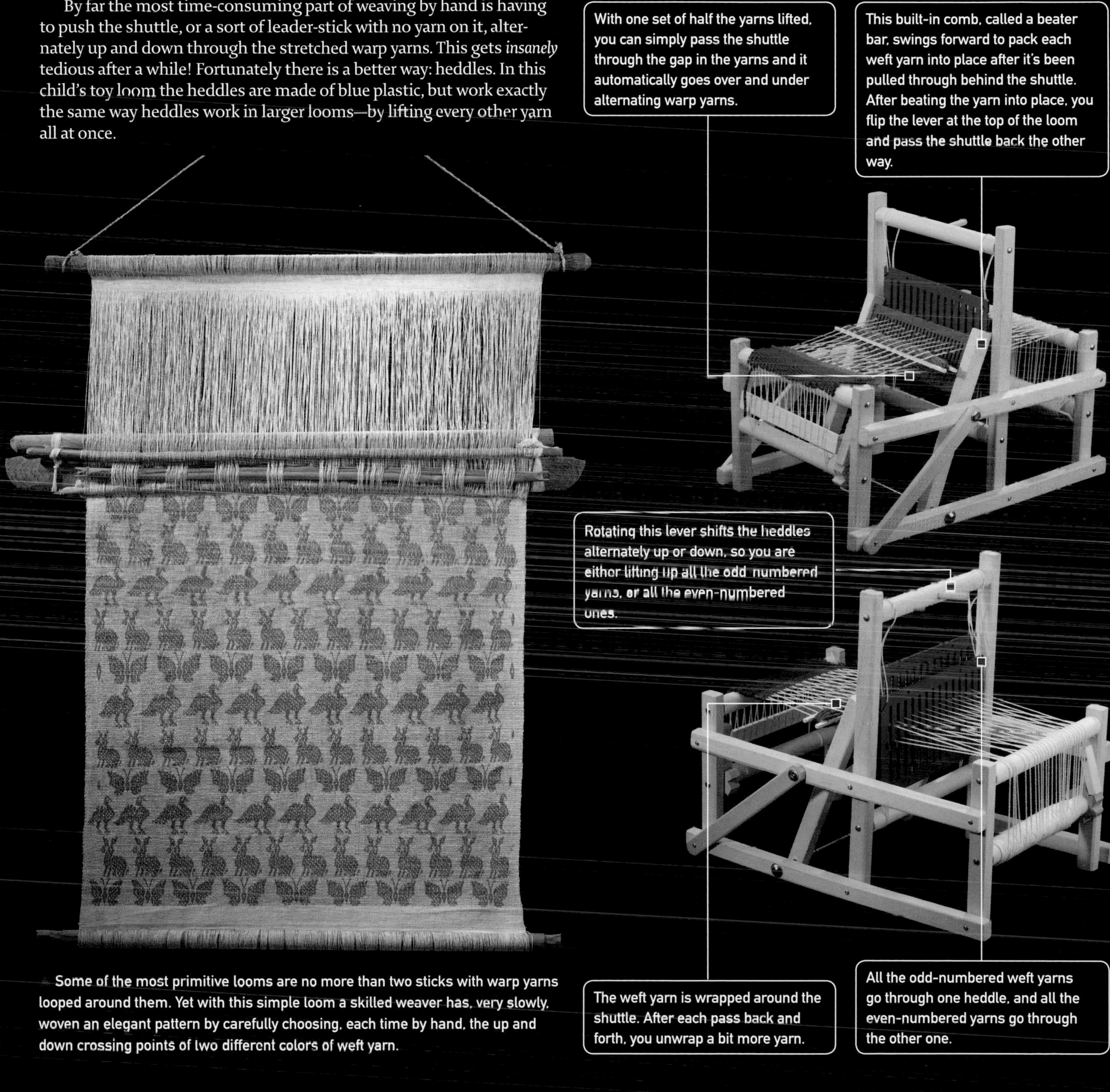

With one set of half the yarns lifted, you can simply pass the shuttle through the gap in the yarns and it automatically goes over and under alternating warp yarns.

This built-in comb, called a beater bar, swings forward to pack each weft yarn into place after it's been pulled through behind the shuttle. After beating the yarn into place, you flip the lever at the top of the loom and pass the shuttle back the other way.

Rotating this lever shifts the heddles alternately up or down, so you are either lifting up all the odd-numbered yarns, or all the even-numbered ones.

The weft yarn is wrapped around the shuttle. After each pass back and forth, you unwrap a bit more yarn.

All the odd-numbered weft yarns go through one heddle, and all the even-numbered yarns go through the other one.

Some of the most primitive looms are no more than two sticks with warp yarns looped around them. Yet with this simple loom a skilled weaver has, very slowly, woven an elegant pattern by carefully choosing, each time by hand, the up and down crossing points of two different colors of weft yarn.

FOOT-PEDAL LOOMS

THE NEXT STEP up in looms is a foot-pedal-operated floor loom. This is a six-heddle Sears model probably from the 1960s. "Six-heddle" means that there are six separate frames that can be raised or lowered, using foot pedals on the floor. When you set up the warp yarns to make a piece of cloth, you can decide which yarns to pass through the eyes on each heddle, depending on the pattern you want to weave. To make a plain weave, you would use just two heddle frames, one for the even yarns and one for the odd yarns.

This bigger loom has basically all the same parts as the toy model, just heavier and wider.

About 40 feet (12 m) of warp yarn is wrapped up around the "warp beam" at the back of the loom. As you weave, you slowly roll the woven fabric up around an uptake beam in the front, while simultaneously unrolling more warp yarn from the beam in the back. The process of "warping" a loom takes many hours—and up to weeks in the case of very large industrial looms—because each of hundreds, thousands, or even tens of thousands of separate warp yarns has to be wrapped onto the warp beam, then individually threaded through the heddle eyes and tied to the uptake beam in the front.

The heddles are wooden frames with wire heddle eyes hanging from a metal bar.

The beater bar is made of metal and weighted to drive the weft yarns home.

Finished fabric is rolled up here.

Pointed ends help the shuttle pass through the shed without snagging any out-of-place warp yarns.

A hole lets the yarn spool out without tangling.

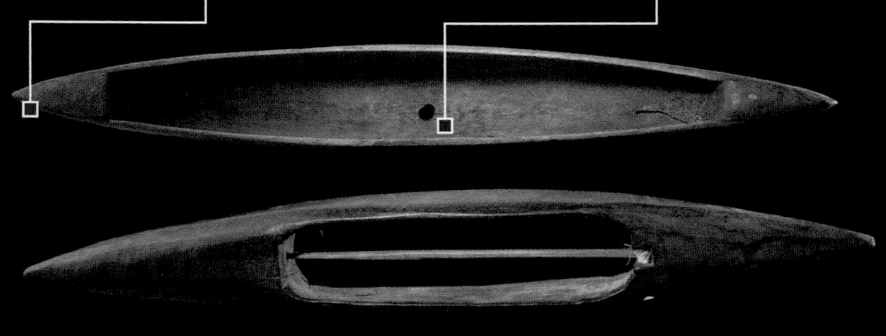

Instead of a stick with yarn wound around it, bigger looms use a "boat shuttle," so named for its shape (though some have open bottoms and would make terrible boats). Inside the boat is a long, thin spool of yarn that unwinds out the end of the boat as it is pulled through the shed.

WHEN IT CAME time to weave my homemade cotton yarn into fabric, I wanted a loom that would waste the absolute minimum amount of hard-won yarn, while allowing me to weave reasonably fast. I decided to make a custom laser-cut acrylic loom to the exact size of the piece of cloth I wanted to make.

My potholder loom is like the simple child's toy loom in that the warp yarns are stretched back and forth between two bars, not wrapped up on a warp beam like in the larger version. But it's like a more sophisticated loom in that it has a mechanical heddle mechanism to automate the up-and-down weaving of the weft yarns.

▲ Close-up of the heddle eyes on my homemade loom

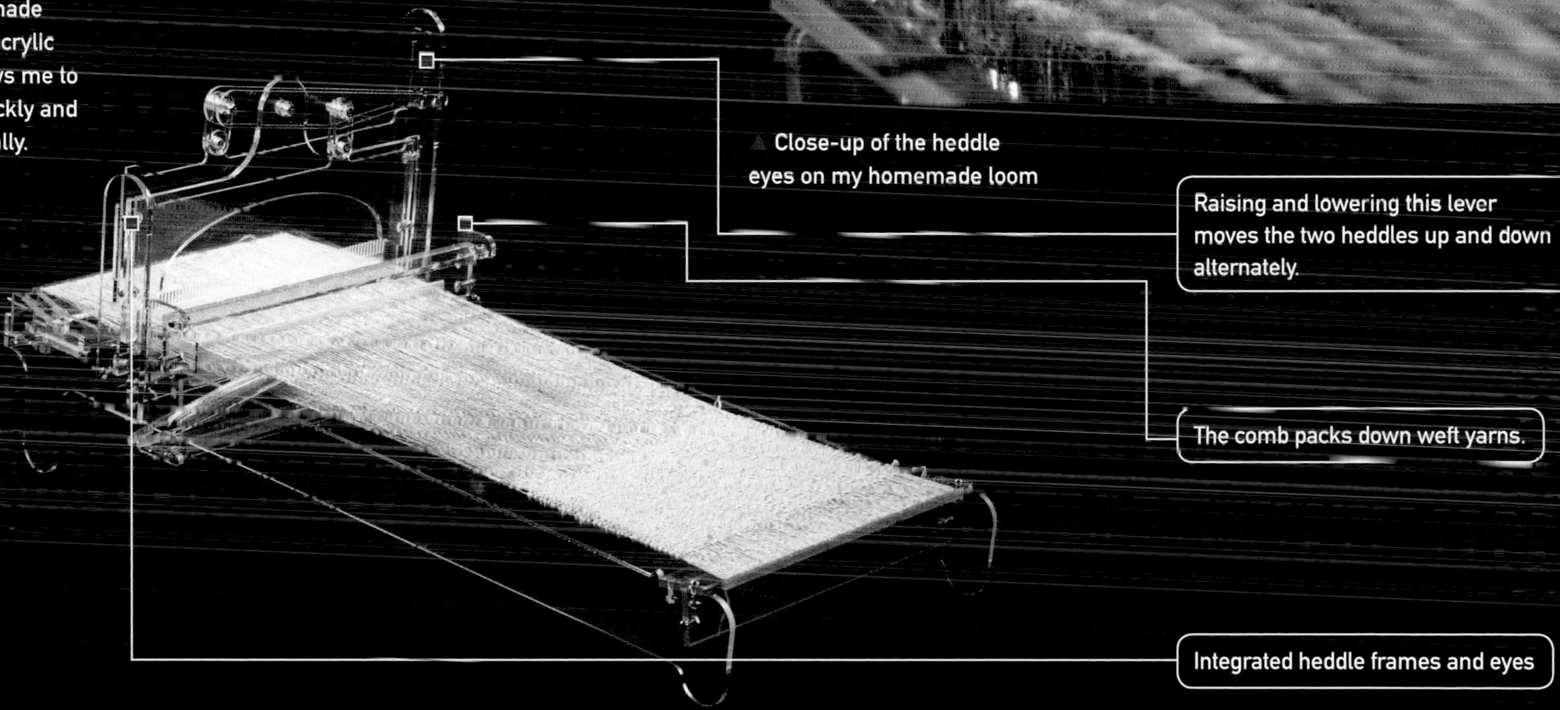

My homemade laser-cut acrylic loom allows me to weave quickly and economically.

Raising and lowering this lever moves the two heddles up and down alternately.

The comb packs down weft yarns.

Integrated heddle frames and eyes

I'm pleased with myself for the design of these acrylic heddle eyes, but for a reason that will probably be appreciated only by people who have been forced to thread hundreds of warp yarns through traditional heddle eyes. My heddles can be slid down like a comb over a set of already-stretched warp yarns, and as long as you interleave the yarns exactly one-to-one with the teeth of the comb, all the even-numbered yarns will automatically end up caught in a sort of yarn trap, while the odd-numbered yarns end up in slots where they can move up and down freely. Then the second heddle frame does the opposite, trapping the odd-numbered yarns and letting the even

This design lets you warp the loom very quickly—in just a few minutes—by looping a single length of yarn back and forth with the whole heddle frame removed. Then you drop the heddle frame in place over the stretched yarns, and you're ready to start weaving.

So far as I know, my potholder loom is the only one in the world with this design, proving that no technology, not even weaving, is so old that you can't come up with an idea that you think is new until a reader sends you the hundred-year-old patent on the same idea.

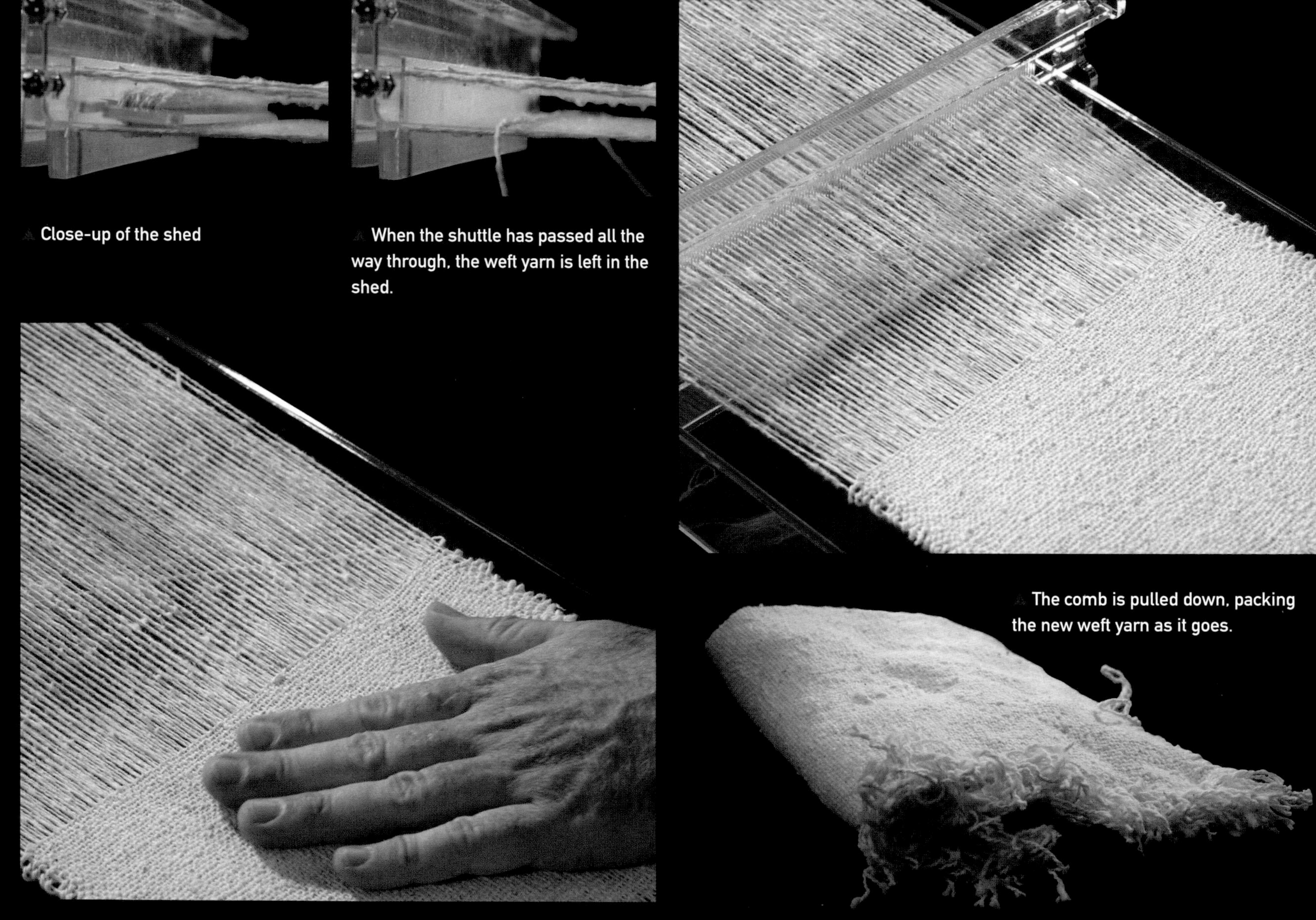

Close-up of the shed

When the shuttle has passed all the way through, the weft yarn is left in the shed.

The comb is pulled down, packing the new weft yarn as it goes.

The new weft yarn should be stretched across the warp yarns at an angle, so when it is packed down against the previous ones, extra length is available for going up and down between the warp yarns. Without this added length, the fabric would get pulled in, becoming narrower and narrower with each new weft yarn, which is a common problem for beginning weavers.

With the weave removed from the loom and with its warp yarns tied off into tassels on both ends, we have achieved a major milestone: cloth! You could actually use this as a potholder as is, but we're going to go further, sewing in batting and quilting it. First, though, we're going to learn about how weaving is done on an industrial scale.

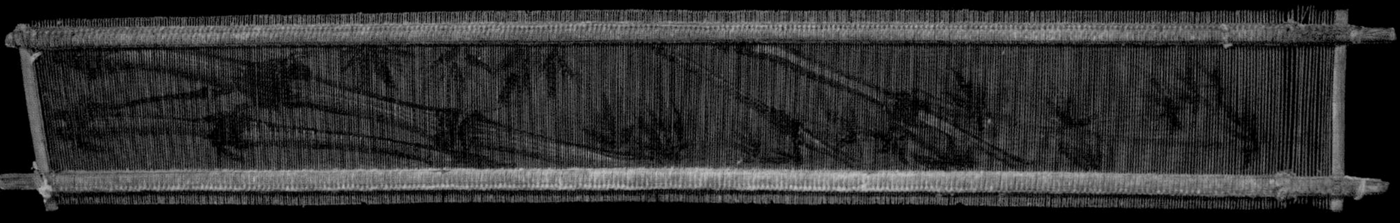

Loom beater bars can be made of split bamboo, as in these old Chinese examples. The result is something worth saving long after the loom is gone. This one has been decorated with traditional painting and is meant to live out its retirement hanging on the wall,

A Visit to a Weaving Factory

NEVER IN MY life have I felt more strongly the beating heart of a place than when I walked onto the weaving floor of this small company in rural Hebei Province, China. A lot of factories are loud, but this one throbs and pulses with an ever-shifting rhythm that made me think of African drums and techno music and Paul Simon. But mere musicians cannot match the raw sonic power of machines like this. They shake the floor and fill the air.

The weaving floor is lit only by the light of its windows, giving the place a depth and texture not found in the sterile, fluorescent factories so common today. The fact that everything not in motion is covered by a hanging moss of cotton fiber adds an otherworldly touch. I wish everyone could visit a place like this while they still exist. They survive now in pockets, but they will soon be gone, even from China, even from India, and someday, even from our memories.

A canvas cloth weaving factory in Hebei Province

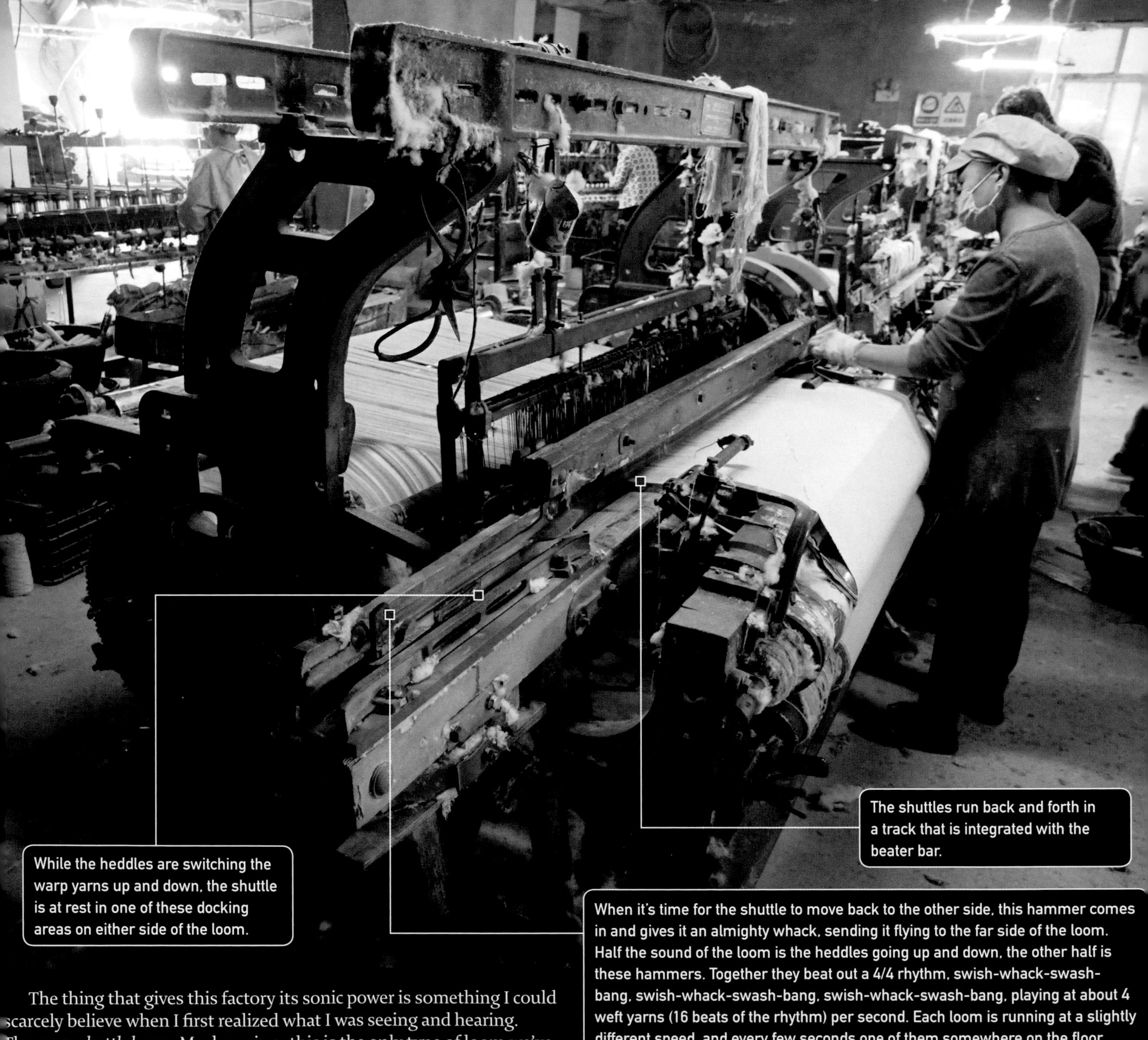

The thing that gives this factory its sonic power is something I could scarcely believe when I first realized what I was seeing and hearing. These are *shuttle looms*. Maybe—since this is the only type of loom we've discussed so far—you're not sure why I should be surprised by this fact? But frankly I was not aware that this type of loom was even still in use anywhere, given its many disadvantages.

Rather than the polished wood shuttles we saw on page 220, the utilitarian boat shuttles in this factory are made of slick plastic, with strong metal points on each end (for when they get whacked by the hammer). They have the same fundamental limitation of all shuttles: you can only put as much yarn onto the spindle as you can comfortably fit between the top and bottom yarns when the shuttle passes through the shed. The maximum gap in the warp yarns squeezes out the potential for continuous operation of any shuttle loom.

At this factory there is one worker standing between each pair of looms facing each other across an aisle. Their job is to keep both their looms running as continuously as possible, by having a fresh shuttle loaded and ready to swap in the instant the old one runs out of yarn. The fact that one worker can handle only two looms is all you need to know about how quickly the machines consume all the yarn on a shuttle.

This loom has a flexible rapier, which wraps around underneath the loom as it comes out of the shed. This saves space on the factory floor compared to versions where a rigid metal rod-of-death, as wide as the whole loom, comes shooting out the side several times per second.

The boat shuttles are made of plastic with strong metal points on each end.

In another room I found this sign of coming modernity: a rapier loom. It's the same in every way as the shuttle looms, except there is no shuttle. Instead there is a flexible metal strip called a rapier that shoots between the upper and lower yarns from one side, grabs the end of a length of weft yarn on the other side, and pulls it through the shed. The spool holding the weft yarn doesn't need to travel through the shed, so it can be as big as you like.

The worker tending this machine only needs to interact with it if something goes wrong: the weft yarn spool lasts all day.

COUNT YOUR THREADS BEFORE YOU BUY

MAKERS OF EXPENSIVE bedsheets often advertise what a tremendously high "thread count" they have. Two hundred threads is decent, 300 is pretty nice, but the "best" brands brag about being 600 or even 1000 thread count. Anything over 220 is almost certainly a lie, even when the advertising swears it's not. Here's why.

Rope model of a plain weave

Rope model of a sateen weave.

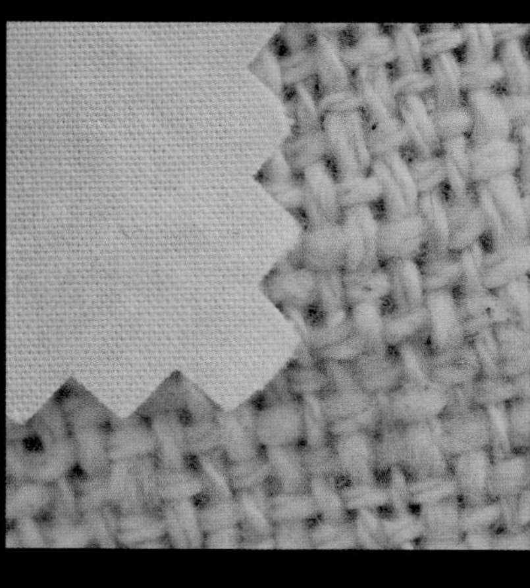

Just look at the difference between my 30-thread-count handmade cloth and this 220-thread-count machine-made fabric!

1000-thread-count sheets are a lie created by the industrial bedding conspiracy.

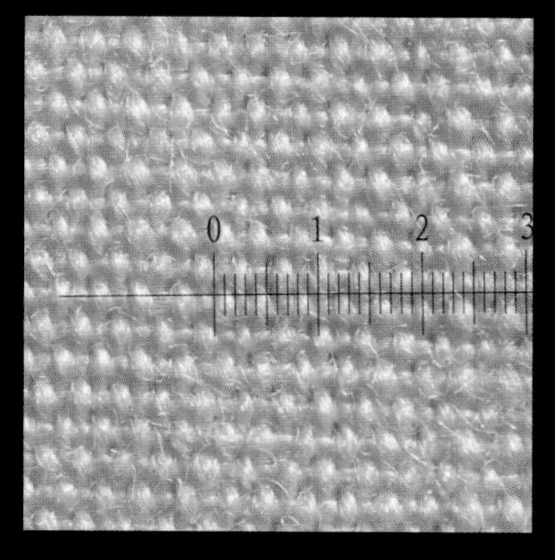

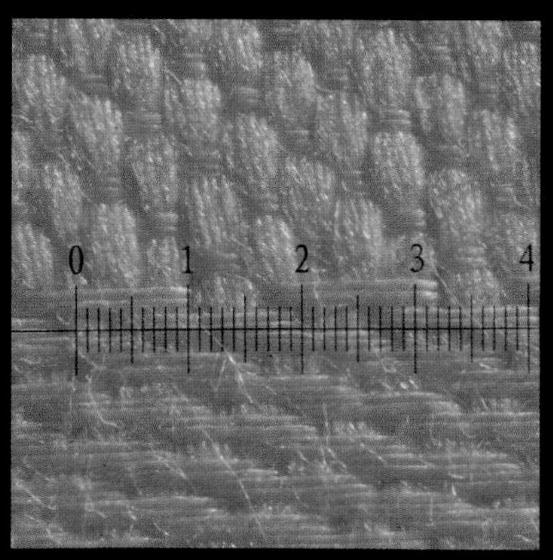

Which of these looks like the more finely woven sheet to you? The one on the top is sold as 220 thread count, and that's exactly what it is, proven by the superimposed micrometer scale. The one on the bottom, which is obviously a much coarser weave, is sold as 1000 thread count! They are counting dozens of individual strands in a satin weave as if they were separately woven threads. It's pure nonsense.

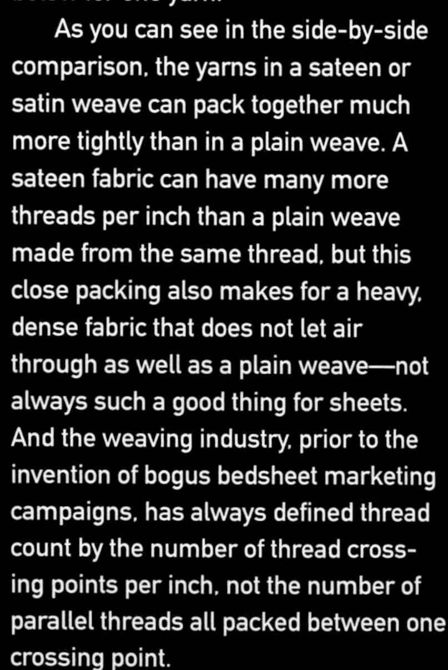

When every yarn goes up and down at every crossing point in both directions, it's called a plain weave. This is both the most common and the strongest, most durable form of weaving. But it's not the only way to do it. Here we see a "sateen" weave. Instead of crossing each time, the weft yarns stay on top for three or four crossings before dipping below for one yarn.

As you can see in the side-by-side comparison, the yarns in a sateen or satin weave can pack together much more tightly than in a plain weave. A sateen fabric can have many more threads per inch than a plain weave made from the same thread, but this close packing also makes for a heavy, dense fabric that does not let air through as well as a plain weave—not always such a good thing for sheets. And the weaving industry, prior to the invention of bogus bedsheet marketing campaigns, has always defined thread count by the number of thread crossing points per inch, not the number of parallel threads all packed between one crossing point.

Two hundred years ago, the one on the right was the standard—the type of cloth people lived in and slept under. The one on the left would have seemed an impossible luxury even for a king. Today, the one on the left is cheap as dirt, but people will pay extravagant amounts for crude handmade fabrics like the one on the right—at least when they're made by people better at it than me.

Machines have made the finest luxury commonplace, but in doing so they have elevated the value of human time and handwork to the point that anything *not* made by machine is rare and expensive.

Weave Me a Shirt of Copper and Steel

MANY OTHER FIBERS can be woven besides cotton, including wool, flax, sisal, hemp, nylon, and any number of other natural and synthetic fibers. Weaving even drifts into "Really, you can do that?" territory, as we will see in this section.

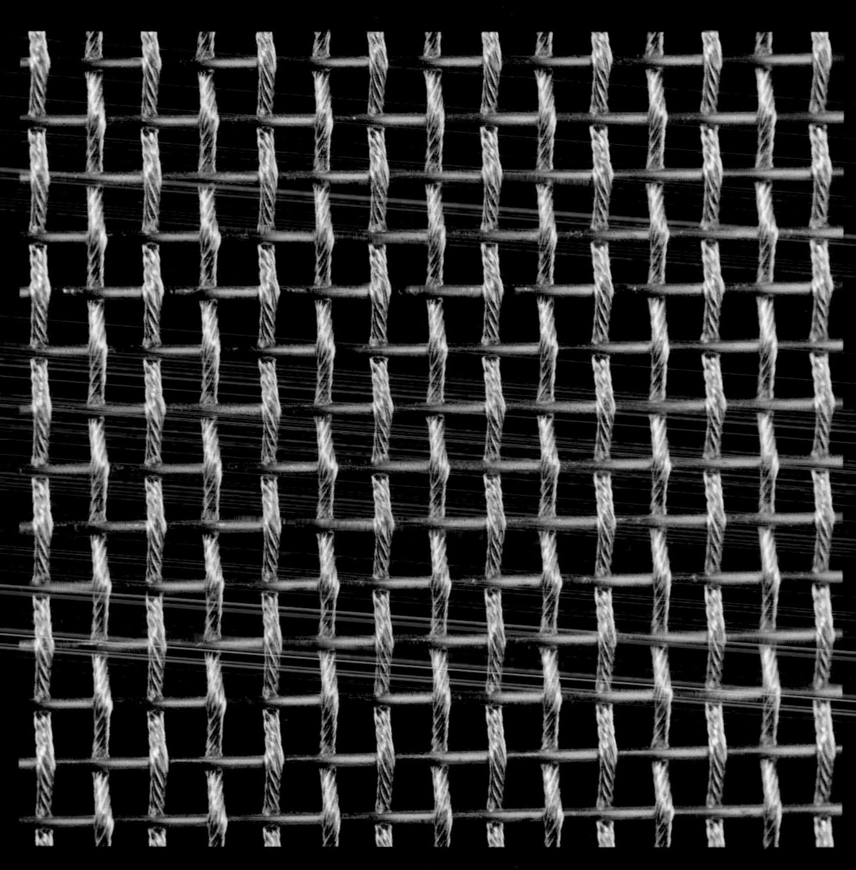

This is exotic woven "cloth" in which one direction of the weave is made of stiff, solid stainless steel wires and the other direction is made of tiny twisted bronze wire cables (bronze is an alloy of copper and tin). The "threads" are made of metal, but are woven just like regular cloth.

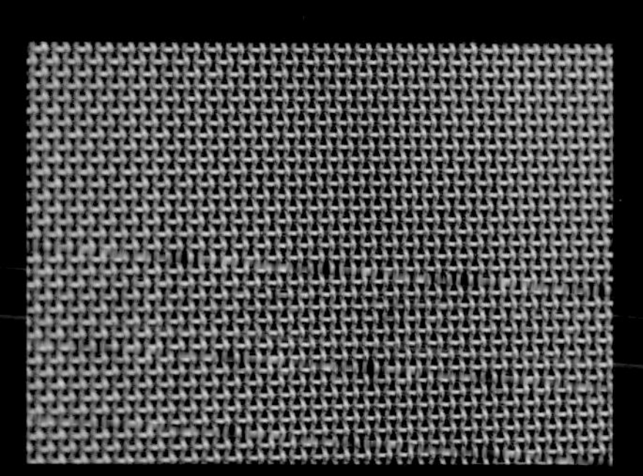

Huge belts of carefully woven fine bronze wire cloth are used in the papermaking industry to press the water out of wood or cotton pulp slurries. The surface of the wire cloth has to be absolutely perfect, because any flaw will be pressed into the surface of the paper it's making, not just once but every time that section of belt comes around.

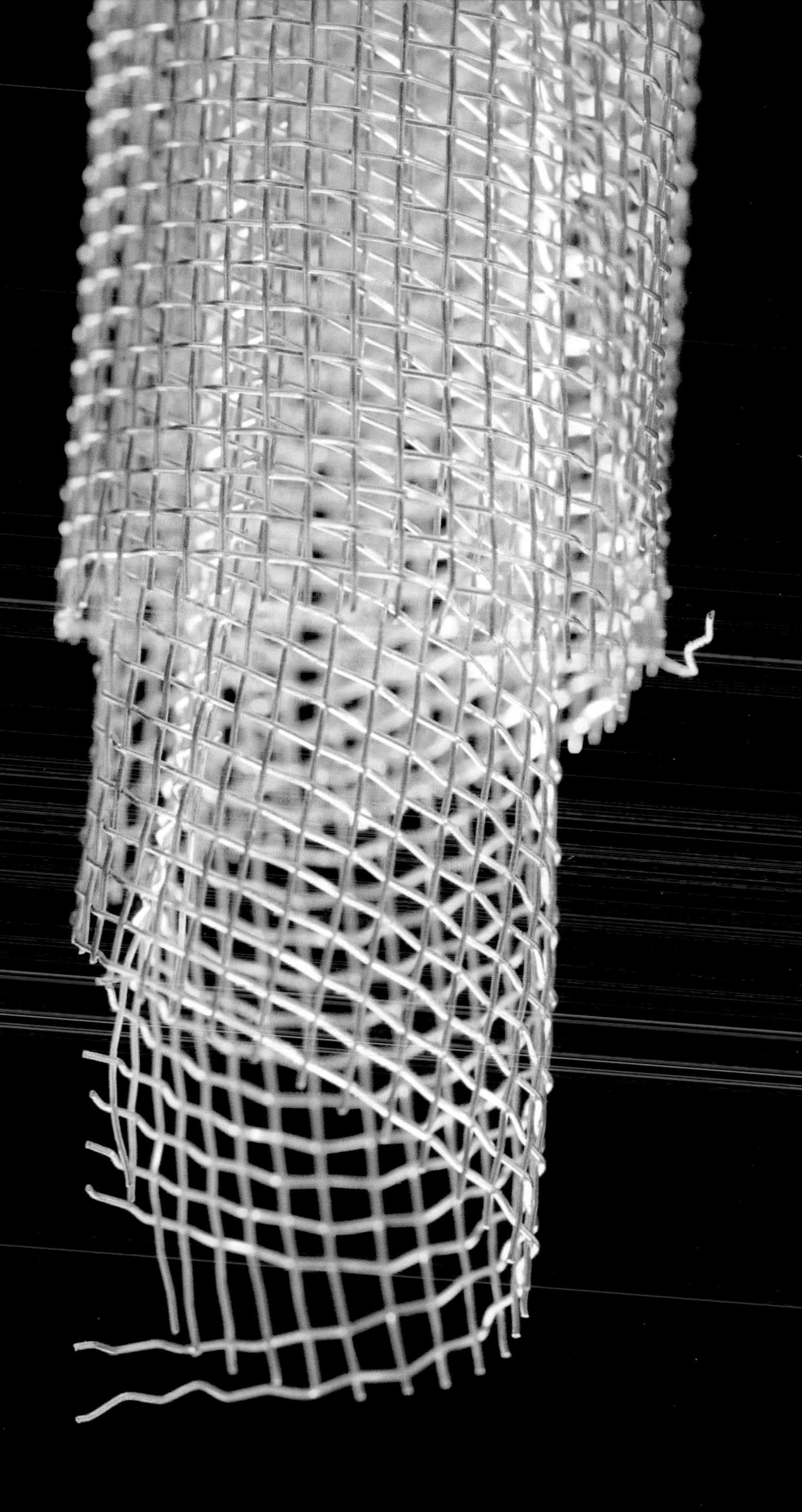

Woven wire mosquito netting keeps people in many parts of the world from being eaten alive.

WEAVING WIRE

WIRE CLOTH CAN be woven more finely than cotton. Unlike cotton, wire and synthetic filament cloth is commercially available up to a true, and truly astonishing, 1270 thread count (meaning 635 wires per inch in both directions).

Weaving wire leads to certain difficulties, which lead to massive machines like this one, a 30-foot-wide (9 m) wire weaving loom in Germany. It weighs about 25 tons—which is a lot of steel. Looms for cotton are big machines, but this one is *much* bigger and *much, much* heavier than anything used to make regular cloth. Why? It's wider because joining together wire cloth is much harder than just sewing together seams of cotton cloth. If you want a 30-foot-wide cotton sheet, you just sew together smaller pieces. If you want 30-foot-wide wire cloth (and people do), it's best to make it that wide to start with. It's *heavier* for a different reason: tension.

Wire doesn't stretch like cotton, and it takes a lot more force to pull a wire straight than it does to keep a cotton thread straight. Imagine you are making wire cloth 30 feet (9 m) wide, and there are 100 wires per inch (40 wires per cm). That's 36,000 wires, each of which needs to be pulled pretty hard to keep it straight. This loom can handle up to 8 tons of force per meter of its width, or 72 tons total. *All* of that force is pulled between the uptake roller in the front and the warp beam in the back. Each one of those 36,000 wires has to be manually threaded from the back of the loom, through the heddle eyes, through the comb, and onto the uptake roller. By hand, one at a time, 36,000 times. This takes a two-person crew (one in back, one in front, handing off wires to each other), working two shifts, 16 hours a day, for 8 days. During that time this very expensive machine is sitting idle, making nothing.

▼ A wire loom at a factory in Germany

▲ Once this machine is all wired up, it runs for about 3 weeks nonstop, day and night, to make around 600 yards (550 m) of very high-quality, flawless wire mesh. This is much, much slower than a cotton loom works. The fastest looms for cotton can insert nearly 30 weft yarns per second, but wire is woven at the more leisurely pace of around one wire per second.

▲ This is the warp beam on the back side of the loom. It's the roll on which the warp wires are spooled before they are fed into the loom. In a cotton loom this beam would be a steel rod about 10 inches (25 cm) in diameter, with cotton thread wrapped around it to a diameter of about 4 feet (120 cm). But in a wire loom, the tension is so high that a far more substantial beam is required. This one is a pipe 32 inches (80 cm) in diameter, with walls made of solid steel 2.4 inches (6 cm) thick. Because the starting diameter is so much larger, not nearly as much wire can be wound on it.

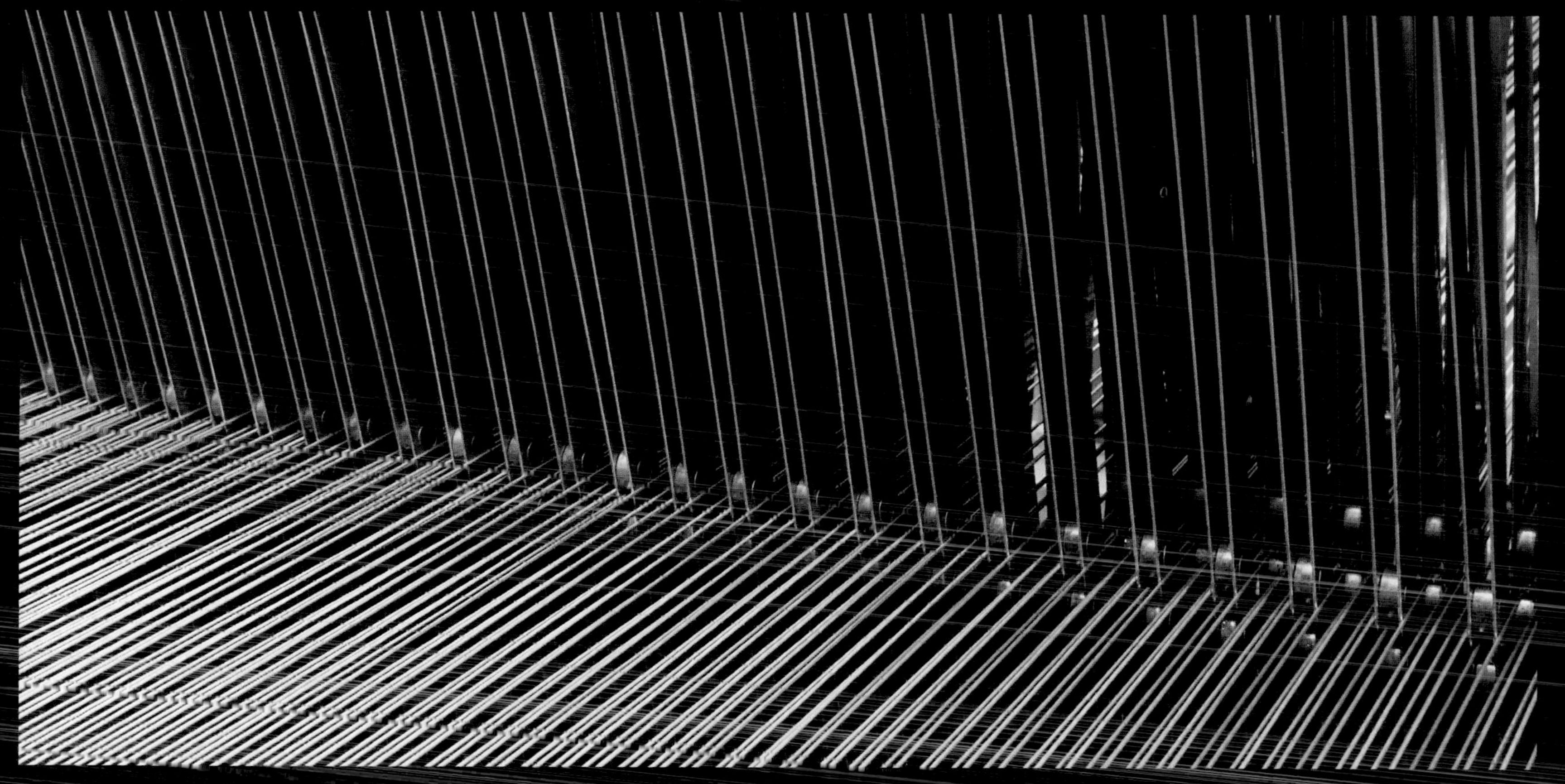

▲ The joy of machines is often in the details. These are heddles, the things that alternately lift and lower the warp yarns (wires) as the cloth is being made. There is one heddle eye for each warp yarn. For thin, smooth wire, or cotton thread, the heddle eyes are simple holes, but that doesn't work if what's going through them is multi-strand twisted steel cable. In that case you need a pair of ball bearings in each and every heddle, to provide smooth passage for the bumpy cable.

▲ As when weaving cotton, there are more options than just plain over/under weaving. Here is what satin weave looks like in stainless steel (front and back).

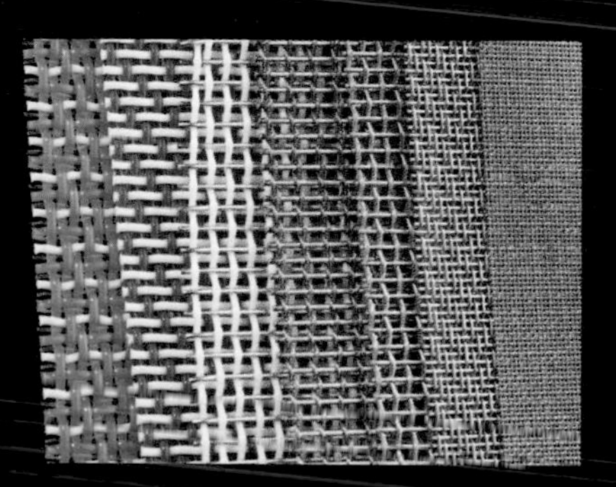

◄ Cloth can be made of mixed materials: stainless steel, brass, or even mixtures of wire and synthetic nylon monofilament. Amusingly, in keeping with weaving industry tradition, no matter what these things are made of, they still call the individual strands "yarns" whether they are twisted thread, multi-ply actual yarn, extruded filament, or drawn wire.

► Nothing beats a nice pair of really big gears. These monsters are what apply the many tons of force needed to keep the warp wires properly tensioned.

Discarded beauty is one of the best things about industrial production. The prize for most beautiful industrial waste goes to these nickel/chromium nodules created (and brutally recycled) by the electroplating industry. But these cut-off edges of woven wire cloth are pretty nice too. Who needs a crown of thorns when you can have a crown of wire cloth scrap?

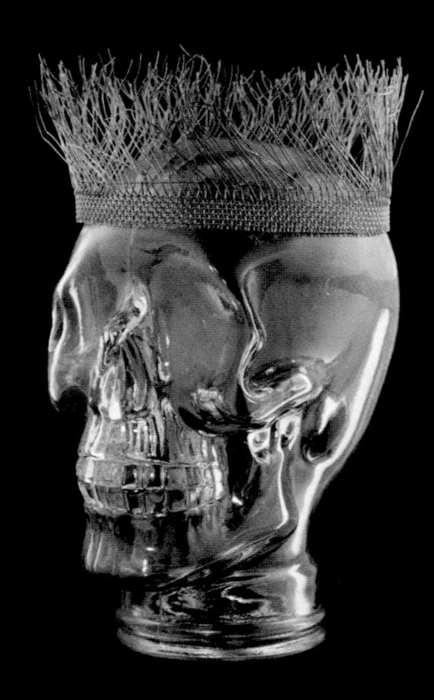

You may have thought the title of this section about making clothes out of wire was a joke. But you can in fact buy pretty much everything—shirts, hats, pants, underwear, socks, and so on—made of metal thread. Actually, most of it is silver-coated nylon or polyester thread, rather than solid wire, but I think that counts. This stuff is made for people who are worried about exposure to electromagnetic waves from power lines, cell phones, aliens, or the government. Personally I'm not worried about any of that, but I do think it's great that these things exist, so I can put them in my book.

Wire is also used to make actually useful protective clothing. This glove is sold for people who chop things up with large knives and prefer not to chop their fingers off.

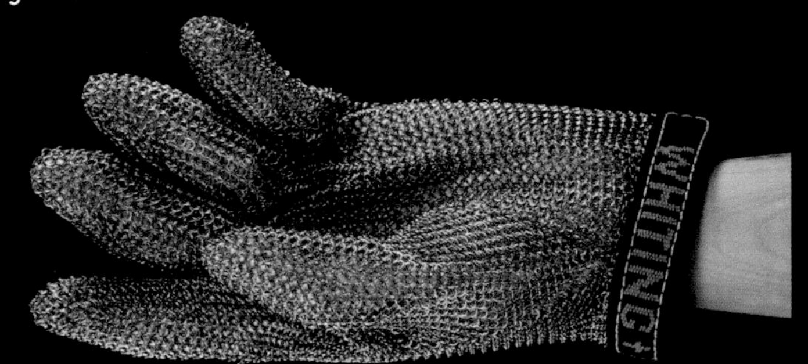

Here's where the pretty scrap edging comes from. After each weft yarn is inserted through the warp by the rapier, it is cut off and the end left long. As the wire cloth is being rolled up, a slitter along the edge cuts off the outer inch or so (a few centimeters) of cloth, which drops into a pile to be recycled. Why waste such a wide strip of expensively produced wire cloth? Because it's imperfect: along the edge the angle of the weft wire may not always be correct.

Woven wire finds itself in some pretty odd places. For example, the left anterior descending artery in my heart. Stents are hollow tubes woven from extremely fine wire. The one you can't see in these pictures (because its wires are so thin they are invisible in the x-ray image) was inserted through a tube from my wrist all the way into my heart, and then allowed to expand in place, unblocking an artery that would otherwise almost certainly have killed me before this book was finished. Thank you, woven wire, x-rays, and skilled doctors.

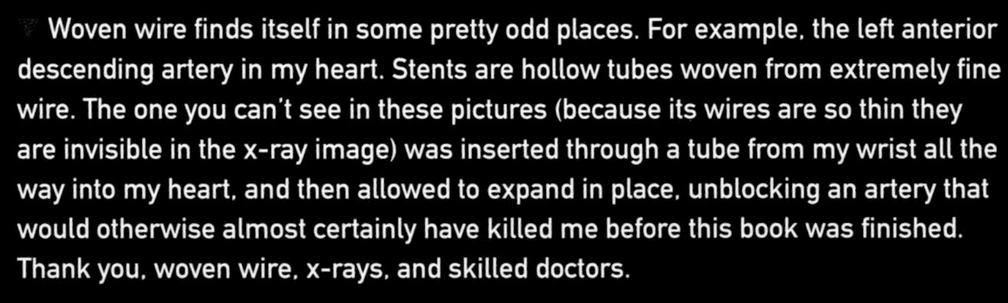

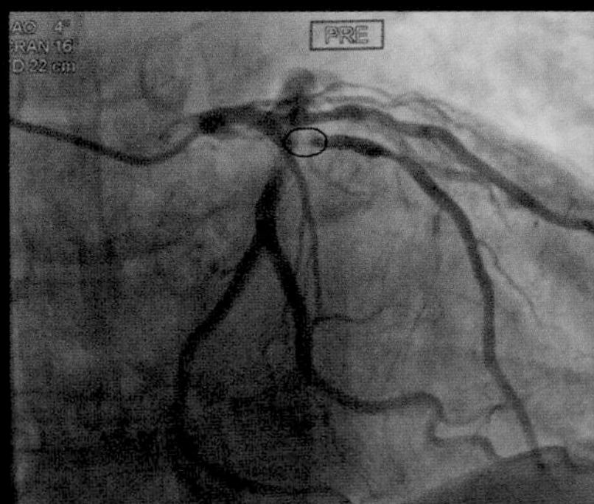

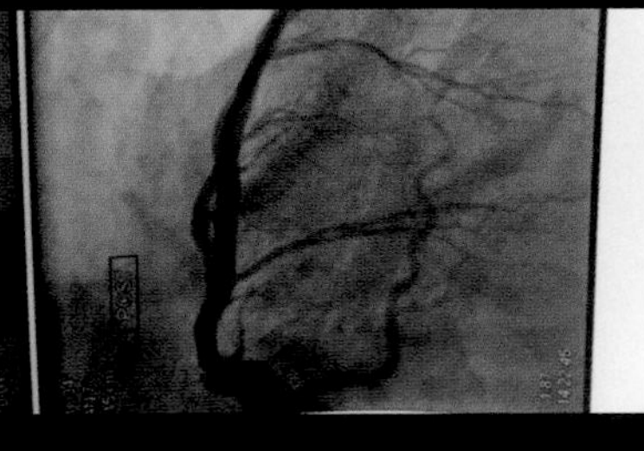

Sewing

WE'RE NOW JUST *one step away* from having a finished potholder: we need to sew the thing together.

Sewing is the process of connecting layers of fabric to each other by passing a thread through and between them. Simple enough, but before we go any further with the potholder, let's understand the topology of sewing and see some of the different ways it can be done.

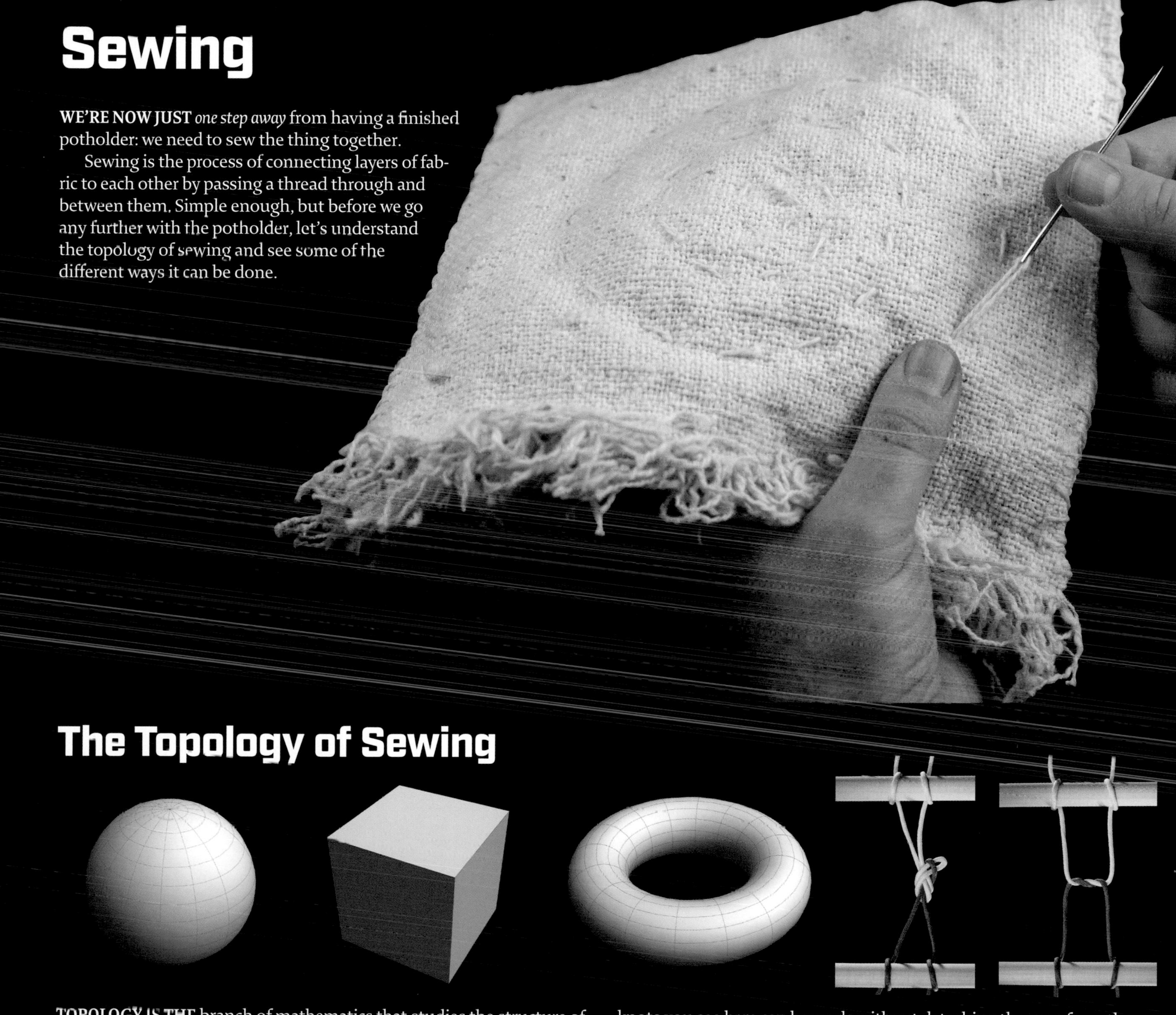

The Topology of Sewing

TOPOLOGY IS THE branch of mathematics that studies the structure of shapes without regard to their details. For example, a sphere is topologically the same as a cube, but different from a donut. Adding the hole makes the donut into a fundamentally different class of shape.

Knots are a classic example of a topological problem. One of the knots you see here can be made without detaching the rope from the post at either end. The other knot can only be made if you have access to one loose end of rope. Analyzing which knots can or cannot be made in a rope fastened at both ends is a topological problem. It's also a problem that is fundamental to sewing.

RUNNING STITCH

THESE CLEAR BLUE acrylic sheets represent layers of fabric. Normally the stitches would be pulled tight, clamping the layers together, but I've left them separated so you can see everything out in the open.

This is how I'm going to sew together my potholder: up and down, pulling the thread all the way through after each stitch. This technique has been practiced for longer than anyone knows. You can recognize handmade "running stitch," as it's called, by the fact that only every other stitch is visible on both sides. This method makes a strong seam, but suffers from the same bottleneck as weaving with a shuttle. Because you have to pull *all* the thread through the workpiece with every stitch, there's a limit to how much stitching you can do before you run out of thread and have to tie on a new length. That limit being, typically, about twice the length of your arm plus the width of your body, because that's how much you can conveniently pull through with the thread doubled over.

The first needles were made of bone. Such needles are still made today, but only for historical re-creation purposes. Modern hand-sewing needles are made of steel, allowing them to be much thinner and sharper. There isn't much more to say about hand-sewing. There are various styles of stitching, but all of them come down to the same thing: push the needle through, then pull the thread after it. Repeat until clothed.

The problem of having to pull all the thread through every time is actually much worse than the shuttle problem in weaving. While inconvenient, mechanized shuttle looms are still in existence. But no machine was ever devised that tried to replicate the hand-sewing process of pulling all the thread through. Instead, there needed to be a method that allowed for arbitrarily long spools of thread to be used (the equivalent of the large spools of weft yarn used in rapier looms).

The earliest sewing needles were made of bone. This one was found in a cave in France and is somewhere between 12,000 and 19,000 years old. Strong evidence exists of similar needles, and the tools used to make them, dating back over 60,000 years.

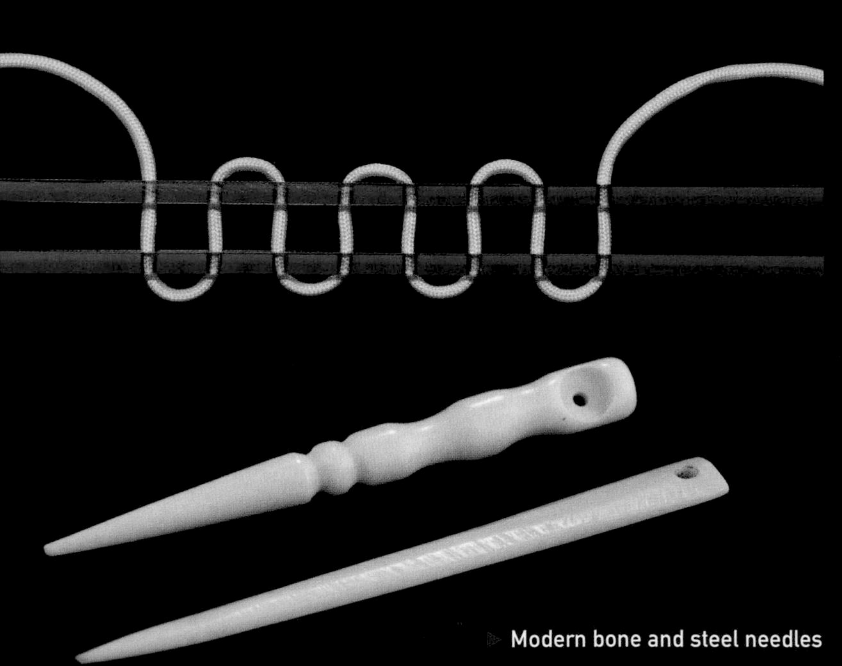

Modern bone and steel needles

CHAIN STITCH

SEWING RESISTED ALL attempts at mechanization until the invention of a radically different method, achieved by putting the eye of the needle at the tip rather than the back of the needle.

It's an unavoidable reality that if you're not going to pass the whole length of the thread through the fabric, then if you poke the thread down through a hole, you've got to bring it back up through the same hole. No matter what you do to the thread underneath, it's still going to form a loop that comes out and goes back in the same hole. That is a topological certainty, and it severely limits the kinds of stitching you can do. (Unless you go around the edge of the fabric, but that's cheating: we are talking about stitches that work in the middle of a sheet of fabric.)

"Chain stitch" is the simplest way to make a seam with a single thread, working from one side only. Each loop on the bottom holds the next one in place, and the thread always enters and exits through the same hole. There's just one tiny little massive flaw with this method.

If you pull on a loose thread in a handmade running-stitch seam, nothing much happens. The thread can't be pulled out unless you alternately pull it out one side, then out the other, and so on, reversing the way in which the seam was originally made. Running-stitch seams are strong.

But if you pull on the end of a chain stitch seam, the whole thing just falls apart. Since the thread always stayed on the same side while the seam was being sewn, there's nothing to stop it from coming right back out again. Because of this flaw, chain stitching is used only for decorative embroidery or easy-to-open bags of grain.

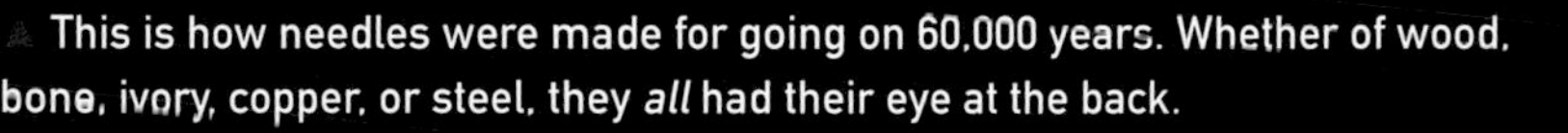

This is how needles were made for going on 60,000 years. Whether of wood, bone, ivory, copper, or steel, they *all* had their eye at the back.

In the 1800s the eye was finally moved to the front, making machine stitching possible for the first time.

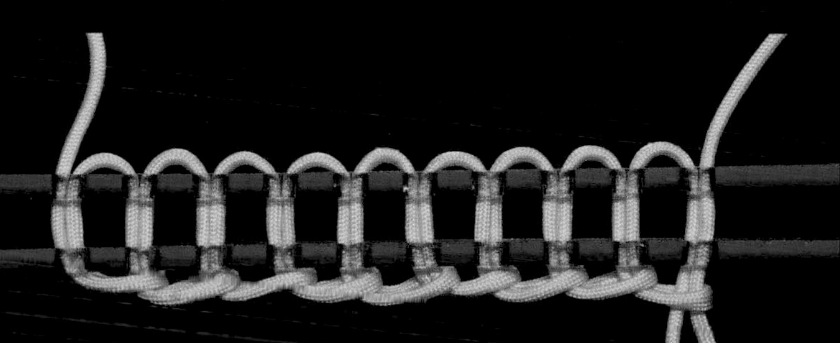

If both ends stay always on top, then the thread must always come back up through the same hole it went down through.

Chain stitch seams are simple, but flawed: they unravel far too easily.

Here's proof that even massively flawed methods can find a way to be useful. Many kinds of grains, rice, animal feeds, fertilizer, etc., are sold in bulk bags that have been sewn shut across the top with a chain stitch. To open them, you just find the right end and start pulling. In a matter of seconds the whole thing unravels and the bag is completely open on top. (This one has two rows of stitching, so it will take a second pull to open.)

As great as this feature of chain stitching is for opening feed sacks, you probably wouldn't want it happening to your pants. A better method is needed.

STRONG STITCHES

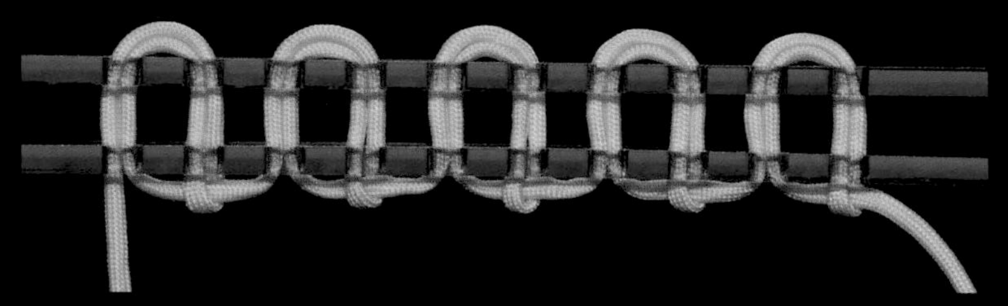

MOCK RUNNING STITCH
This interesting variation on single-thread machine stitching is called a "mock running stitch" because from above it looks just like a handmade running-stitch seam where the thread has been pulled through with every stitch, leaving only every other stitch visible. But in fact it's machine-made with the thread fed always from below. There is exactly one commercially available machine that does this, called the Baby Lock Sashiko. I'd tell you how it works, but it makes my head hurt thinking about it. (OK, I admit, I don't understand it, and since this form of stitching suffers from the same problems as the far more common lock stitching described next, we're not going to dwell on it.)

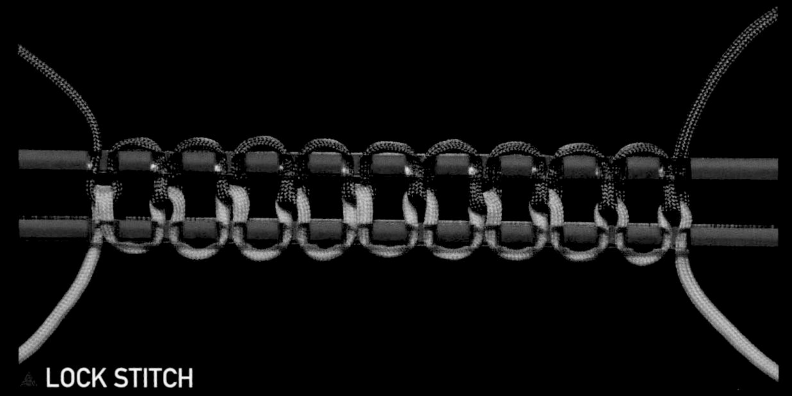

LOCK STITCH
The "lock stitch" is how nearly all serious, strong machine seam stitching is done. Notice that there are *two different* threads involved, one that always stays on the top of the fabric and one that stays on the bottom. At each stitch, they are looped around each other, always in the same direction.
Lock stitching is strong. If you start pulling on one of the threads, a few stitches may pull out, but fairly quickly the top and bottom threads will tangle with each other, preventing any more stitches from pulling through. That happens because the threads are continually twisted around each other, one twist per stitch. Even if the fabric suddenly vanished, the two threads would not come apart, because they are spiraled together, like strands of a rope.

I chickened out on trying to explain the Sashiko machine, but lockstitch sewing machines are too important to gloss over. The system I describe here is universal: all standard sewing machines work this way. In the many years since the advent of mechanized sewing no one has come up with a better type of stitch, or even a plausible alternative for all but the most specialized uses. After a hundred years of trying to do better, this is just the way it's going to be.

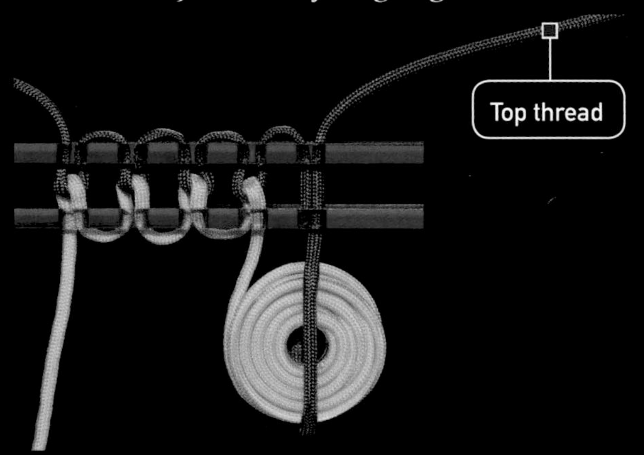

If the bottom thread could also come from a large spool somewhere away from the action, that would be ideal. The machine could keep sewing for miles without stopping. Sadly, that is fundamentally, topologically impossible. If the bottom thread were also coming from a distant source, it would be impossible to twist the two threads around each other. To achieve the strength of a continuously twisted lock stitch, the bottom thread *must* come from a spool small enough to allow the top thread to be passed entirely around it on each stitch.

The top thread is coming from a large spool somewhere away from the action. In practice, commercial sewing machines typically use spools of thread about the size of a soda can holding about 6,000 yards (3.4 miles or 5.5 km) of thread. But there's no limit to the size of spool you could use, because it doesn't come into play in the sewing operation. In this sense it's like the weft yarn spool in a rapier loom.

So this is how lock stitches are formed. The top thread is pushed down through the fabric, then pulled out in a loop, which is then passed entirely around a "bobbin" containing the bottom thread and pulled back up toward the top until the new stitch is tight. Passing the top thread around the bobbin is what twists the two threads together. If you repeat the process, moving the fabric a small step each time, you create a lockstitch seam.

Now the question is, how does the machine make this black magic happen? In particular, how does it deal with the fact that the top thread has to pass entirely around the bobbin on all sides—front, back, and all around—while still somehow holding the bobbin in place? To see that, we need to add some more parts to our model.

The mechanism wrapped around the bobbin took a *long* time to work out, as reflected in the range of solutions invented and then rejected. This model replicates the modern system, minus a few details that we'll see later in real machines. There are two motions in a sewing machine: the needle goes up and down in a smooth harmonic oscillation motion, and the "hook assembly" that surrounds the bobbin rotates counter-clockwise at a constant speed. Nothing is starting and stopping abruptly, which allows the machine to run smoothly, with minimal vibration, at very high speeds—60 stitches per second for the fastest commercial sewing machines, 30 for most high-speed industrial machines, and around 10 or less for common home machines.

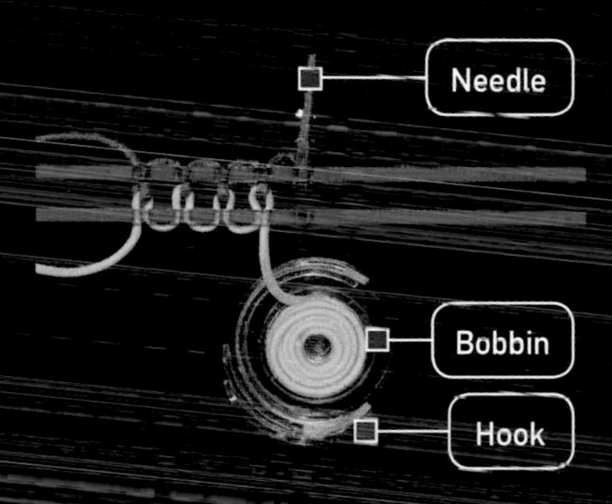

▲ We start the cycle with the needle on its way down through the fabric, and the hook starting to come up on the right side.

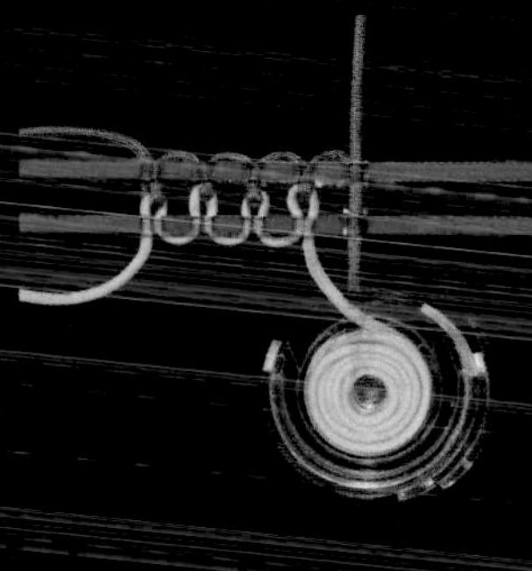

▲ Notice that the needle is bent to form a slight indent toward us, just above the eye hole.

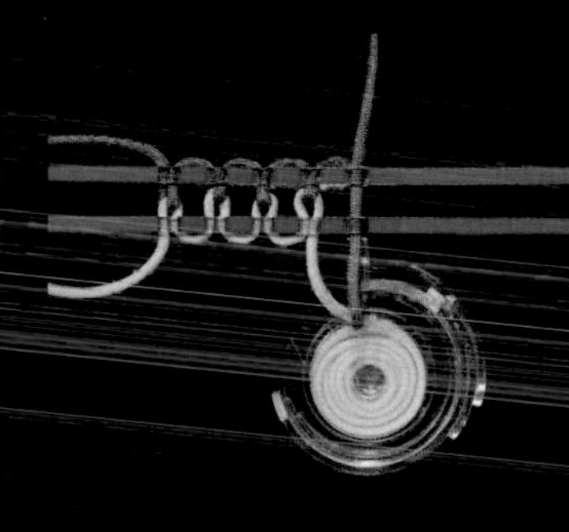

▲ The hook has a sharp angled point, which allows it to slip between the needle and the top thread right at the bend in the needle.

▲ As the hook continues to rotate around the bobbin, it pulls the top thread into a loop. Notice that both sides of the top thread are still on the same side of the bobbin thread; no twist has been formed yet.

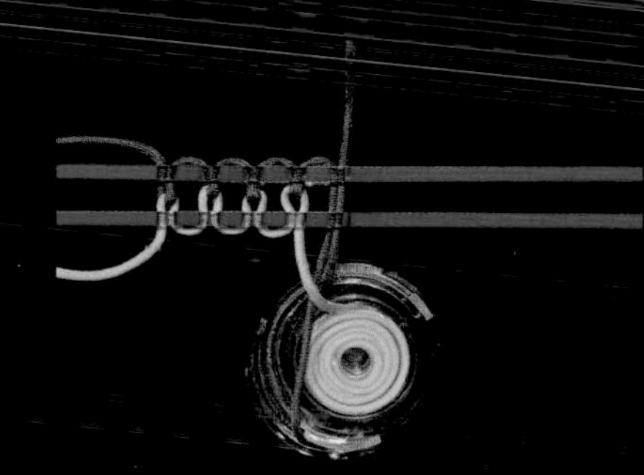

▲ The angled slot behind the hook makes one side of the loop travel backward, behind the bobbin, while the other side stays in front.

▲ The loop is now at its maximum size and completely surrounds the bobbin holder, which is floating in a cage, but has enough space all around it for the top thread to pass by on all sides.

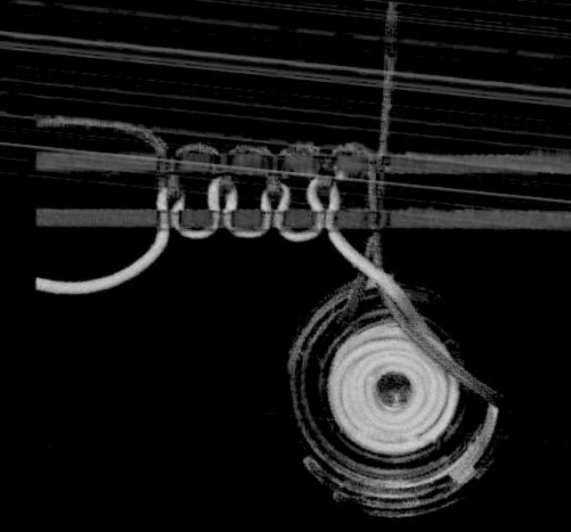

▶ As soon as the thread clears the bobbin, it is jerked back up by the take-up lever at the top of the machine, tightening the loop and completing the cycle.

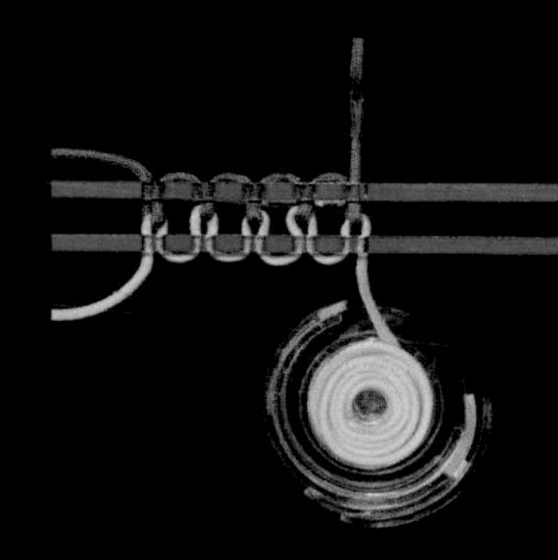

Sewing Machines

IT'S ESSENTIALLY USELESS to photograph the inside of a real sewing machine in action, both because the opaque metal obscures everything that's going on, and because the thread is so thin you can't easily see it. But these views of the bobbin, the bobbin case, and the hook assembly give some idea of how the parts fit together.

The bobbin case was not shown in the model on the previous page. It fits around the bobbin, inside the rotating hook assembly. The loop of top thread goes around the entire bobbin case.

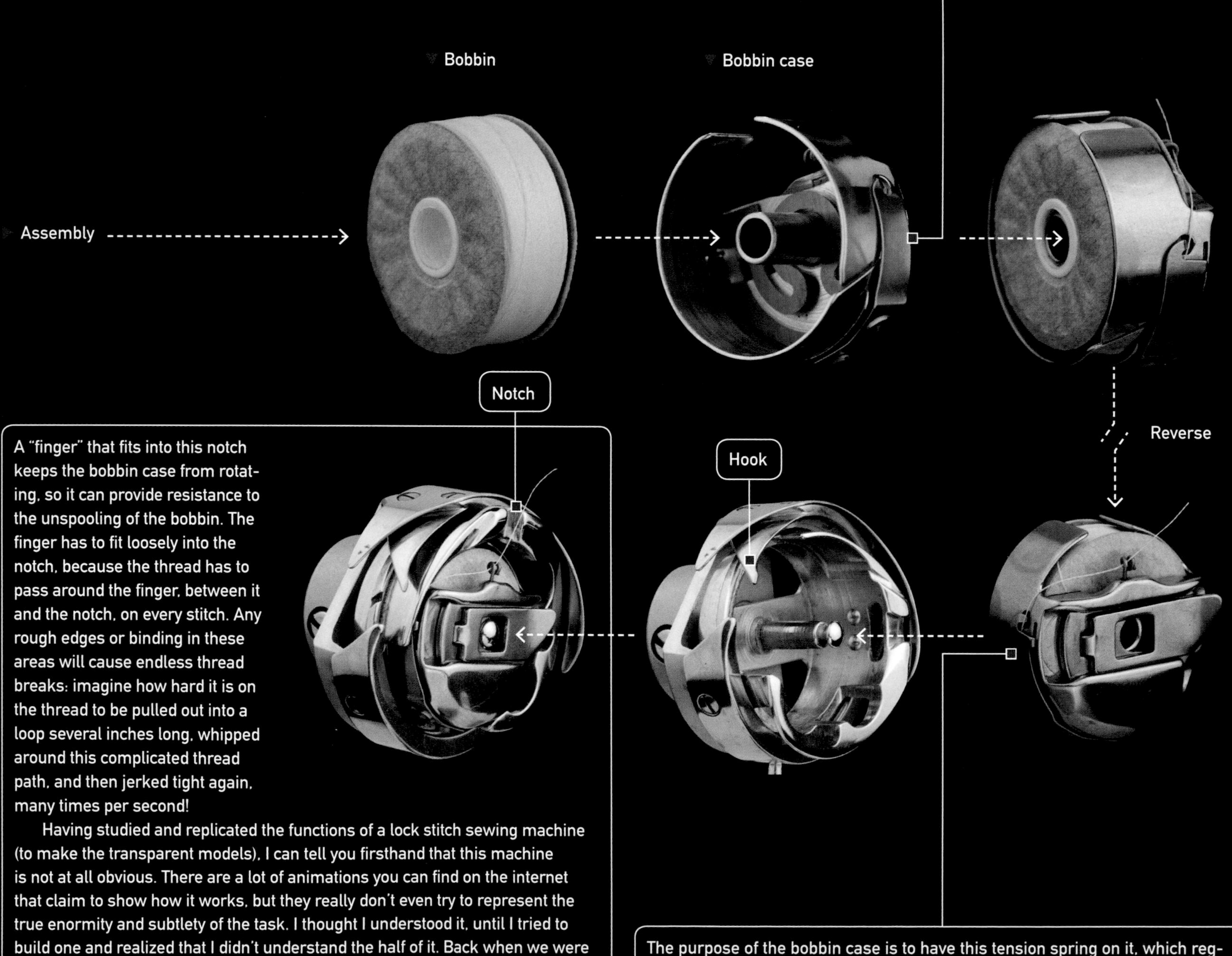

A "finger" that fits into this notch keeps the bobbin case from rotating, so it can provide resistance to the unspooling of the bobbin. The finger has to fit loosely into the notch, because the thread has to pass around the finger, between it and the notch, on every stitch. Any rough edges or binding in these areas will cause endless thread breaks: imagine how hard it is on the thread to be pulled out into a loop several inches long, whipped around this complicated thread path, and then jerked tight again, many times per second!

Having studied and replicated the functions of a lock stitch sewing machine (to make the transparent models), I can tell you firsthand that this machine is not at all obvious. There are a lot of animations you can find on the internet that claim to show how it works, but they really don't even try to represent the true enormity and subtlety of the task. I thought I understood it, until I tried to build one and realized that I didn't understand the half of it. Back when we were talking about the cotton gin, I said I thought it could have been invented a thousand years ago. This machine I'm frankly surprised was ever invented at all. It is deeply impressive.

The purpose of the bobbin case is to have this tension spring on it, which regulates how much force it takes to pull thread off the bobbin. If the bobbin could just spin freely (as it could in the model), the stitches would come out way too loose.

There are a lot of different sizes of bobbins used in home and commercial sewing machines, but all of them are pretty small. The biggest ones are found only in relatively slow machines used to sew with very thick thread (more like string than thread). All high-speed sewing machines, even the most powerful industrial models, use bobbins less than an inch (2.5 cm) in diameter.

This is a serious problem for operators of these machines: these bobbins typically hold about 220 yards (200 m) of thread, which at full speed will last less than 10 minutes. After that the machine has to be stopped and an operator has to put in a new bobbin. In a factory production environment that's a major hassle.

So for heaven's sake why don't they just make the bobbins bigger? Well, in the early days a few manufacturers tried making machines with unusually large bobbins, but they didn't work very well. The problem is that with a bigger bobbin, you have to pull down a bigger loop of top thread to get around it. That slows down the maximum speed, and it puts more stress and wear on the thread. Suppose you are making 1/8-inch (3 mm) stitches, and you need to pull a 5-inch (125 mm) loop to get around a large bobbin. Each section of top thread will end up being pulled around the bobbin case *forty times* before it's finally laid to rest in a stitch.

The industry settled on the sizes of bobbins in use today after generations of experience showed that you just don't want to go any bigger.

As we saw in the model on page 235, the alignment between needle and bobbin hook is critical. The hook has to *just* touch the needle, passing through the tiny space between the needle and the top thread, and it has to do this at *exactly* the right time, as the needle is starting its movement back up through the fabric. Every sewing machine has to have some way of assuring this alignment: usually it's done with a sturdy metal frame and gears or toothed belts, but there are other ways, as we will see.

TWO-SPOOL MACHINE

THIS "TWO-SPOOL" MACHINE was an experiment that did not catch on. It lets you use a full standard home-style spool of thread as the bobbin. It has an extra-large take-up lever on top, which needs to travel much farther than normal to pull up the large loops needed to get around an entire spool of thread. With this machine you can sew for hours without running out of bobbin thread, but the whole time you're wishing you had a smaller bobbin.

A Menagerie of Machines

SEWING MACHINES COME in a wide range of shapes and sizes. They started out strong and heavy, made of thick cast iron and lasting nearly forever. I've restored a few that are over a hundred years old and needed only a bit of oil and some adjusting. In the present day they have morphed into cheap plastic models that only last a couple of years. But on the plus side, they can do a lot more complicated things, and they don't weigh nearly as much.

This machine is not much bigger than a toy, but it is a *lot* heavier, and it's definitely a serious machine, made for the ages of solid cast iron. These Singer "Featherweight" machines are much prized by sewing enthusiasts for their smooth operation, compact size, and solid quality. They were made in America for over thirty years, ending in 1969.

This is a tiny Russian-made chain stitch machines from the Soviet period, just a few inches high. Simple, not particularly well made, but meant to be used.

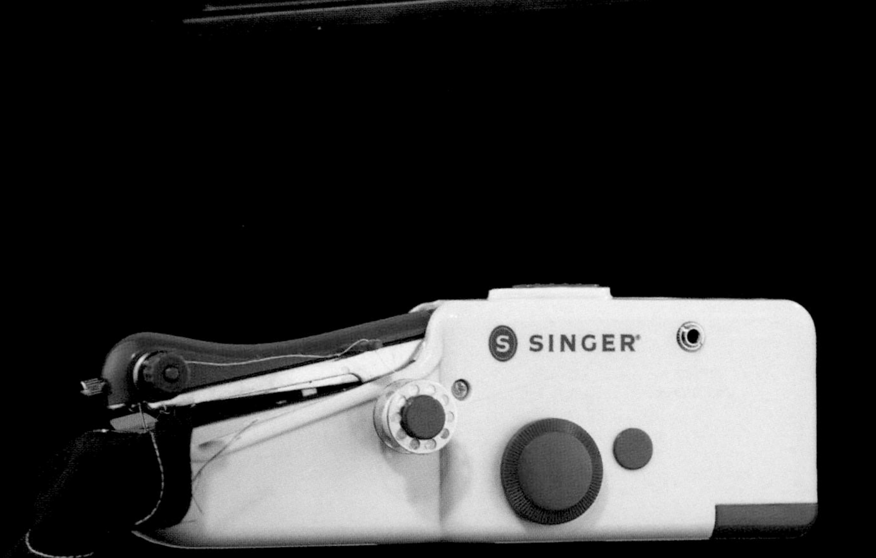

The modern equivalent of the tiny Russian chain stitch machine is this cheap plastic thing, one step above toy-grade. Because it makes a chain stitch, it's best for temporary repairs.

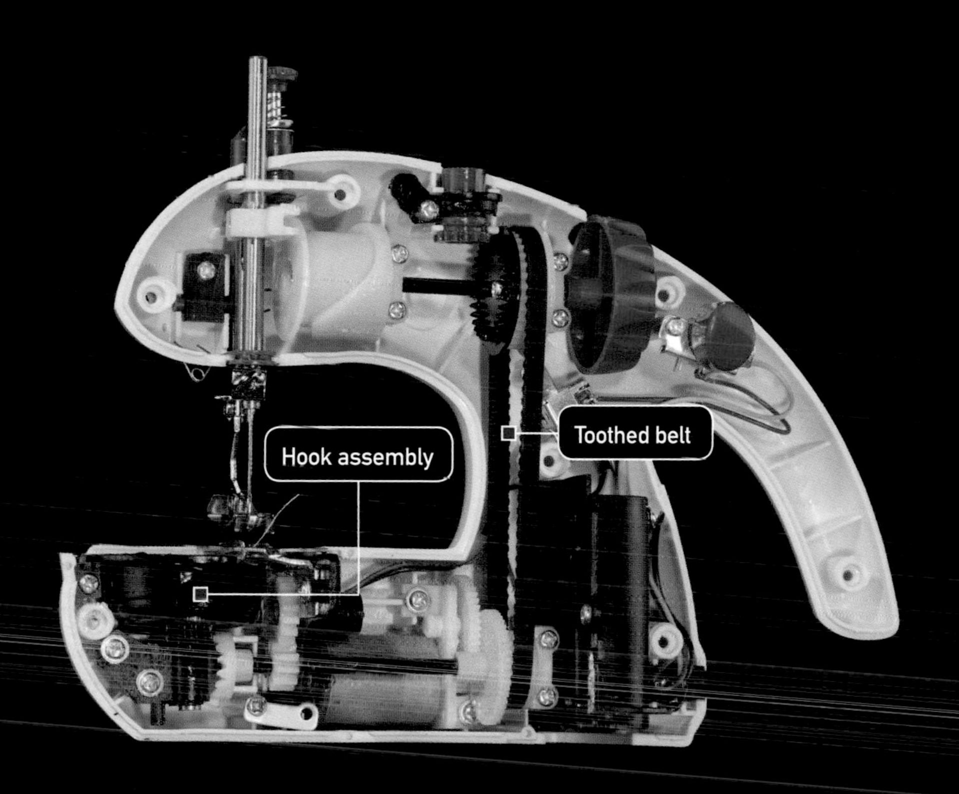

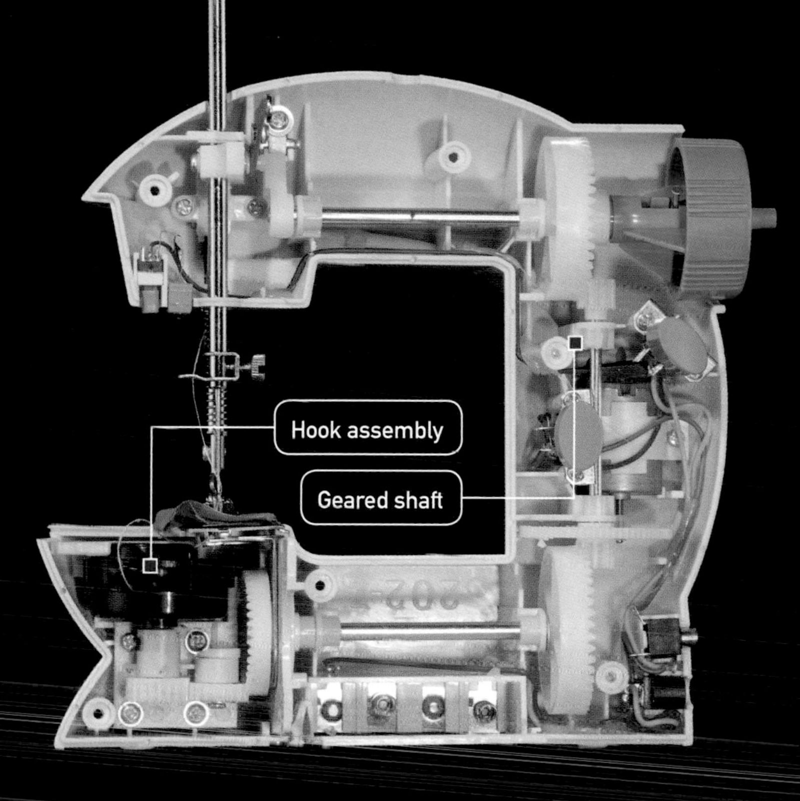

These two, very much in the toy size and price category (under $20) are real lock stitch sewing machines. Just not very…serious.

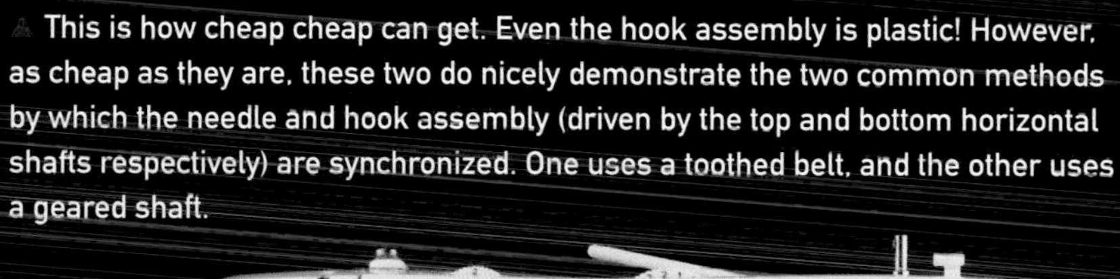

This is how cheap cheap can get. Even the hook assembly is plastic! However, as cheap as they are, these two do nicely demonstrate the two common methods by which the needle and hook assembly (driven by the top and bottom horizontal shafts respectively) are synchronized. One uses a toothed belt, and the other uses a geared shaft.

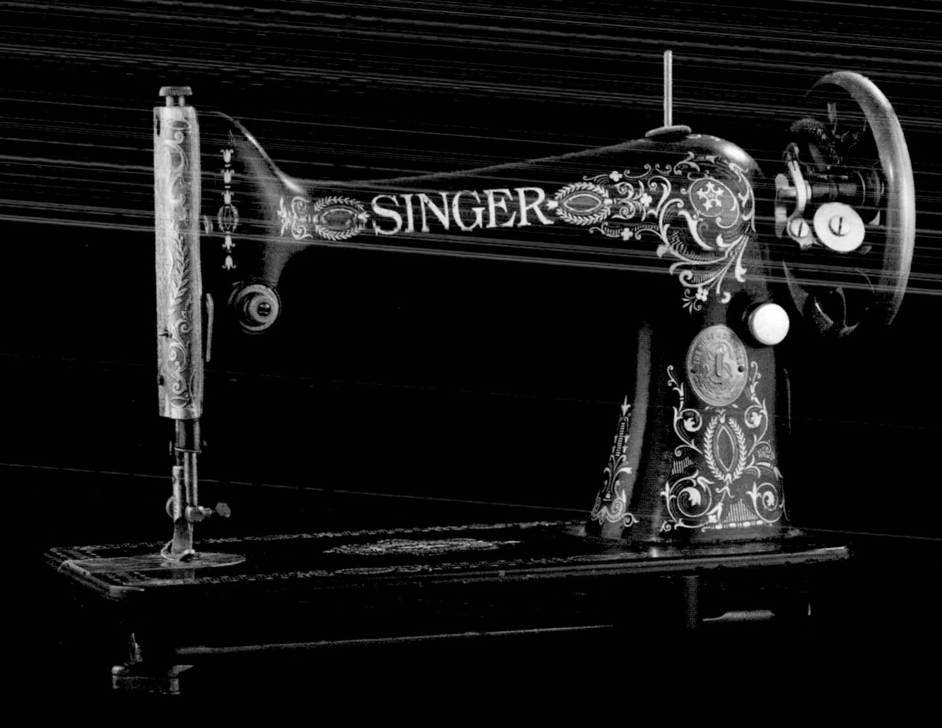

For about a hundred years this is what home sewing machines looked like. They were basically the same size and identical in function to the modern version, just much simpler (no fancy zigzag stitches) and much more durable.

Sometime in the 1960s, sewing machine companies decided that people were keeping their machines too long (i.e., forever), cutting into the market for new ones. So they started building crappy ones literally designed to break after a few years. This practice, essentially invented by sewing machine companies, has a name: planned obsolescence. It is claimed that some companies even went around buying up their own everlasting machines and destroying them because their existence was ruining the market for crappy machines.

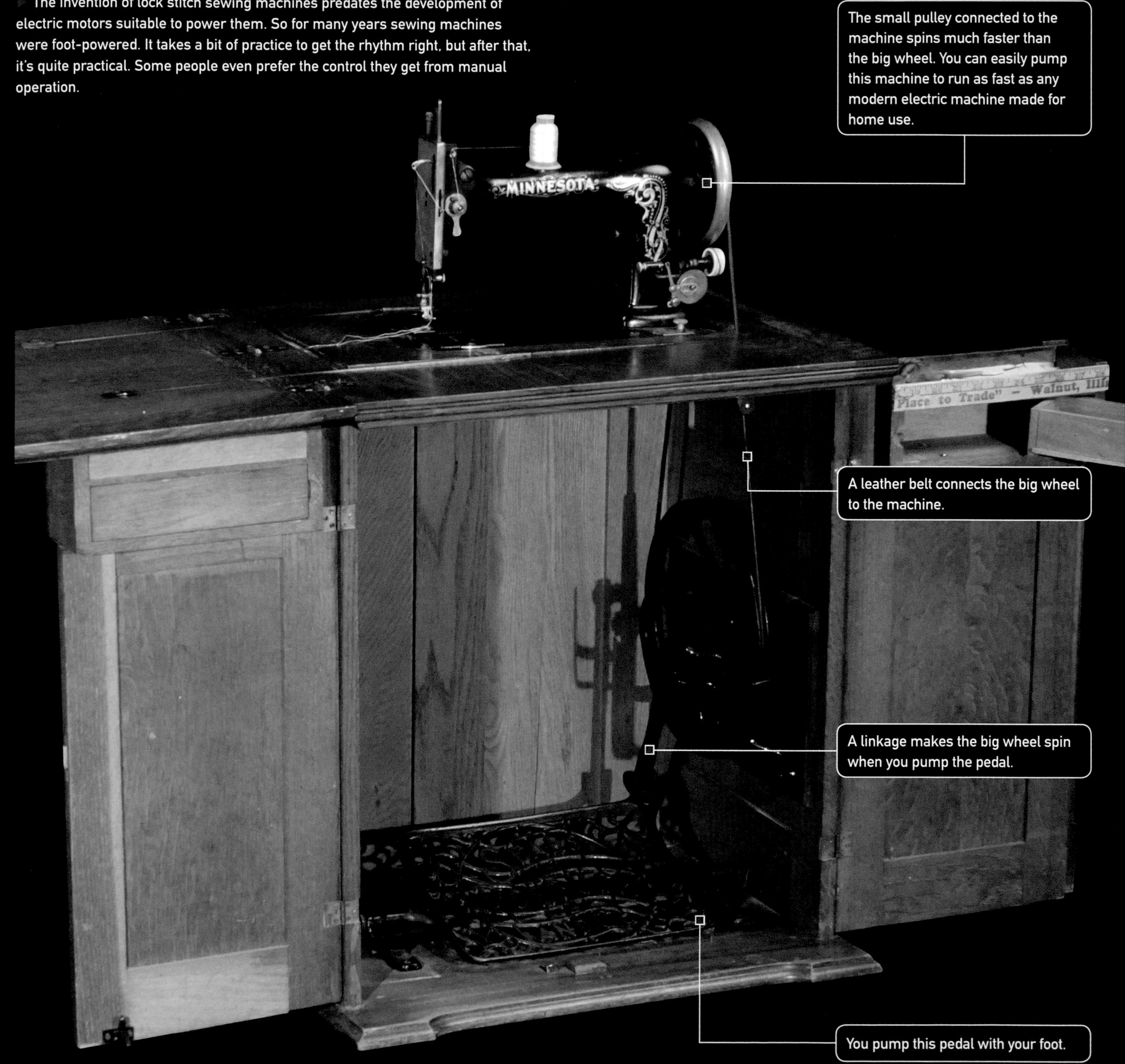

The invention of lock stitch sewing machines predates the development of electric motors suitable to power them. So for many years sewing machines were foot-powered. It takes a bit of practice to get the rhythm right, but after that, it's quite practical. Some people even prefer the control they get from manual operation.

▲ Inside this mid-range home machine we find a decently solid die-cast aluminum frame, with a lot of plastic and metal gears, levers, cams, and clever little mechanisms—many ideally suited to breaking after a fairly predictable length of time. It's nice to have a machine that will last forever, but then again, that durability is expensive and you're stuck with the same machine forever. Cheap machines that break quickly can be replaced with new cheap machines that have clever new features, like computer-controlled fancy stitching—built to work for a couple of years.

▲ There's a whole zoo of special-purpose sewing machines in the world. This is a fine old German-made shoe-making machine, still in use at a cobbler shop in Switzerland. Its job is to sew the tops to the soles of shoes. That means it has to be able to stitch very thick thread through multiple layers of literal shoe leather, and do that in very tight spaces around the inside and outside corners of a shoe.

▲ The needle is super thick and curved so it can sneak in from the side (it travels in a circular arc instead of straight up and down as in a normal machine). Underneath is a larger-than-normal bobbin assembly. This machine stitches very slowly, usually hand-cranked because you need to position each stitch carefully.

▲ It's impossible today to buy a robust, lasts-forever style of sewing machine meant for home use. But planned obsolescence and cheap plastic are not acceptable in the commercial market, where machines are expected to earn their keep by working hard every day for a long time. So commercial sewing machines are every bit as solid and heavy as they ever were. This cast-iron machine has a very deep "throat" and can make zigzag stitches through a dozen layers of tough synthetic outdoor fabric. It's intended for making sails for boats, but can be used the same way any home sewing machine would be. It's just a lot stronger. As for the price, adjusted for inflation it's probably similar in price to a last-forever home model of the past. It seems expensive (several thousand dollars) by the standards of today's sewing machines only if you compare it to the price of the cheap plastic alternatives.

▼ This modern Chinese shoe-making machine wouldn't cut it in a factory production environment, but for occasional shoe repair, it's fine. (Who even repairs shoes anymore?) It's solid metal, not cheap plastic, but unadorned and without any luxury features: all business, no polish. Solid value. I like this kind of machine, whether it's a $100 shoe-making machine like this one, an $80 drill press, or a $500 milling machine. Very cheap industrial machines are stripped down to their bare essence, not made flimsy, but made raw (and usually in need of a bit of grinding before they run smoothly).

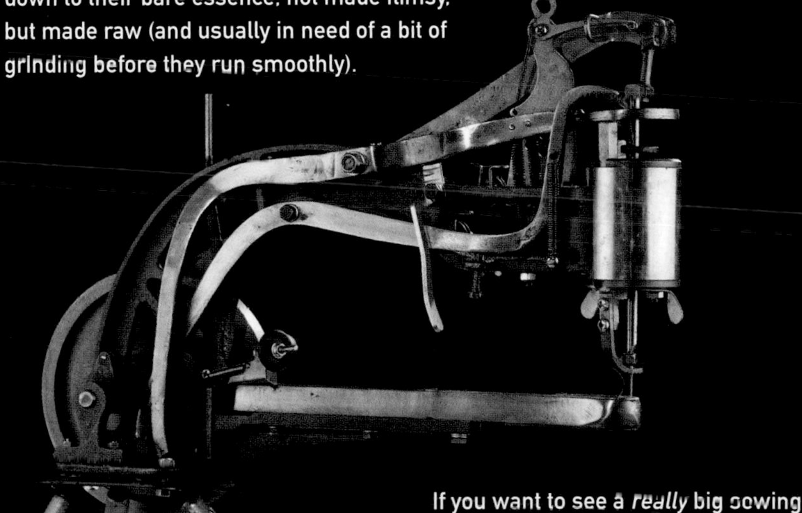

If you want to see a *really* big sewing machine, turn the page....

QUILTING MACHINE

THIS IS CLOSE to the biggest sewing machine I know of, and for complicated relationship reasons, it's in my studio. (Technically it belongs to an ex-girlfriend, but it was too big to move out when she did.) I have a mix of feelings about this thing, which is also a mix of two machines: a high-speed commercial sewing machine, and an X-Y plotter. A normal plotter would have a pen on the head and draw ink lines on paper. This one has a sewing machine instead of a pen, and it draws lines of thread on layers of cloth and batting. It's pretty fast: 2,000 stitches per minute, over 30 per second, on a straight or gently curved section. (When it needs to turn a sharp corner it has to slow down a lot: the crossbeam weighs a ton—literally—and can't just start and stop instantly.)

At the heart of this monster is a seemingly ordinary sewing machine, with a couple of differences that make me marvel that it works at all, let alone that it stitches reliably for hours on end at very high speed.

As we saw in the previous pages, nearly all sewing machines, from the cheapest to the most industrial, have two things in common: a sturdy metal frame linking the needle and bobbin assembles to keep them perfectly aligned, and a pair of drive shafts linked with gears to keep them perfectly in sync. This machine has neither.

Instead, the needle and bobbin are mounted on completely separate traveling heads, each with its own toothed belt to move it sideways. And they are driven by their own separate servomotors, electronically synchronized in a sort of "fly by wire" system.

My initial reaction on reading about this design, before we got the machine, was that it would never work. Once I saw the sheer size and weight of the crossbeam and the precision of the linear bearings that the two heads ride on, I thought "ah, so that's how it's done." I am in awe of this machine's ability to maintain millisecond timing and tenth-of-a-millimeter alignment as it flies across a wide expanse of fabric.

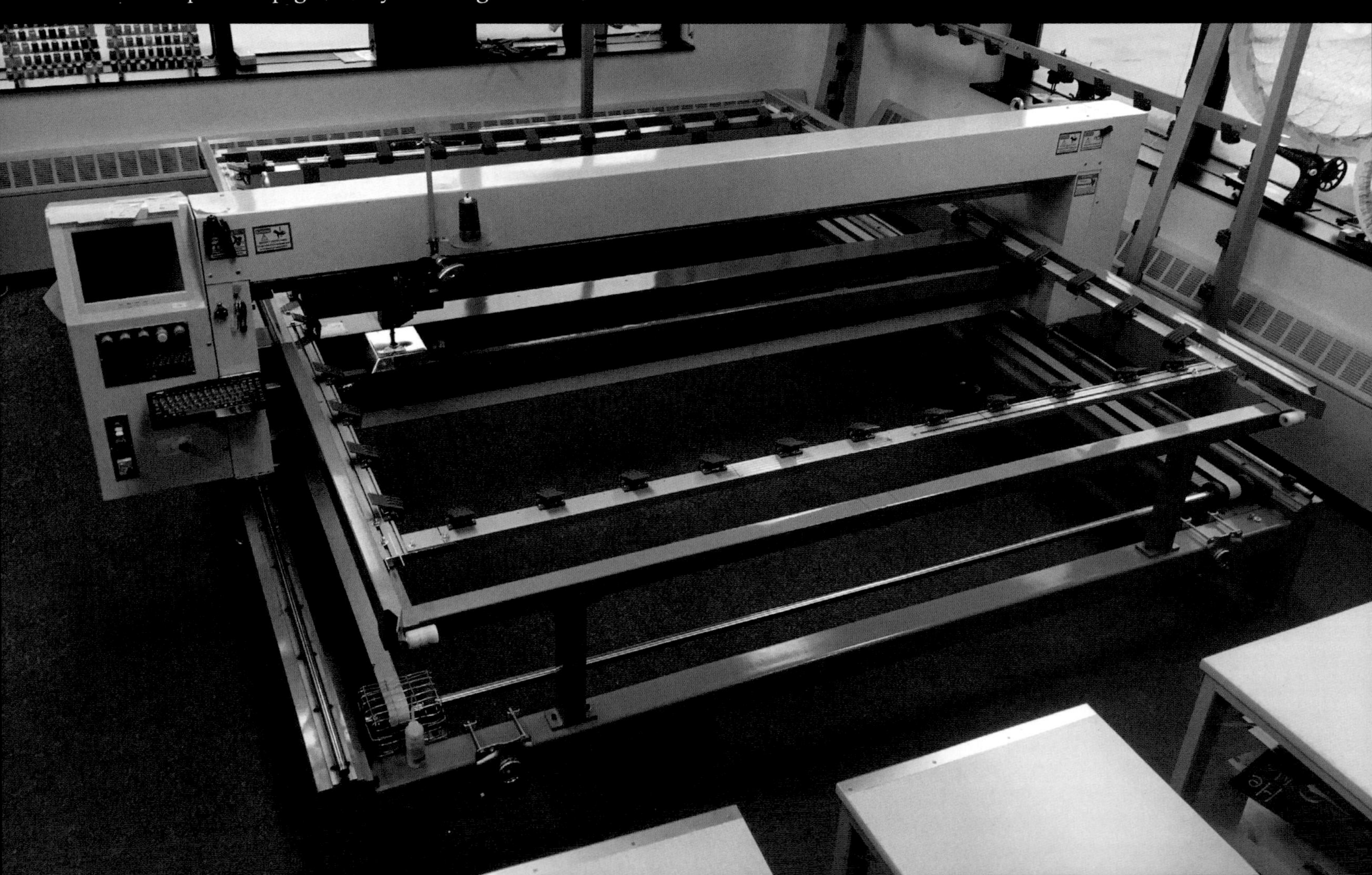

▲ One secret to the astonishing precision of the machine is that what appear to be rubber-toothed belts are actually made of a continuous stack of side-by-side steel aircraft cables embedded in a thin layer of rubber. These belts are stretched incredibly tightly, and they have *zero* give or stretch to them.

I use the machine to make quilts with various nontraditional designs, like a periodic table quilt, or this molecule quilt that shows the immunosuppressant drug cyclosporin. It takes between one and a few hours to make these designs, mainly because of all the sharp corners in the stitching, which slow the machine down.

▶ I'm sort of misusing the machine, having it take hours to make one quilt. Its true calling is making a *lot* of quilts very fast. This factory in China has five of them and makes, if you can believe it, *several thousand quilts a day*. That's a fantastic amount of material to go through in one day, and it's possible only because, when it's making a simple pattern with smooth curves and long straight stretches, it can make a whole quilt in just a few minutes. If you buy a cheap bedspread or quilt at Walmart (or an expensive one at a department store), chances are it was made on a machine very much like this one.

The Two-Thousand-Dollar One-Dollar Potholder

AFTER THAT BREATHTAKING example of mass-production sewing, here we are back at the potholder, which I have painstakingly finished sewing together by hand. Between the two layers of fabric is some loose cotton fiber called batting, which adds insulation value. Interior stitching over the whole surface (the quilting) keeps the batting in place.

Potholders are for holding hot pots, and that's exactly what I did with this example. Once, to get me the picture at the beginning of this chapter. Then I carefully put it away in a safe place to make sure nothing ever happens to this ridiculously overblown year and a half of my life.

AT MANY POINTS along the story of my potholder we've seen the contrast between doing something on a small scale, and doing it on a vast industrial scale. This is a *tremendously* important difference.

If you add up all the money I put into making this potholder, it's at least a thousand dollars. Actually, in line with the excellent country song "Thousand Dollar Car," it's probably closer to two thousand dollars. It all adds up: renting a rototiller, paying a couple guys to help till, rake, and plant the field, realizing that my irrigation pump has irretrievably sunk to the bottom of the lake, that extra 300 feet of water line to reach the field, sprinklers, a walk-behind cultivator to clean up between rows so the pictures come out pretty....It just keeps going, and that's just to get a few pounds of raw cotton. Ginning, carding, spinning, and weaving just kept piling on the costs.

It cost me thousands of dollars to make one potholder that is worse in every objective way than one I could buy at Walmart for a dollar or two. Commercial agriculture and industrial production systems are *fiercely* efficient. They can create beautiful finished products at preposterously low costs with very little human effort.

Is this a good thing? I think the answer is yes, and I'll tell you why.

I heard a talk once by a fashion designer who was proud of the fact that her shirts were entirely handmade of cloth that had been woven on hand looms, from handspun yarn that was dyed with natural indigo dye. It took two months of hand labor by workers in a small village to make one of her shirts.

Think about what that means. Her business is based on us living in a world where there's one group of people who can afford to spend more money on one shirt than another group of people have to live on for two months. That's not a world I want to live in.

This designer wanted the people who make her shirts to keep making them by hand. That's another way of saying she wanted to keep them poor. The village of Nan Tang Tuan Cun took a different path: they invested in large-scale automation, greatly increasing their ability to produce high-quality goods for sale. Today they have automated planting, ginning, and spinning. Someday perhaps they will automate harvesting too. With each step, individuals become more productive, able to earn a better living, and able to bring more resources back into their community.

Look, fundamentally I'm an optimist. Yes, people are displaced when machines take away their jobs. But over time, in every country on every continent, when that has happened, the result has been new jobs for those people, better education, and a better life for everyone. It's not easy, and there have always been great hardships on the way. But it's always proven worthwhile. In recent years my optimism has been severely tested as our world has slipped backward in many ways, with hardworking people finding their income and standard of living slipping backward as they watch the rich grow ever richer. This is dangerous, and it must stop.

I believe it will stop, and that we will once again move into a time of ever-increasing opportunities for all. I have to believe, because the alternative is too horrible to contemplate. So I choose hope over despair, and trust that there will continue to be a steady march of productive improvements in machines and technology. I hope for this future, if not for myself, then for my children, and for the children of Nan Tang Tuan Cun as they and all the children of the world weave their futures together.

▲ Children from 南塘疃村 fly the drone I used to photograph their parents' cotton field.

▲ OK, that's all, folks, end of the line, you can go home now: the potholder is finished.

AFTERWORD:

THE THING WHISPERER

AN EX-GIRLFRIEND ONCE described me as a thing whisperer. This is a bit like being a horse whisperer, except that instead of being able to form a deep emotional bond with a majestic animal, I can fix things. The work is quite similar.

It might seem odd to say that understanding the inner life of a horse is like fixing things, but when done right, fixing horses and fixing things involve the same kind of deep empathic bond. That's the only way to bring the full capacity of the human brain to bear on the problem.

Think about learning to do a magic trick, or throw a ball, or pronounce a difficult word. You watch someone do it, and then try to mimic them. First you imagine your body going through the motions, then you do it, then you watch again to see what you messed up, and so on. Imitation requires forming a mental connection to the person you're watching, imagining that you are them, that their body is your body. Your brain internally goes through the motions while you are watching, priming circuits to carry out the action for real. There are special "mirror neurons" in the brain for exactly this purpose.

A horse whisperer, I'm sure, is using the same parts of their brain, except instead of mirroring another person, they are mapping their own identity onto that of the horse. They are becoming one with the horse, feeling what it feels, sensing its motivations, and learning how to communicate with it.

Thing whispering is no different, except it is done with mechanical, electrical, or computerized devices. You have to form a bond before you can truly understand a thing. You have to map your identity and your body onto the parts of the device and make it part of yourself.

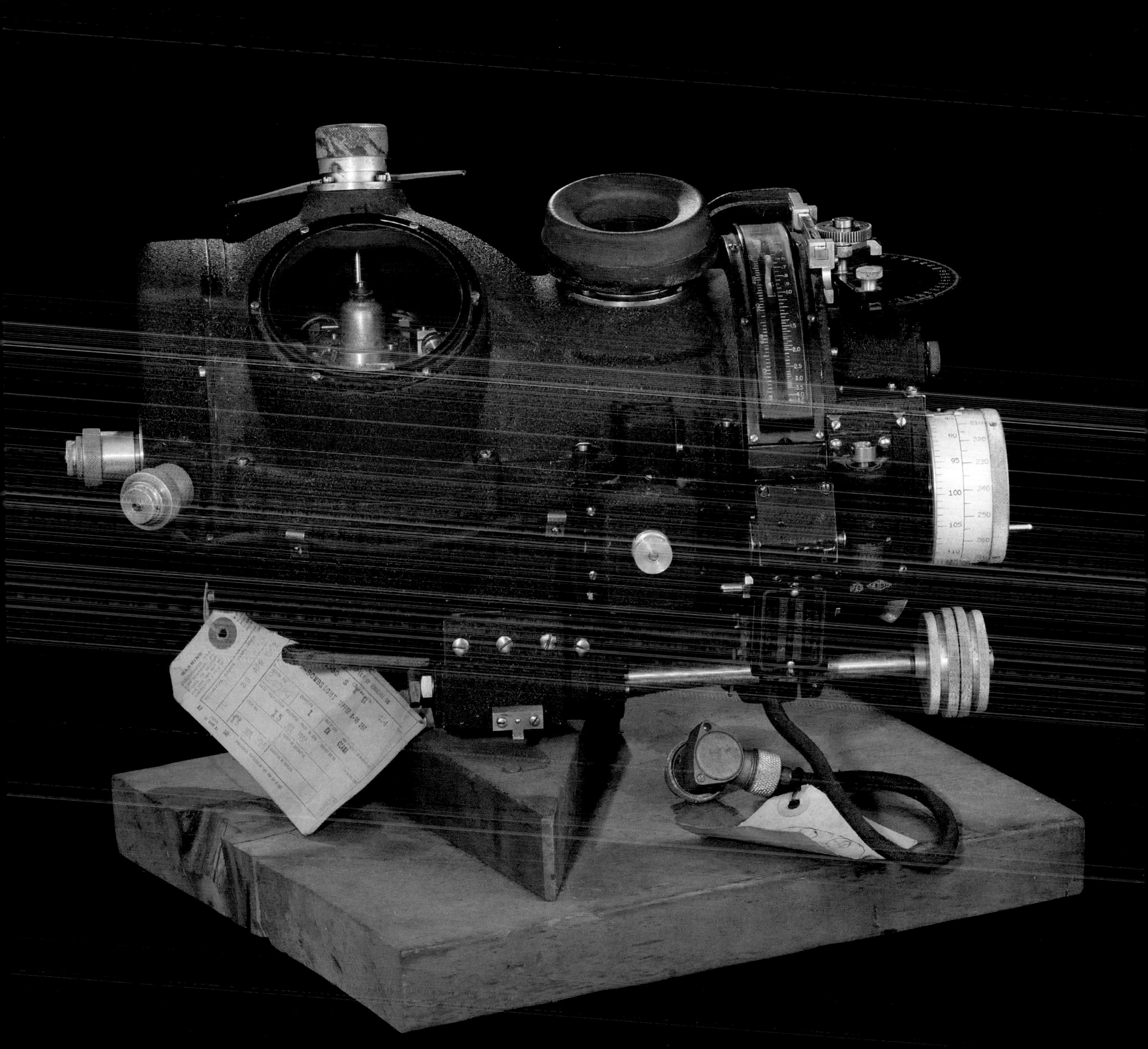

LEARNING TO DRIVE a car and to ride a bicycle are perfect examples of what I'm talking about. Once you can do either of these things well, you have incorporated the mechanical device, the car or bicycle, into a mental picture of your body. You "feel" its extent and movement just as you feel your own body. You don't have to calculate or think about where you are in the lane, or how far to lean into a turn.

Extending this ability to a wider range of mechanisms is just a matter of practice. Say you want to know what will happen when a certain lever moves against another one. You can, subconsciously, project that lever onto your own arm, and then imagine how your arm would move. Your brain is *really good* at imagining and predicting the behavior of your body. This kind of projection is a far more powerful way of understanding devices than any sort of intellectual understanding of the thing.

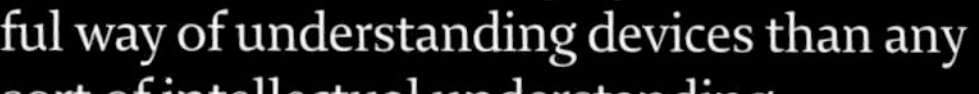

Reaching out and touching something with the tip of your finger is something you learn to do as a baby. By the time you're a few years old your finger will touch exactly where you want it to *every time*, without fail and without thinking. When an expert carpenter is at work, the hammer becomes a part of their extended body image. They are no more likely to miss the nail with their hammer than you would be to miss tapping it with the tip of your finger.

This idea of internalizing the external is the same for anyone who's an expert at what they do. A painter's brush or a surgeon's knife becomes a part of their extended body, controlled not by the intellect, but by the lower, animal parts of the brain, the same parts that let us run and jump or balance on one foot. Guided by higher thought, yes, but executed deep in the unconscious wiring of the machine that is our brain.

Done at the highest level, any of these activities is, by necessity, effortless. If you have to think hard when you're riding a bicycle, you're doing it wrong. You have to feel it, not think about it. That's obvious, but maybe it's less obvious that exactly the same is true for the guy cutting open your chest to rewire your heart. If he's thinking about it too much, well, that's not good. Hopefully he's just a student and there's a more experienced surgeon watching

Nick Mann on his bike

Great skill is a thing of beauty to watch, no matter the field. Our minds recognize perfection. The efficient, effortless motion of a practiced hand looks good, because it is good. We subconsciously know these motions to be the way they should be.

over his shoulder. It's the surgeon who *isn't* thinking too much about the craft who is less likely to make a mistake.

It is a testament to the power and flexibility of the human brain that you can teach yourself to do this kind of projection even onto things that are completely unlike your own body. What's even more remarkable is that your brain can project itself even onto abstract ideas. Mathematicians or computer programmers spend a good bit of time not "thinking" in the way one does when taking a math test. Instead they are imagining strange shapes transforming, or patterns of light and dark. If you watch them working on a tricky problem, you'll often see them moving their hands, or shifting their fingers back and forth in complicated patterns. These movements are leaking from the intense activity of the very same brain systems that guide the carpenter's hammer, or the surgeon's knife. They are using the bond-forming and action-predicting parts of their brain to figure out something about an algorithm or mathematical proof.

Richard Feynman wrote in his book *"Surely You're Joking, Mr. Feynman!"*: "The mathematicians would come in with a terrific theorem, and they're all excited. As they're telling me the conditions of the theorem, I construct something which fits all the conditions. You know, you have a set (one ball)—disjoint (two balls). Then the balls turn colors, grow hairs, or whatever, in my head as they put more conditions on. Finally they state the theorem, which is some dumb thing about the ball which isn't true for my hairy green ball thing, so I say 'False!'"

Of course doing any of these things—programming, surgery, playing music, carpentry—requires a great deal of concentration. It can be exhausting. But it doesn't feel "hard" in the way it is hard to do something you're not very good at. Instead, in the best of moments, it feels *good*. It is satisfying in a far deeper way than anything else I know of.

As much as Nick and I would love to take credit for this remarkable image, it was in fact taken by the doctor-photographer Max Aguilera-Hellweg, and is included in expression of the highest achievement of which a human being is capable. It is the embodiment of the other, the foreign, into our own consciousness. It is the feeling of making the thing a part of yourself, and thus enlarging yourself to include that thing outside the confines of your body.

It's said that the ultimate goal of meditation is to extend your mind to encompass the whole of existence—to become one with the world, just as the carpenter becomes one with the hammer. I don't know if this is true. I have not—at least not yet—ever felt one with the world. But I have felt one with a few things—a hammer, a bicycle, a couple of computer languages, and just once, for an instant, the whole of a house with all its plumbing and electrical systems. It felt good and right, like something I'd like to do more of. Maybe someday I'll expand to include the whole of Cleveland. (If I lived in Cleveland, which I don't. I'm just using Cleveland as an example because it's the only city, other than Bielefeld, Germany, that is ridiculous on its face.)

I tried playing a few instruments, but I never got good enough at any of them to feel the flow of playing them well. But I liked the idea of making music, so sometime in high school I built this thing. It's a touch-sensitive keyboard.

Designing the circuit took weeks, maybe months, I'm not sure anymore. There is a separate channel for each of the twenty-five keys that implements an attack, decay, sustain, and release curve that is used as an envelope for the audio waveform for that key. The individual frequencies for each key are created by a CD4040 frequency divider chip, a nifty little integrated circuit still available today.

I don't remember all the details of how it works anymore, but I do remember the feeling of flow while I was designing and building it. I also remember feeling, a few years later, that I had lost the ability to feel that way. I'm not sure why: I just couldn't immerse myself as completely in a project anymore. Maybe it was college, maybe it was (the lack of) girls, I don't know. Fortunately it was temporary, because the feeling has come back many times since, and I expect it will again. I hope everyone can feel this way sometimes.

I made these circuit boards using a resist-ink pen to hand-draw the contact patterns on blank printed circuit boards (fiberglass boards coated with sheets of copper). After drawing out the pattern, you soak the boards in a solution of ferric chloride, which dissolves all the copper not covered with resist ink. Then you solder the various components down to the copper traces.

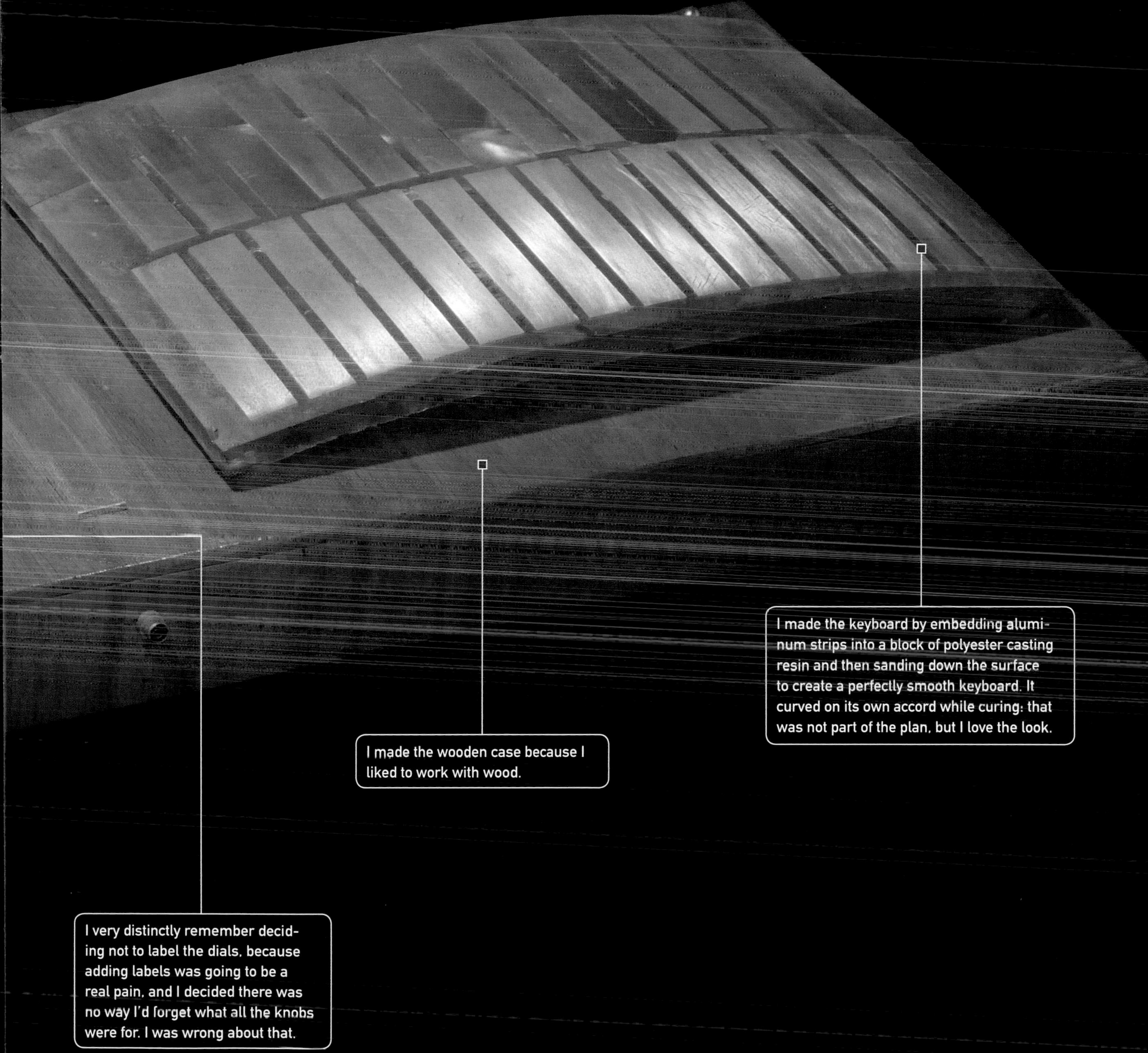
I made the keyboard by embedding aluminum strips into a block of polyester casting resin and then sanding down the surface to create a perfectly smooth keyboard. It curved on its own accord while curing: that was not part of the plan, but I love the look.
I made the wooden case because I liked to work with wood.
I very distinctly remember deciding not to label the dials, because adding labels was going to be a real pain, and I decided there was no way I'd forget what all the knobs were for. I was wrong about that.

Are You a Thing Whisperer?

EVERYONE IS GOOD at something, and everyone has worked at getting good at that thing. For example, some people are good at being popular at school. It turns out that this is actually a lot of work. Of course you don't notice, because part of the work is making it all seem effortless. What you're not seeing is the many hours the popular kids spend studying trends—to decide what to wear so that they will have exactly the right look, without looking like they are trying to have the right look.

There's nothing wrong with being popular, but I definitely was not. I wish someone had told me back then that this was because I wasn't trying hard enough. Not because it would have made me try harder to be popular: quite the opposite! Instead it would have reinforced my view that being popular was not something worth spending my time on. Being popular in high school is temporary; it's a skill that doesn't translate well into real life.

Instead I spent my time building things, learning to fix them, programming computers, making gunpowder, and a lot of other things that I don't regret for a minute. If you, too, find yourself called to a life of things—if you are a thing whisperer—then be proud as you go out into the world. You will find your people, and they will know you, and welcome you.

Acknowledgments

THIS BOOK, LIKE all my books, is a collaboration between me and the life and people around me. More than anything that means my longtime photographer Nick Mann, who took nearly all the pictures in the book (except those from China and Louisiana, which are generally not as good because I had to take them myself).

The book is also a collaboration between myself and my editor Becky Koh and designer Matt Cokeley. They bring great skill to the process, without which the book would just be lonely Word files on my computer. Also in that category is my long-suffering assistant Gretchen, without whom I would probably be in jail for some kind of criminal tax negligence.

Beyond these central figures is a zoo of individuals who sustain me, break up with me, give me purpose in life, and just generally keep me on my toes. We won't get into the exes, but in the sustaining category are my kids, Addie, Connor, and Emma (listed either alphabetically or by height, depending on your point of view). Then there's the new people in my life, Maribel and Toby, and the long term people, Bobbie, Tristan, Koatie, Alexus, Brianna, and Quinton (definitely by height this time). And even though we are no longer officially in the app business together, Max and Fiona continue to be a source of support, advice, and inspiration.

New to the list of people to thank for this book is a contingent of friends in China. I have been visiting quite frequently, often as a guest of Xu Jizhe of the Newton Project and/or Professor Ben Koo of Tsinghua University. I am very grateful to them for indirectly sponsoring an important part of the research for the book: These visits allowed me to make detours to assorted farms, factories, and markets, which were the sources of many of the stories and objects in this book.

I am also tremendously thankful for my outstanding Chinese guide and translator 王 · 苍 (Debbie Wang), who took me all over the cities and countryside, and whose parents and siblings, in a remarkable stroke of luck, are in the cotton-farming business, giving me a delightfully personal look into the making of cotton products in China. I am also grateful for the time and attention of 周宁 (Zhou Ning) of 河北昱昌古典钟表有限公司 (the Hebei Yuchang Antique Clock Company), who spent several days taking me around to the far-flung outposts of his company's clockmaking empire.

Also in connection with cotton I thank Logan Farms and the Gilliam Gin Company of Caddo Parish, Louisiana, for allowing me to film their operations. Sue of the Champaign-Urbana Spinners and Weavers Guild very kindly tried to teach me to spin cotton, and when that didn't work, spun the needed yarn for me. Toby and Alexus helped with weaving, and Quinton spent many hours engaged in probably illegal child labor to run my hand-crank cotton gin.

At the very start of the process, I thank Koatie, Bobby, Qwantrell, Quinton, Brianna, and Alexus for the backbreaking labor of tilling and planting my cotton field. During the growing season Jason and Cindy kept an eye on the field, watered it when necessary, and warned me of impending insect catastrophes.

Koatie gets a special mention for introducing me to the whole world of prison-based clear objects, without which the first chapter would not have come about. I thank Gabby for her patience in standing on the big Toledo scale until we got the pictures just right.

Bruce Hannon provided invaluable insight into the operation and origin of the clocks in this book. Igor Pudiszak is the source of the excellent Toledo scale, and of the grace to release me from the burden of a rusty bearing.

Daniel and Eric of the GU Eagle laser cutter company were tremendously supportive in getting me the very fine laser cutter used to make all the acrylic models you see in these pages (and which you can get yourself at Mechanicalgifs.com, made in my studio and lovingly packaged by Koatie, Bobbie, and Gretchen).

Finally I would like to thank all the companies that make this wonderful stuff, and all the people at those companies who design, manufacture, and market the products. I'd particularly like to thank my local Bergner's for going out of business so I could get their mannequins at fire sale prices. You have no idea how many pictures of mannequins did not make the cut because they were just too creepy, so enjoy the ones that did make it into the book.

Image Credits

You can get build-your-own versions of several of the models shown in this book at mechanicalgifs.com. Having the physical object in your hand gives you a whole new perspective on how it works, and these clear acrylic models let you see everything in action. They make everything transparently obvious.

Unless otherwise noted below, all photographs by Nick Mann or Theodore Gray and copyright © 2019 Theodore W. Gray.

P31: Glass Frog by Geoff Gallice, used under Creative Commons Attribution license.

P37: Glass table and chairs courtesy of the Corning Museum of Glass, used with permission: 2004.2.13, Collection of the Corning Museum of Glass, Corning, NY; 2014.2.5, Collection of the Corning Museum of Glass, Corning, NY.

P75: Taken from Wolfram MathWorld by Eric Weisstein, used with permission.

P84: Engraving by Francis Place after Robert Thacker c. 1676, courtesy of the Royal Observatory Greenwich. Image is in the public domain.

P85: The time ball atop the Royal Observatory, Greenwich, London, courtesy of User Repton1x at Wikimedia Commons, used under Creative Commons Share-Alike license.

P85: The Troughton 10-foot Transit Instrument. Drawn by J. Farey and engraved by T. Bradley. Plate 16 from Pearson's *An introduction to practical astronomy* (London, 1829). Image courtesy of Robert B. Ariail Collection of Historical Astronomy, Irvin Department of Rare Books and Special Collections, University of South Carolina Libraries, via the Royal Observatory Greenwich. Image is in the public domain.

P85: Antikythera Mechanism copyright © Tony Freeth PhD, Images First Ltd., used with permission.

P93: John Harrison's marine timekeeper H1, item L5695-002. Copyright © National Maritime Museum, Greenwich, London, Ministry of Defence Art Collection, used with permission

P131: Louis Essen and J. V. L. Parry standing next to the world's first cesium atomic clock, developed at the UK National Physical Laboratory in 1955. Courtesy of the UK National Physical Laboratory. Image is in the public domain.

P132: Graph of day-length variations created by the International Earth Rotation and Reference Systems Service. Image is in the public domain.

P133: Cesium fountain clock courtesy of the UK National Physical Laboratory. Image is in the public domain.

P140: Segment of the Papyrus of Ani from the collection of the British Museum. Image is in the public domain as a faithful reproduction.

P143: Statute of the UK Parliament, 25 Edward III st. 5 c. 9 (1350).

P163: Photo of Armin Wirth, photographer unknown.

PP172–173: Copyright © 2018 Brian Resnick, Courtesy Vox.com and Vox Media, Inc.

P178: Ship disaster, copyright DHA-Depo Photos via AP, used with permission.

P179: NASA Science Officer Bill McArthur on board the International Space Station, image courtesy of NASA. Image is in the public domain.

P200: Slave shackles: Collection of the Smithsonian National Museum of African American History and Culture, Object Number 2008.10.4. Image is in the public domain.

P207: Machine hall of the manufacturer Richard Hartmann in Chemnitz, artist unknown, image is in the public domain.

P249: Copyright © 2019 Max Aguilera-Hellweg

Illustrations are indicated in **bold**.